T0133244

Canada and Ballistic Missile Defence, 1954-2009

Studies in Canadian Military History
Series editor: Dean F. Oliver, Canadian War Museum

The Canadian War Museum, Canada's national museum of military history, has a threefold mandate: to remember, to preserve, and to educate. Studies in Canadian Military History, published by UBC Press in association with the Museum, extends this mandate by presenting the best of contemporary scholarship to provide new insights into all aspects of Canadian military history, from earliest times to recent events. The work of a new generation of scholars is especially encouraged, and the books employ a variety of approaches – cultural, social, intellectual, economic, political, and comparative – to investigate gaps in the existing historiography. The books in the series feed immediately into future exhibitions, programs, and outreach efforts by the Canadian War Museum.

John Griffith Armstrong, *The Halifax Explosion and the Royal Canadian Navy: Inquiry and Intrigue*
Andrew Richter, *Avoiding Armageddon: Canadian Military Strategy and Nuclear Weapons, 1950-63*
William Johnston, *A War of Patrols: Canadian Army Operations in Korea*
Julian Gwyn, *Frigates and Foremasts: The North American Squadron in Nova Scotia Waters, 1745-1815*
Jeffrey A. Keshen, *Saints, Sinners, and Soldiers: Canada's Second World War*
Desmond Morton, *Fight or Pay: Soldiers' Families in the Great War*
Douglas E. Delaney, *The Soldiers' General: Bert Hoffmeister at War*
Michael Whitby, ed., *Commanding Canadians: The Second World War Diaries of A.F.C. Layard*
Martin Auger, *Prisoners of the Home Front: German POWs and "Enemy Aliens" in Southern Quebec, 1940-46*
Tim Cook, *Clio's Warriors: Canadian Historians and the Writing of the World Wars*
Serge Marc Durflinger, *Fighting from Home: The Second World War in Verdun, Quebec*
Richard O. Mayne, *Betrayed: Scandal, Politics, and Canadian Naval Leadership*
P. Whitney Lackenbauer, *Battle Grounds: The Canadian Military and Aboriginal Lands*
Cynthia Toman, *An Officer and a Lady: Canadian Military Nursing and the Second World War*
Michael Petrou, *Renegades: Canadians in the Spanish Civil War*
Amy J. Shaw, *Crisis of Conscience: Conscientious Objection in Canada during the First World War*
Serge Marc Durflinger, *Veterans with a Vision: Canada's War Blinded in Peace and War*
Benjamin Isitt, *From Victoria to Vladivostok: Canada's Siberian Expedition, 1917-19*
James Wood, *Militia Myths: Ideas of the Canadian Citizen Soldier, 1896-1921*

Canada and Ballistic Missile Defence, 1954-2009: Déjà Vu All Over Again

James G. Fergusson

UBCPress · Vancouver · Toronto

20 19 18 17 16 15 14 13 12 11 10 5 4 3 2 1

Printed in Canada on FSC-certified ancient-forest-free paper (100% post-consumer recycled) that is processed chlorine- and acid-free.

Library and Archives Canada Cataloguing in Publication

Fergusson, James G. (James Gordon), 1954-
Canada and ballistic missile defence : déjà vu all over again / James G. Fergusson.

(Studies in Canadian military history)
Includes bibliographical references and index.
ISBN 978-0-7748-1750-9

1. Ballistic missile defenses – Canada – History. 2. Canada – Military policy. 3. National security – Canada – History. 4. Canada – Politics and government – 20th century. 5. Canada – Politics and government – 21st century 6. Canada – Military relations – United States. 7. United States – Military relations – Canada. I. Title. II. Series: Studies in Canadian military history

UG745.C3F47 2010 358.1'740971 C2010-901123-6

Canadä

UBC Press gratefully acknowledges the financial support for our publishing program of the Government of Canada (through the Canada Book Fund), the Canada Council for the Arts, and the British Columbia Arts Council.

This book has been published with the help of a grant from the Canadian Federation for the Humanities and Social Sciences, through the Aid to Scholarly Publications Programme, using funds provided by the Social Sciences and Humanities Research Council of Canada.

Publication of this book has been financially supported by the Canadian War Museum.

UBC Press
The University of British Columbia
2029 West Mall
Vancouver, BC V6T 1Z2
www.ubcpress.ca

For Betty and Jack McGeown
whose library
and endless discussions with a young man
created a career

and

H. John Skynner
friend, brother, and patriot

Contents

Illustrations

Figures

Tables

Photos

Preface

The immediate origins of this study of Canadian policy on ballistic missile defence and its various iterations over time date back to the release of the 1994 Defence White Paper. Having participated in the Joint Parliamentary Committee hearings on Canadian Defence Policy in 1993, I was naturally interested in comparing the committee's findings with the government's subsequent White Paper. In so doing, the findings, not surprisingly, were very similar, except for one glaring exception. Whereas the parliamentary report largely ignored the missile defence question, and its companion, the minority report of the Bloc Québécois committee members, fully condemned missile defence, the government moved in the other direction. It did an about-face by opening the door to Canadian participation in the United States' efforts, thereby reversing the previous policy of the Mulroney government that formally rejected Canadian government participation in the context at that time of the Strategic Defense Initiative (SDI) research program. Naturally, the government, or perhaps more accurately the White Paper drafters in the Department of National Defence, placed the usual caveats around this policy reversal in emphasizing interest in obtaining a better understanding of the US programs – then in the transition from George H. Bush's Global Protection against Limited Strikes (GPALS) architecture to Bill Clinton's National Missile Defense (NMD) – referencing consultation with allies and placing firm limits on possible areas of Canadian participation. Regardless, a major shift in policy had occurred, even though a close friend in National Defence suggested that the new policy was not a significant reversal.

On the surface, his suggestion may have simply reflected the bureaucratic view that academics read more into policy statements than warranted. Alternatively, it could be understood as an attempt to keep this contentious issue off the public agenda for as long as possible. Given the public furor surrounding SDI in 1985, playing down the significance of the missile defence parts of the White Paper might enable the government, and National Defence, to engage quietly the issue with the United States as a preliminary to developing Canadian policy options if or when the United States proceeded to deployment. The White Paper was simply to set the stage and protect the government from any charges of duplicity and at the same time signal the United States that Canada was open to discussions. Regardless, the new policy implied that there had been some developments in Canadian policy thinking with regard to missile defence long before the words appeared in the White Paper. Recalling the end of SDI when missile defence disappeared from the Canadian public agenda, the deafening silence in Canada that had followed the GPALS announcement, and the lack of attention paid to the initial steps taken by the Clinton administration on missile defence, perhaps there was much more than met the eye behind the White Paper's missile defence referents. Indeed, the missile defence question may have been much more prominent in policy-making circles between 1985 and 1993 than known in the public domain.

Regardless, the 1994 White Paper reawakened my interest in ballistic missile defence, whose roots date back to my first exposure to the issues surrounding nuclear deterrence and strategic defence as an undergraduate student in a seminar on strategic studies taught by my future colleague Paul Buteux. The 1994 White Paper also reminded me of my attendance at a seminar in Vancouver in 1985 by a representative of the High Frontier – a pro-SDI lobby group – on the suggestion of my fellow graduate student and colleague Alex Moens. The representative's passionate articulation of the case for SDI stood in stark contrast to the dominant academic perspective, still with us today, that missile defence undermined a stable nuclear deterrent relationship between the United States and the Soviet Union and thus was a threat to peace. Unfortunately, my renewed interest in the debate between the pro- and anti-SDI arguments, generated by the presentation, was placed on the back burner with my doctoral research focused on an entirely different area of study in international relations.

Beginning with my op-ed article on ballistic missile defence in 1995, my personal research agenda became almost solely devoted to the issue of ballistic missile defence in general and Canadian policy in particular. What was initially going to be a study on contemporary policy issues facing the Chrétien government surrounding the issue of potential Canadian participation in what would become NMD in 1996 expanded into the much larger study of the history of Canadian policy. The beginnings of the public debate in 1997 not surprisingly framed the issue of Canadian participation around the Cold War arguments of SDI. It seemed clear that any understanding of the contemporary policy issues and the policy process within the Canadian government on the NMD question would be based upon the issues and process that had led to Mulroney's 1985 policy on SDI. From there, understanding SDI necessitated an understanding of its predecessor, the late-1960s anti-ballistic missile (ABM) question, which had resulted in the ABM exclusion clause in the 1968 North American Air Defense Command (NORAD) renewal agreement – and this policy decision naturally drew upon the issues and process that began in the early 1960s with the first proposed US missile defence program – Nike-X – and the initial stages of research and development that began in their infancy in the final stages of the Second World War following the first German V-2 attacks on England.

Reinforcing the logic of a historical study was the simple reality that the issue of Canada and ballistic missile defence had not been examined in any detail whatsoever, which reflects the paucity of systematic, historical studies of Canadian defence policy. Certainly, there were many contemporary analyses of the missile defence question each time the issue emerged after a US program announcement, with the notable exception in Canada of GPALS. Although these were valuable to provide context, each missile defence iteration was largely treated in isolation from its predecessors, providing a snapshot rather than a systematic evaluation of Canadian thinking and the Canadian policy process. Of course, any contemporary study faces the difficulty of obtaining access to primary government documents vital to a full understanding of any policy debate, and this study is no different in examining the last decade or so of the issue in Canada.

Indicative were the results of my formal access to information requests. The request for historical documents from the Privy Council Office (PCO) produced a wealth of primary data, especially on the SDI question. Even though large sections of the key 1985 Kroeger Report commissioned by the prime minister were deleted, other released documents concerning the report provided clear evidence of its contents. In contrast, a similar request for Privy Council documents of the contemporary period yielded only letters from politicians and interest groups to the prime minister, the government's generic responses, and forty-two pages of exemptions with no referent dates. Even so, it was possible to draw limited conclusions from pages of exemptions. Thus, for example, the relatively small number of pages exempt from release covering the last fourteen years confirmed information from other sources that cabinet had been rarely involved, and even though document dates are now also withheld, the most recent first eighteen pages of exemptions was likely the 2003 memorandum to cabinet, which led to the beginning of formal discussions on Canadian participation with the United States. Regardless, a detailed full representation of the contemporary policy process awaits the passing of time and governments.

As such, this study of the Canadian policy process on ballistic missile defence can be seen in some sense as two studies in one. The first is the historical study that begins in the early 1950s, when the chief of the air staff identified missile defence as a priority, and concludes with the election of the Chrétien government in the fall of 1993. It is based on documents from the National Archives, the PCO access request, and the valuable cooperation of the departments of Foreign Affairs and National Defence. In particular, the Foreign Affairs academic research program proved invaluable in providing access to its documents on missile defence from the 1960s to the early 1990s. As part of the agreement governing access, academics are able to take notes from screened Foreign Affairs documents with two caveats. First, photocopying of the documents is prohibited. Second, Foreign Affairs holds a veto on citations, but not substance, within the relevant chapters of any book based on material supplied under the agreement. In this case, Foreign Affairs did not delete any citations of the submitted chapters. After lengthy discussion, National Defence agreed to a similar arrangement. The archival research was supplemented by a series of confidential interviews with government officials who had worked on the file over time.

Turning to the contemporary era with the election of the Chrétien government, access to documents, as noted above, was severely limited. Like any contemporary policy study, I had little choice but to rely upon alternative research material. In this regard, the research benefited from several factors. My involvement through the University of Manitoba Centre for Defence and Security Studies with the Department of National Defence Security and Defence Forum created a very positive yet arm's-length relationship with department officials and facilitated my understanding of the policy process and my ability to conduct wide-ranging confidential interviews with National Defence officials. Similarly, my work with the Department of Foreign Affairs Verification Research Unit in the 1990s, which included the writing of two reports on ballistic missile defence, generated a similar positive relationship with Foreign Affairs officials involved with the missile defence file. Along with these interviews, I was very fortunate in being able to conduct confidential interviews with a wide range of other government officials in Canada, including many of the political actors involved, as well as officials in the United States.

Along with the use of standard secondary research material from academic and journalism sources, the research for this book also benefited greatly from my direct involvement in the public debate from the mid-1990s onward. As a result of my extensive publications on the question of ballistic missile defence, I acquired the reputation as one of Canada's leading academic experts, and, for good or ill, the most prominent academic proponent of Canadian participation. As one colleague put it, I was and remain Mr. BMD in Canada. This reputation resulted in my involvement in nearly all the public fora on the missile defence question over the past decade or so, as well as invitations to testify to the standing committees on foreign affairs and national defence. These appearances not only put me in contact with most, if not all, of the individuals and organizations engaged with the missile defence question inside and outside government but also provided this study with a better understanding of the arguments on both sides of the missile defence divide. Of course, as a proponent of Canadian participation, my analysis is not entirely neutral, as no study can be. Nonetheless, to the best degree possible, this book offers a balanced account of the debates inside and outside government over time.

Despite the differences in research material for the historical and contemporary parts of this study, evidence from the past and present clearly indicates that the contemporary debate and policy process on Canada and ballistic missile defence have not wandered far, if at all, from the historical debate and process. From my first foray into the historical documents, it is striking how the contemporary debate and process mirrored the past. It is indeed déjà vu all over again, even though the world changed with the end of the Cold War, and missile defence technology advanced significantly. Certainly, the new post-Cold War strategic rationale for missile defence is markedly different from the Cold War case. Although many members of the bureaucracy understood the significance of the new strategic environment, overall the issues and problems as seen through the lens of successive Canadian governments in the 1990s and 2000s remained firmly entrenched in the past, even though there is little evidence to indicate that the policy analysts in Foreign Affairs and National Defence tapped institutional memory from earlier experiences.

In the end, this study is about much more than just the question of Canada and ballistic missile defence. It is really about a nation and its governments and bureaucracy over time trapped within a fixed mindset about its place in the world and the means through which to manage foreign and defence policy requirements in response to the actions of others. It is about a national groupthink that affects and determines the manner in which policy is conceptualized and processed. Neither Conservative nor Liberal governments, nor the bureaucracy that transcends both, have escaped from this national groupthink that affects Canadian foreign and defence policy deliberations. Even twenty years after the end of the Cold War, and fifty-four years after the end of the Second World War, when the fundamental principles of this national groupthink were established, nothing has shattered this mindset.

All things being equal, the future is likely to remain déjà vu all over again. The ballistic missile defence question will likely return to the public agenda in the near future, especially if NATO proceeds forward with a strategic defence for Europe. When it does, the government will likely behave as its predecessors have done in seeking to keep a low profile and hope that the inherent contradiction between being a participant in an alliance with a ballistic missile defence capability and being a non-participant in a ballistic missile defence capability for the defence of North America and Canada simply fades from public view.

Completing this study would not have been possible without the support and assistance of many people. I am particularly indebted to Francis Furtado, who provided extremely valuable comments and editorial suggestions on each of the draft chapters. Paul Buteux's numerous hours of conversation on strategic questions, his in-depth understanding of the logic and practice of nuclear deterrence and helpful suggestions on each chapter, assisted greatly. Hector Mackenzie and Greg Donaghy of the Historical Section of the Department of Foreign Affairs provided valuable assistance in the context of the department's academic access research program. I also greatly appreciate the many government officials who agreed to be interviewed confidentially and discussed the issue of ballistic missile defence informally with me on many occasions over the past decade, without whose support the second half of this book in particular would not have been possible. In addition, Aaron Hywarren and the entire Security and Defence Forum community played a significant role in the development and execution of this study, not least of all through their patience on innumerable occasions listening to me drone on about ballistic missiles and through their valuable comments in response to my many presentations. I am indebted to the numerous archivists at the National Archives and National Library for facilitating, many times on short notice, my research. I would also like to thank the two anonymous reviewers of the manuscript for their helpful suggestions and Emily Andrew, Melissa Pitts, and Laraine Coates of UBC Press for their work in support of this effort. Finally, and most importantly, this study would not have been possible without Laura and my family, who never wavered in their support and patience. As always, any errors or omissions are my responsibility.

Abbreviations and Terms

ABM	anti-ballistic missile defence
ALCM	air-launched cruise missile
ASAT	anti-satellite weapon
BMDO	Ballistic Missile Defense Organization
DRB	Defence Research Board
DSP	Defense Support Program (of early warning infrared sensors in geosynchronous orbit)
GMD	Ground-based Midcourse Defense
GPALS	Global Protection against Limited Strikes
ICBM	intercontinental ballistic missile
INF	intermediate-range nuclear forces
IRBM	intermediate-range ballistic missile (1,500 to 3,000 kilometres)
JROC	(US) Joint Requirements Oversight Council (within the Joint Chiefs of Staff)
MAD	mutual assured destruction
MDA	Missile Defense Agency
MIRV	multiple independently targetable re-entry vehicle
NATO	North Atlantic Treaty Organization
NDP	New Democratic Party
NIE	US National Intelligence Estimate
Nike-X	modified high-altitude surface-to-air system
NMD	National Missile Defense
NORAD	North American Air (Aerospace in 1981) Defense Command
NORTHCOM	US Northern Command
NWS	North Warning System

PAC-2	Patriot Advanced Capability – 2 (proximity explosion)
PAC-3	Patriot Advanced Capability – 3 (kinetic kill)
Patriot	Phased Array Tracking to Intercept of Target (tactical anti-ballistic missile system)
PCO	Privy Council Office
PJBD	Permanent Joint Board on Defence
Safeguard	Nixon-era ABM architecture
SALT	Strategic Arms Limitation Talks/Treaty
SAM	surface-to-air missile
SDA	Strategic Defense Architecture
SDI	Strategic Defense Initiative
SDIO	Strategic Defense Initiative Organization
Sentinel	Johnson-era ABM architecture
SLBM	submarine-launched ballistic missile
SORT	Strategic Offensive Arms Reduction Talks/Treaty
SRBM	short-range ballistic missile (up to 500 kilometres)
START	Strategic Arms Reduction Talks/Treaty
STSS	Space Tracking and Surveillance System
TMD	theatre missile defence
UCP	US Unified Command Plan
USAF	United States Air Force
V-2	Vergeltungswaffe or revenge weapon (first operational ballistic missile employed by Germany in the Second World War)
WMD	weapons of mass destruction

Canada and Ballistic Missile Defence, 1954-2009

Prologue
What's with Defence?

On September 8, 1944, the first German V-2 rockets armed with conventional explosives struck London. Eleven months later, the United States detonated the first atomic weapons over the cities of Hiroshima and Nagasaki. So was born the ballistic missile and nuclear age, whose marriage first with short-range missiles in the 1950s and then with long-range, intercontinental missiles in the early 1960s became a strategic benchmark of the world as we know it. Even though this benchmark receded from prominence in the post-Cold War era of peace operations and the post-9/11 war on terror, it has always remained close to the surface of strategic concerns. As part of the natural dynamic and logic of weapons and war, the V-2 attacks also signalled the beginning of attempts to counter or defend against ballistic missiles. Indeed, both the United States' and the Soviet Union's missile research and development programs on ballistic missiles included research on the means to counter them – ballistic missile defence. Yet, by the 1960s, the idea of defending against long-range ballistic missiles armed with nuclear weapons had become increasingly viewed as dangerous and illegitimate.

As this new world emerged in the wake of the Second World War, a dramatic, if not revolutionary, idea became codified in international law: that the only legitimate use of armed force was in self-defence. Thus, the great irony and paradox was born. Self-defence became equated with offensive preparations, a situation entirely at odds with the dangers of offensive military preparations, which many believed had played the key part in the outbreak of war in August 1914. The difference, of course, was nuclear weapons and the belief that no rational or sane decision maker would take a nation to war knowing full well that the result would be suicide. Ensuring the prospect of suicide became central to self-defence in the nuclear age. Defending against the prospect of suicide became an anathema to self-defence.

This is a story about fundamental ideas and beliefs about war and peace since the end of the Second World War, when nuclear weapons turned the world upside down, when defence became simultaneously legitimate and illegitimate, when hawks and doves became strange bedfellows, and when one nation – Canada – grappled with the practical political problems that followed from this upside-down world, made worse by its geostrategic location beside one of the two Cold War protagonists, and the post-Cold War's sole dominant power – the United States. No other nation faced the conundrum that Canada did, and in many ways still does, concerning preparations for war in order to sustain peace in the post-Second World War era. Yet this is more than a story about Canadian policy toward self-defence as told through its five-decade-long confrontation with ballistic missile defence in a world of Cold War nuclear deterrence and post-Cold War proliferation concerns. Despite the uniqueness of the Canadian confrontation, it is also a story about secondary states dealing with Great Powers and the much deeper and broader realities of such a nation's foreign, defence, and security policy as told through the lens of successive governments, Liberal and Conservative, faced with a superpower neighbour and ally slowly but inevitably moving forward to acquire a defence against ballistic missile attack.

Canada's missile defence conundrum, and thus its much broader and deeper defence relationship with the United States in the post-Second World War era, comprise a story played out in five acts, each initiated by its friend and ally: the anti-ballistic missile (ABM) era consisting of McNamara's 1967 Sentinel program and Nixon's 1969 Safeguard revision; Reagan's 1983 Strategic Defense Initiative (SDI), better known as Star Wars; George H. Bush's 1991 Global Protection against Limited Strikes (GPALS) architecture; Clinton's 1996 National Missile Defense (NMD) idea; and finally, George W. Bush's Ground-based Midcourse Defense (GMD) system. In each case, successive Canadian governments feared they would have little choice but to join in because each American initiative had direct implications for fundamental Canadian defence interests, of which the cooperative binational North American Aerospace Defense Command (NORAD) was key. In each case, Canadian governments preferred the initiative would disappear and after each occurrence (except for the recent case) hoped it would not return. In each case, Canadian governments looked to avoid making a decision in order to avoid two perceived undesirable outcomes: an alienated ally or a divided nation and government. In each case, Canadian governments looked south for a means to escape making a decision, and the United States formally or informally to provide such a means, even in the case of the final act when Prime Minister Martin (or more accurately his foreign minister, Pierre Pettigrew) on February 24, 2005, said no to Canadian participation – a decision announced when the United States neither expected nor sought one.

Despite the notion that the recent no is inconsequential, it has proven pivotal in a dramatic, yet ignored, transformation in Canada's defence relationship with the United States, with major implications for Canadian strategic defence interests. But then Canadian strategic defence interests never amounted to much in all the debates, despite the best attempts by National Defence to insert them into the decision-making process. Instead, each debate was conducted around the larger global issues of war, peace, and stability under the implicit assumption that Canadian interests were (and are) synonymous with global or superpower interests. Certainly, sandwiched between the two superpowers during the Cold War, Canadian elites had no choice but to be concerned about the implications of missile defence for the stability of the deterrence relationship between the United States and the Soviet Union. But the idea that distinct Canadian defence interests should drive Canadian policy was largely peripheral in public debates, and this extended even after the end of the Cold War when the relevance of deterrence stability arguments simply evaporated overnight. Overall, it was as if Canada was a Great Power itself instead of a secondary power reacting to decisions and events out of its hands. Canadians mimicked the superpower (American) debates long after they were politically over, except for the case of SDI. Here in a strange twist, the Canadian debate would conclude before the most substantive and politically significant part of the US debate really began and, of course, would be missed within Canada.

The Great Power overtones embedded in each Canadian debate reflect a Canadian hubris – naive beliefs that Canada could truly influence the decisions of its superpower friend and ally, if not the global community, by saying either yes or no. Alongside this hubris, of course, is the presence in each act of a large dose of the Canadian independence and identity question, where pro and con constructed their arguments in terms of Canadian sovereignty as if a decision would really add to, or subtract from, Canadian sovereignty. It was also the question that made Canadian decision makers believe that any missile defence decision, regardless of direction, would have major domestic political implications. In this regard, each turn of the missile defence question exhibited a clear lack of leadership by national decision makers in allowing the issue to fester, sending mixed signals at home and abroad. This was compounded as decision making over time became increasingly centralized in the Prime Minister's Office, with cabinet and the bureaucracy being on the outside.

The lack of leadership is also reflected in the relationship between the two primary governmental departments at play – External/Foreign Affairs and National Defence – representing different organizational interests and responsibilities. It comes as no surprise that, in the absence of clear political leadership, the two departments largely stood in opposite camps on many, but not all, occasions – the former opposed to and the latter in favour of Canadian participation. Yet, when some degree of leadership was present, both departments overcame their inherent differences. In this sense, five decades of missile defence are also about the difficulty confronted by a bureaucracy trying to manage an important policy question, and thus defend Canadian interests, in the absence of any clear direction from above. It is in many ways a five-decade-long stall to avoid a decision, in which governments and bureaucracies simultaneously feared getting a US invitation to participate and not getting an invitation.

As the issue first unfolded in Canada in the mid-1960s, the patterns were set for the subsequent missile defence iterations – the United States leaving the door open for Canada to make an initiative and decide how it might wish to participate while proceeding on a unilateral basis and, in turn, Canadian officials seeking to obtain detailed classified information in order to make a decision without making any commitment. The resulting dilemma, or Canada-United States dance, became the relationship between the United States' willingness to provide such information in the absence of a Canadian commitment and the difficulty Canada faced in making a commitment in the absence of such information. All this took place again and again in an environment of uncertainty about whether missile defence would proceed and what the implications of being on the outside might really be for Canada, despite occasions of US reassurance. This reality about the relationship between sovereign states took place against the repeated backdrop of supposed but mythical US covert and overt pressure on Canada to participate.

The story of Canada's ballistic missile defence policy, to borrow from Yogi Berra, is truly déjà vu all over again as each act plays out a similar story. Perhaps this is not surprising, as missile defence is missile defence regardless of the iterations the issue went through over five decades. However, each iteration had enough substantive difference from its predecessor to change the policy parameters: ABM was a limited system initially designed (Sentinel) to defend against the growing Chinese threat and an accidental Soviet launch and then designed as "Safeguard" to protect the American land-based intercontinental ballistic missile fleet from a disarming first strike; SDI was a massive research and development program sold in Hollywood terms; GPALS a limited land- and space-based system to deal with growing proliferation challenges and potentially as part of a United States-Russian global protection system (GPS); NMD an even smaller land-based research, development, testing, and evaluation program; and, finally, GMD the first operational element of a larger layered defence against proliferators (and possibly the Chinese).

Above all else, the world transformed when the Berlin Wall fell in 1989, ending the Cold War, and the political conditions that had informed the strategic debate around missile defence disappeared. Parts of the debate in Canada on missile defence also changed, especially for advocates. Even so, the change was not transformative in any substantive sense. The parameters set in the ABM debate remained in place as if the world and technology had remained static. But then the fundamental premises underpinning Canadian foreign and defence policy that emerged out of the Second World War didn't transform with the end of the Cold War, despite the rhetoric of the 1994 Defence White Paper, the 1995 government Foreign Policy Statement, and the more recent (2005) international and defence policy statements. It remains to be seen whether the current Harper minority government's policy will be any different. If these premises, with the popular public myths informing them, didn't change after 1989, why would one expect Canadian policy on any issue, including missile defence, to change now? Missile defence could have been the occasion for a fundamental debate about the premises of Canada's worldview and its place in the world. However, this was not to be, as none of the major actors, pro or con, truly challenged the fundamentals. As such, déjà vu is also the story of national groupthink, even among the supporters and opponents of Canadian participation.

In this sense, the foundation of national groupthink is laid in Canadian elite reaction to the failure of appeasement and isolationism in the interwar period. The resulting post-Second World War policy of internationalism emerges and becomes by the 1960s an unquestionable article of faith among Canadian elites and the Canadian public. However, not only is the success of internationalism a function of its ability to offer something to everyone regardless of their ideological or strategic predispositions, it is also a straightjacket from which decision makers cannot escape. It contains two fundamental propositions: the linkage of Canadian defence and security to international peace and security, and the linkage of Canadian defence and security to American defence and security. As long as the two propositions can be sustained together, such that international peace and security thinking and interests are perceived to be consistent with American defence and security thinking and interests, then Canadian internationalism is at equilibrium. If this linkage is disrupted, as is increasingly the case from the 1960s onward, coinciding with the emergence of missile defence onto the international and bilateral agenda, then Canadian decision makers perceive themselves to be confronted on every issue with a choice between the world and the United States. To choose the world is to reject Canada's most important friend, ally, and trading partner, with fears of significant consequences to follow. To choose the United States is to reject the world, with fears of significant domestic consequences of being seen as subservient to the United States.

It is for this reason that Canada's NATO allies play a seminal role in Canada's missile defence debate and much larger defence and security post-Second World War story. This is not the counterweight idea per se, in which NATO Europe provides a balance to American superiority. Rather, it is the NATO allies potentially providing legitimacy and cover for Canadian policy makers. Being in step with the allies provides a means to disguise a bilateral decision as a multilateral one, thereby defusing the Canadian subservience argument. Unfortunately, much of the missile defence story is one in which the NATO allies, or at least significant European members, are out of step with US missile defence policy – fearing in their case the decoupling of the United States from their defence and security during the Cold War and the return of a hostile Russia after the Cold War. Nonetheless, the ups and downs of Canadian missile defence policy, in which governments appear to be leaning one way or another, are closely related to NATO Europe, and recent missile defence initiatives within the alliance are setting the stage for a possible revision to Canadian policy on missile defence.

Before the 1960s, the inherent conflict or contradiction within Canadian internationalism doesn't exist. From the Second World War until then, Canadians generally saw international peace and security as synonymous with American defence and security, with the Suez Crisis of 1956 an exemplar. The United States has not yet become fully recognized by Canadians as a dominant force in Canadian political life, with Great Britain and the Commonwealth link still remaining relatively prominent. Nuclear weapons and nuclear deterrence are yet to dominate strategic thinking as informed by the Cold War geopolitical situation and with it the self-defence–no defence paradox. Yet the origins of the paradox roughly date back to a decade before the first use of ballistic missiles – the German V-2 rocket attacks on England beginning in September 1944 – and the first and only use of nuclear (then known as atomic) weapons – the American bombing of Hiroshima and Nagasaki.

Informed by the disaster of the First World War and the complicated political forces at play in the interwar period (1919-39), the seeds of the no defence–self-defence paradox lie in the development of the heavy bomber and ideas that airpower in the form of the bomber would transform warfare and avoid a repetition of the carnage of the First World War. Airpower enthusiasts, led by the vision of General Giulio Douhet, among others, posited a strategic vision in which airpower (heavy bombers) could provide a rapid knockout blow against an adversary, thereby avoiding a repetition of the First World War stalemate.[1] Victory through airpower would be quicker and less costly in men and material than war on the ground.

Underpinning this new vision was the belief that the bomber would always get through, with three important implications. The first was the implicit futility of defence against the bomber. Second was the vulnerability of modern industrial societies, which collocated industrial production with large concentrations of civilians. While future debates occurred around whether one should target industrial capacity, in which civilians would be unintended collateral damage, or target civilians to destroy the national will on the presumption of their weakness, reality, given the technology and demography of the day, made the issue moot. To bomb one was to bomb the other. Finally, this situation suggested a strategic environment favouring the offence and a decision to strike first.

Of course, the reality of the war that followed was markedly different. Decision makers did fear the implications of bombing on a weak civilian population, but the result was a tacit agreement until the second stage of the Battle of Britain not to strike at cities.[2] The war did not begin with a massive strategic air attack, what would be known today as a pre-emptive strike. When strategic bombing did result, civilian populations on all sides proved remarkably resilient. As for the belief of defeating the enemy by destroying its capacity to wage war, the predictions of airpower strategic bombing advocates were found to be wanting, though one cannot disregard the losses that did result from the bombing campaign.[3] Finally, all the nations devoted significant resources to air defence. In the case of Great Britain, this included the development of the Spitfire and Hurricane fighter interceptors, anti-aircraft batteries, and radars within an integrated air defence network. Germany followed suit, and according to Richard Overy the greatest value of the allied strategic bombing campaign was the diversion of large-scale German resources to air defence and away from the eastern and western fronts.[4]

The ideas underpinning the case for strategic bombing, which Canada participated in through the Royal Air Force's 6 Bomber Group during the Second World War, would be the forerunner of the nuclear age, in which nuclear deterrence and mutual assured destruction (MAD) would become the guarantors of peace – the final realization of the old Roman axiom, if you wish peace, prepare for war. However, before this realization, nuclear weapons were simply seen as providing the means to bring the vision of interwar airpower enthusiasts into reality. Except for the voice of Bernard Brodie, considered the "father" of nuclear deterrance thought, atomic and then nuclear weapons following the first thermonuclear test in 1953 were simply seen as another weapon of war – artillery on a much larger, cheaper, and more efficient scale.[5] Developing a defence against nuclear weapons was seen as no different from developing a defence against any other weapon of war entailing two elements, which came to be defined as passive and active defence in the nuclear age.

Passive defence is designed to negate or prevent the effect of the weapon itself. In the medieval era, for example, armour and chain mail were designed to defend a knight against a sword or arrow. During the First World War, it consisted of digging deep trenches to protect soldiers from artillery, machine gun, and rifle fire. In the Second World War, passive air defence for Londoners was the underground subway or tube providing bomb shelters. Confronting nuclear weapons in passive defence terms simply meant more, deeper, stronger, and better provisioned shelters à la the major US public civil defence bomb-shelter exercise, or the extensive Soviet effort, which included the Moscow subway.

Active defence is intended to prevent the use of the weapon, usually by intercepting or destroying the delivery system. If one could locate and destroy an opponent's archers, a delivery system very vulnerable without infantry protection, then one could eliminate the weapon itself – the arrow – thereby defending one's troops. In modern artillery parlance, this is the idea of counter-battery fire intended to destroy an adversary's capacity to fire high explosives at one's forces. Active defence may also take the form of pre-emptive or disarming first strikes against an adversary's offensive military capabilities. During the 1991 Gulf War, the coalition's Scud hunting campaign, albeit largely unsuccessfully, sought to destroy the Iraqi mobile missile launchers dispersed in the desert before they could release their missiles or afterward to prevent reloading. It is also the military side of counter-proliferation as practised, for example, in the Israeli strike against the Iraqi nuclear reactor at Osiriq in 1985. These latter two examples represent the modern offensive element of defence in which a defensive act is carried out offensively.

Active air defence as traditionally understood consists of intercepting and destroying bombers or strike aircraft before they reach their target and deliver their weapons (bombs and, later, air-to-surface cruise missiles) through fighter interceptors, anti-aircraft artillery, or surface-to-air (SAM) missile batteries. Active missile defence substitutes missiles, the buses, or their warheads for aircraft as their targets. Each of these targets gets progressively smaller and faster from launch to target, making such defence extremely challenging technologically.

Missile defence in the Canadian context almost exclusively refers to one class of ballistic missiles: strategic. This class as originally defined during the Cold War generally consists of two types: land-based intercontinental ballistic missiles (ICBMs) with a range of five thousand or more kilometres and sea-launched ballistic missiles (SLBMs) carried by nuclear-powered ballistic missile submarines (sub-surface ballistic nuclears – SSBNs) capable of relatively short- and long-range attacks. These two types of ballistic missiles, as well as long-range bombers, would become the focus of strategic arms control negotiations and agreements between the superpowers. As far as the five missile defence iterations, ICBMs would be the primary concern.

In terms of active missile defence, the three possible targets – missiles, buses, and warheads – correlate roughly to the three stages of flight for an ICBM. The first, or boost-phase, is the period from missile launch until the final burn of the third stage of the rocket booster. Its conclusion roughly coincides with the exit or release of the bus carrying one or more warheads and possibly decoys. During the boost phase, the missile is moving at its slowest as it ascends and provides the largest target for intercept. Depending upon the distance from launch to target, this stage lasts roughly about three minutes. Interception, while technologically the easiest during this phase, requires the deployment of an interceptor close to the target – on a forward-deployed ground or naval platform, aircraft, or potentially on-orbit in space. A variety of air- or space-launched missile interceptors can be deployed, as well as potentially exotic intercept technology, such as a laser. This exotic technology is currently being developed in the form of the US Airborne Laser program.[6] Politically, however, it is the space-based idea that has proven most contentious in the Canadian debate, even though there is no formal space-based missile defence development program in the United States.

Once the final-stage rocket burn is completed, the missile – or, more accurately, the nose cone or bus with its warhead(s) – enters the second phase of flight, the mid-course phase. This is the longest phase of flight, with a duration of twenty-five to thirty minutes, depending again upon the distance involved, where the bus/warhead reaches speeds of about eight kilometres per second in free flight (unpowered) on its ballistic trajectory. Depending upon the sophistication of the attacker, the bus may carry only a single warhead or re-entry vehicle, multiple independently targetable re-entry vehicle (MIRV) warheads, or decoys. At some point during the flight, the bus releases the warheads and decoys if available.

Interception attempts can be made at the bus before it releases its cargo or the warheads during the mid-course phase. As this phase for ICBMs occurs in the vacuum of outer space, a conventional explosive intercept is largely impossible. Instead, intercept must rely either upon a kinetic kill, where an interceptor warhead collides with the bus or incoming warhead, or on a nuclear warhead, which generates X-rays and electromagnetic pulse capable of disabling the warhead and preventing its detonation. The first-generation US and Soviet missile defence systems employed nuclear warheads because the technology required to employ a kinetic kill was not available. Current US missile defence programs all employ kinetic kill, including the Ground-based Midcourse Defense (GMD) system deployed at Fort Greely, Alaska, and Vandenberg Air Force Base, California. Space-based mid-course intercept is also possible employing kinetic or exotic technology, and this form was the centrepiece of the proposed SDI architecture known as Brilliant Pebbles. Nuclear weapons could be employed, but their deployment in space is banned by the 1967 Outer Space Treaty.

The final phase is the terminal one, when the warhead re-enters the atmosphere and falls to target. Decoys generally burn up on re-entry, making tracking and target identification much easier. However, the flight time in this period is extremely short (a minute or less), the warhead is very small and moving very fast, and the attacker can employ atmospheric drag and other penetration aids to defeat an intercept attempt.[7] During this phase, a nuclear, conventional, or kinetic-kill intercept can be employed, but the defender is limited to one attempt only. In the case of one-stage short-range missiles, which do not leave the atmosphere and have a flight time of several minutes, it is possible to take several shots. Currently, the United States has not developed a terminal-phase system for strategic defence, though it may be possible for the GMD system to shoot at an incoming warhead in the terminal phase. However, except for an attack on San Francisco or Fairbanks, Alaska, the GMD system is not deployed in a location to take a terminal shot.

Almost immediately after the first German V-2 attack on London, attention began to be paid to developing an active defence capability, especially in the United States. However, the United States and its allies lacked a ballistic missile capability, which at a minimum was needed for any defence. The speed of the V-2 made anti-aircraft artillery ineffective. As such, the allies' only option was to identify and destroy the V-2s on the ground. Research on missile defence did not end with the defeat of Nazi Germany. Following the American-Russian race to capture German rocket scientists, of whom Werner von Braun would become the most famous, the United States and the Soviet Union both embarked upon ballistic missile and missile defence development programs.

The near-simultaneous pursuit of a defence against a new weapon system is a long-standing dynamic in the weapon's development process. Identifying and developing possible defences or counters to a new weapon are a natural response to the expectation that the technology will eventually diffuse and an adversary will either develop the same weapon and/or methods to counter it. Thus, simultaneous defence developments are designed to ensure the effectiveness of the new weapon system and prepare for the day when it may be employed against oneself. In the case of the ballistic missile, defence after over a decade of research (see Act 1) because of the speeds involved came to rely upon the same delivery system as the offence – ballistic missiles armed with nuclear warheads to attack and ballistic missile interceptors armed with nuclear weapons to defend (and also ballistic missiles to launch satellites into orbit that would increasingly become important to the strategic equation).

Although short-range single-stage missiles armed with nuclear warheads would be deployed in the 1950s, ICBMs did not come into operation until the early 1960s. Until then, strategic nuclear delivery systems designed to strike at the national homelands of the two major protagonists were exclusively the domain of long-range strategic bombers. As the Soviet Union and the United States developed large strategic bomber fleets, both invested significant resources in strategic air defence. For the Soviet Union, these defences were very large because of the size of the American bomber fleet and the need to cover nearly 360 degrees for the defence of the Soviet Union. The American strategic bomber fleet as a function of US bomber bases on American or allied territory encircled the Soviet Union and its ally, Communist China (until the Sino-Soviet split in the late 1960s and American rapprochement beginning with Nixon's 1971 visit to China). The United States' geostrategic position made the over-the-pole route the most direct means to strike the Soviet Union, especially after the adoption of a policy of immediate strategic strikes on the Soviet Union in the case of war, embodied in Eisenhower's New Look strategy of massive retaliation. Dispersed overseas bases were made possible by the globalization of containment through the creation of a series of interlocking military alliances, from Europe (NATO) to the Middle East and Central Asia (the Central Treaty Organization – CENTO) to Southeast Asia via the SouthEast Asia Treaty Organization – SEATO) and finally Japan through the bilateral Mutual Defence Pact.

For the United States, strategic defence was much simpler and more straightforward. The only feasible avenue of attack by Soviet strategic bomber forces was across the Arctic, from north to south, which meant overflying Canada to targets in the continental United States. By virtue of geography, American strategic air defence involved Canada, and Canada was clearly an important piece of strategic real estate. Theoretically, Canada had three basic choices. It could ally with the Soviet Union, which, of course, made no political or moral sense at all. It could opt for neutrality, knowing full well that in the case of war both the United States and the Soviet Union would violate its neutrality unless Canada developed a large, sophisticated, and very expensive independent air defence capability to deter both. Finally, it could ally with the United States and cooperate in the air defence of North America against the Soviet Union.

The choice to ally and cooperate with the United States was a given, and effectively little if any consideration was given to the option of neutrality. It would not be until the 1960s that this option would be seriously broached by a national political party – the New Democratic Party (NDP) – and critics of the policy of neutrality would point directly to the costs of neutrality in terms of defence requirements and the inconsistency of the policy relative to Canada's place in the world as a liberal democracy economically integrated with the United States.[8] Regardless, in the late 1940s, with fears of the development of a long-range

Soviet bomber threat, which would manifest itself in the famous 1955 bomber gap debate in the United States, cooperation began between the United States and Canada in the air defence of North America.

Air defence cooperation followed politically from the evolving bilateral defence relationship that had predated the Second World War. The roots of continental defence cooperation went back to the 1938 unilateral Roosevelt pledge to defend Canada if attacked by a third party and Mackenzie King's response that Canada would not permit its territory to be used as a base from which to attack the United States. On paper, the King response looked no more than a pledge of neutrality. However, in the air defence world, bombers in sovereign Canadian airspace made King's response more than just about neutrality. Soviet bombers entering Canadian airspace without permission was an act of war, dictating a Canadian military response. Such an act would also bring Roosevelt's pledge into operation. The formal mutual defence commitment or alliance between Canada and the United States would be embedded in Article 5 of the 1949 Treaty of Washington establishing the North Atlantic Treaty Organization (NATO), even though both parties for all intents and purposes actually kept the mutual defence of North America beyond European influence.

The two pre-Second World War commitments would be formalized with the 1940 Ogdensburg Agreement on mutual defence, which created the Permanent Joint Board on Defence, followed by the 1941 Hyde Park Agreement on defence economic cooperation and the development of the bilateral annual Joint Defence Plan. Building on Canada-US multidimensional military cooperation in the Second World War, with the primary threat to North America an air-breathing one, direct cooperation between the newly created United States Air Force and the then Royal Canadian Air Force began.[9]

For the two air forces, strategic air defence cooperation made simple functional sense, reinforced by the natural ties that exist via organizational culture between functionally identical military services. Similarly, the Soviet strategic bomber forces as the primary direct threat to North America gave both services a legitimate claim on national resources, especially for the fighter-interceptor community as the most effective air defence resource. Having observed the lessons of air defence in the Second World War, intercepting attacking bombers hundreds if not thousands of kilometres from their targets with fighter-interceptors rather than waiting until they were close to their targets increased the probability of the successful defence of North American cities and of industrial and transportation centres. Forward air defence gave the defender many chances to intercept bombers, and successful intercepts would down the bombers with their nuclear payloads far away from any major and minor targets of value.

Effectiveness, in turn, would be increased by the use of nuclear weapons in an air defence role, either as an air-to-air weapon on forward-deployed interceptors or as a last line of ground-based SAM defences. Today, most Canadians and Americans are unaware of the extensive air defence network that existed in North America during the 1950s.

The logic of Canadian defence cooperation with the United States was perhaps most clearly laid out over four decades ago by Carleton University professor Peyton Lyon: "Even if we tried to remain neutral, Canada is virtually certain to be mauled in any assault on the United States. It is, therefore, in our vital interests that the United States not only be invulnerable, but appear invulnerable. If the United States were in danger of attack, and Canada were able to make a significant contribution, anything but our maximum effort could be folly."[10] By the 1960s, however, one would be hard-pressed to see Canada's effort at anywhere near maximum, and Canada's most significant contribution was in reality the provision of territory as Canadian defence spending declined precipitously.

One might have expected that the new threat posed by ICBMs would have occasioned a cooperative Canadian effort similar to that of the 1950s. This was not to be the case. For many reasons, successive Canadian governments would look to avoid investing anything on ballistic missile defence, even fearing that any agreement with the United States on Canadian involvement would come with a large price tag. Implicitly, they hoped that the possible provision of territory, with the United States providing the bulk of infrastructure money, might suffice to keep the vital bilateral defence relationship healthy and to protect vital Canadian strategic interests. Even then, the idea of Canada providing territory for US missile defence interceptors became a non-starter early on and long before the 1972 Anti-ballistic Missile Treaty (ABM Treaty), which placed prohibitions on third-party involvement. As the issue unfolded, the meaning of participation for Canada came to be seen strictly in passive terms – the possible provision of sites for tracking radar and Canadian involvement in the command and control process, both consistent with the existing NORAD air defence mission. Indeed, missile defence for Canada came to be almost exclusively about the future of NORAD as the institutional key in the relationship and the meeting of Canadian strategic defence interests.

Although rarely if ever fully articulated by Canadian governments, in an environment dominated by aerospace threats to the nation (bombers and then missiles) at least until the end of the Cold War or 9/11, NORAD was the key to Canada's five core interrelated strategic defence interests as they concerned North America and the US relationship – the expropriation of US defence resources for the defence of Canada, access to US defence thinking and planning, influence opportunities to ensure Canadian defence interests were taken into account in US defence plans, protection of Canadian sovereignty from US unilateral defence undertakings, and reassuring the United States of Canada's commitment to continental defence. In addition, NORAD gave Canada a window onto the global strategic world, which it could not hope to obtain on its own. Finally, NORAD had intangible international political value in representing Canada's unique and special relationship with the United States – one that no other ally possessed. Only Canada worked with the United States in a formal binational command relationship in which a foreigner could command US forces.

All these strategic interests were at play in the United States' six-decade-long quest to develop and deploy a missile defence capability for the defence of North America. All remain at issue following the Martin decision to say no publicly to Canadian participation beyond early warning of ballistic missile attack. It all began, at least publicly, in the fall of 1967 when the US secretary of defense, Robert McNamara, reluctantly announced that the United States would proceed to deploy a limited ground-based anti-ballistic missile system – Sentinel – in the continental United States to defend against Communist China and an accidental ballistic missile launch.

Act 1
Anti-Ballistic Missiles: Don't Worry, Be Happy (1954-71)

On September 18, 1967, in a speech to the United Press International editors in San Francisco, Secretary of Defense Robert McNamara announced that the United States would deploy a limited anti-ballistic missile (ABM) shield, code-named Sentinel, around the continental United States. This thin ABM system was designed to defend America against an accidental attack and the growing nuclear and ballistic missile threat of Communist China – a country portrayed in ways similar to that of rogue states today. On March 14, 1969, President Nixon renamed the program Safeguard, downsized the shield, and shifted the primary rationale from the defence of cities to that of land-based intercontinental ballistic missiles (ICBMs). A year and a half later, the United States and the Soviet Union signed the ABM Treaty. It limited both parties to two ABM sites, around the national capital and/or an ICBM field(s), with each site to consist of no more than a hundred land-based interceptors. In 1974, the protocol to the treaty limited each side to one site only. The next year, the United States declared operational the Safeguard site for the defence of the Grand Forks ICBM field, located at Cavalier, North Dakota.[1] Six months later, on October 2, Congress voted to close the site, and in February 1976 dismantlement of the system began with the removal of the warheads and missiles.

So ended the ABM era – one that actually began with a multidimensional ABM research and development program and called for an extremely large or thick ABM system capable of defending against the growing Soviet ICBM threat. Today, all that physically remains of Safeguard is the Cavalier phased array radar as part of the ballistic missile early warning system and space detection and tracking network linked to NORAD. Yet, unbeknownst to the future players, the ABM debates would continue to resonate – conceptually, politically, and psychologically – through each successive missile defence era. It would truly be déjà vu all over again every time missile defence emerged out of the shadows over the next five decades, even though ABM itself was long forgotten.

As the US ABM debates in the 1960s proceeded in two stages – Sentinel and Safeguard – so did the ABM question in Canada. During the first stage, ABM was framed almost exclusively around the issue of the renewal of NORAD, set to elapse in May 1968. Government fears that NORAD renewal could not be

Stanley R. Mickelson Safeguard ABM Complex outside Cavalier, North Dakota. *Courtesy US Army*

separated from ABM were allayed five months before the Sentinel announcement, thanks to McNamara. On April 5, 1967, McNamara informed Paul Hellyer, the Canadian minister of national defence, that Canada could remain outside ABM without affecting NORAD or the bilateral defence relationship. He even offered Canada the opportunity to acquire ABM at cost if interested. Above all else, McNamara offered the inclusion of a formal public clause excluding ABM from NORAD if Canada so desired. A year later, in May 1968, the clause appeared in the renewed NORAD agreement.

Canadian participation in the ABM system was at the centre of the second stage. For Canadian officials, the exclusion clause did not amount to an actual decision on ABM participation. Instead, it only prevented an automatic commitment through NORAD. The immediate issues at hand were two elements of the ABM mission: early warning of a ballistic missile attack and operational command and control. In the end, foreshadowing the missed window of the 1990s, the government did and said nothing one way or the other. Yet NORAD acquired the early warning mission for Safeguard with no public fanfare. Without a Canadian commitment, command and control were assigned to the US Continental Air Defense Command. By default, the meaning of participation, and thus the parameters of future debates in Canada, came to entail three parts: early warning as non-participation, direct involvement in the actual operational side of interceptors, likely to entail the deployment of interceptors on Canadian soil, as full participation, and operational command and control as somewhere in between.

Despite the angst that ABM generated in Canada, missile defence had not always been problematic. In November 1954, well before nuclear deterrence became the overarching strategy of the West and the United States, MAD emerged as the accepted strategic situation in East-West relations, and missile defence was viewed as a threat to international peace, security, and stability, the chief of the air staff identified missile defence as a future requirement for the defence of North America and Canada.[2] Drawing on intelligence estimates, such a defence was essential to deal with an expected Soviet force of 100 ICBMs carrying between 3,000- and 7,000-pound warheads with a range of 5,500 miles, travelling at 15,000 miles per hour, with an accuracy of approximately one mile. Accordingly, the Royal Canadian Air Force "requires the development of an effective defence system against high supersonic surface-to-surface ballistic missiles fitted with nuclear warheads. Studies and research should commence at once and every possible means of combating this threat should be investigated in order to find an effective defence before 1960." The requirement was divided into five components: early warning, target identification, target acquisition and tracking, weapon assignment (essentially command and control), and interception and destruction.

In January 1955, discussions began with the United States and the United Kingdom to develop a coordinated missile defence research and development effort.[3] The task of coordinating this effort led to the creation of the Tripartite Committee on Defence against Ballistic Missiles and the US-UK-Canadian Technical Sub-Group for Defence against Ballistic Missiles. Three additional subcommittees, labelled D, E, and M, were established to coordinate research and development for the warning and tracking, weapon assignment, and interception components respectively. Along with this tripartite effort, Canada's defence research establishment engaged in direct bilateral cooperation with the United States. The Defence Research Board (DRB)/Defence Research Telecommunications Establishment worked with the US Air Force on a range of basic scientific research, including the complicated problem of missile tracking. The DRB/Canadian Armament Research and Development Establishment engaged in supporting research by the US Army Rocket and Guided Missile Agency for the Nike-Zeus surface-to-air missile system – the first proposed interceptor.

The primary focus of initial Canadian efforts concerned the development of supporting science and technology for the larger US effort. These included radar propagation and echo effects, infrared sensors, rocket propulsion, astrophysics, and systems analysis. In addition, the Canadian Armament Research and Development Establishment began to undertake research through Project Helmut to employ long-range artillery in an ABM role.[4] The initial intercept approach planned to employ a 240mm smooth-bore gun firing a projectile containing 1.2 tons of pellets. The pellets were to be released at a high altitude outside the atmosphere to collide with the incoming warhead travelling roughly at twenty-four thousand feet per second, generating sufficient kinetic energy to destroy the warhead either by destabilizing the warhead's trajectory and causing it to burn upon re-entry or by destroying the fusing and arming mechanism. It was estimated that 120 tons of pellets were needed for a 95 percent kill probability. The system was seen as a potential complement to the US Army Nike-Zeus system.

Nike-Zeus was originally a high-altitude air defence interceptor, which the US Army sought to modify for an ABM role. The United States initially planned to deploy an operational system by 1963.[5] The DRB estimated that at least four batteries at a cost of $200 million were required to defend a large city such as Montreal. Even then, Nike-Zeus had no capacity to deal with submarine-launched missiles. The board also estimated that, at a minimum, ten interceptors were needed to kill one incoming warhead.[6] Moreover, a $1,000 defence investment was needed to counter a $1.00 investment in offence, and it was unlikely that technology in the foreseeable future would narrow this ratio.[7] The DRB concluded that Canada's effort should concentrate upon passive defence and ballistic missile early warning.

The early warning effort, which eventually led to the US radar sites at Fylingdales in the United Kingdom; Thule, Greenland; and Clear, Alaska – with supporting radars in the continental United States – and which became known as the Ballistic Missile Early Warning System, was the responsibility of the United States Air Force.[8] In addition to early warning of a ballistic missile attack to ensure that the US ICBM fleet was not vulnerable to destruction by a surprise attack by providing enough time for a launch decision, the primary radars were to provide long-range warhead tracking for the missile defence radars of Nike-Zeus and all its successors. In 1958, the DRB suggested that the United States would deploy an early warning radar in Thule, Greenland, with a Nike-Zeus site by the mid-1960s.[9] The limits of the initial radar were expected to require gap-fillers at two Canadian sites, Norman Wells and Frobisher Bay. Eight months earlier, the Canadian Chiefs of Staff Committee had recommended Canadian support for the United States to begin site surveys for a communications link through Canada in the context of a recognition of the magnitude of the overall missile defence project: "If an active defence against ballistic missiles is to be developed and placed in operational use, the magnitude of any significant

First-generation ballistic missile early warning radar, Clear, Alaska.
NORAD Neg# 69-669, Courtesy Canadian War Museum

contribution could transcend that of any previous Canadian military program. The task would not only be beyond the capacity of any single Canadian service, but more over would be beyond the capacity of Canada."[10]

With the missile age rapidly approaching, the chiefs of staff established the Joint Ballistic Missile Defence Staff in May 1959, which became the Directorate of Strategic Studies in July 1961. The staff was to liaise with the Joint Intelligence Committee, the Joint Special Weapons Policy Committee, the services, and the DRB and to report to the Chiefs of Staff Committee. Its work on missile defence immediately recognized that Canadian participation would be limited, though ballistic missiles would influence the future of Canadian foreign and defence policy. Recognizing strategic, technical, and cost barriers, the Joint Ballistic Missile Defence Staff also cast serious doubts on the effectiveness of missile defence.[11] Furthermore, the most important analyst within the staff, and perhaps the only true strategic thinker within Canadian defence circles at the time, R.J. Sutherland, came to the conclusion that missile defence was an anathema to a stable nuclear relationship based upon deterrence and the threat of an invulnerable nuclear retaliatory capability.[12]

By the early 1960s, Canadian research concluded that intercepting warheads either outside the atmosphere (exo-atmospheric) during the mid-course phase of a missile's or warhead's flight or inside the atmosphere (endo-atmospheric) during the warhead's final descent or terminal phase was beyond current technology. With warheads transiting through space at speeds around 8 kilometres per second (roughly 28,800 kilometres per hour), the precise timing required for a launch and intercept was simply beyond the capabilities of the day if a shotgun-type kinetic-kill or conventional explosives were concerned. Nuclear weapons, however, were a different thing. Their large blast radius, which outside the atmosphere entailed intercept via the generation of X- and gamma rays, could make up for the lack of precision. Even then, effectiveness was limited. Warheads could be hardened. Nike-Zeus could detect, track, and intercept a single object at ballistic missile speeds, but there was no economical solution to a saturation attack.[13] Certainly, one could ring North American cities with hundreds or thousands of nuclear-tipped anti-ballistic missile interceptors. But if nuclear arsenals grew as expected and missile defences followed apace, the spectre of intercepting incoming nuclear warheads with massive nuclear detonations at high levels in the atmosphere above Canadian and American cities made the cure appear politically worse than the disease, as McNamara found out with the emergence of a wave of public opposition to Sentinel.

Even though a formal cabinet decision to disengage from missile defence research did not take place, all signs pointed defence officials in a direction away from missile defence. Research indicated that the technology of the day was unlikely to provide a cost- and military-effective defence system. Sutherland, along with a growing body of US experts, communicated the precepts of mutual nuclear deterrence and the danger defences posed to these precepts. Above all else, significant political constraints existed. Roughly at the same time the above conclusions about missile defence were reached, the Diefenbaker government faced the contentious issue of acquiring nuclear warheads for the new Bomarc-B surface-to-air missile purchase, as well as the acquisition of the Genie nuclear-tipped air-to-air missile and the public recognition of a nuclear role for Canadian forces in NATO-Europe.[14] Even more, these were not isolated events but, rather, part of a string of earlier defence decisions by the Diefenbaker government, accompanied by the deterioration of personal relations between the new president, John Kennedy, and the prime minister.

RCAF Bomarc-B nuclear-tipped ground-to-air missile interceptor.
NORAD Neg# 69-114, Courtesy Canadian War Museum

From the unexpected political fallout surrounding Diefenbaker's official signing of the NORAD agreement in 1958, the cancellation of the Avro *Arrow*, the Cuban Missile Crisis, the existing animosity between President Kennedy and the prime minister to Bomarc-B (which eventually brought down the government), defence officials easily concluded that a Conservative government would not appreciate another American defence initiative, especially one that entailed nuclear weapons.[15] It was no time politically to push the ABM file, even if strategy, bilateral relations, technology, and cost, or some combination therein, had warranted forward movement. The political environment dictated a risk-avoidance strategy to manage any further possible damage to the vital defence relationship with the United States. The strategy then, and repeated until 2003, was as best as possible to keep missile defence off the political agenda.

Nor would the time be right when Pearson and the Liberals took office. Pearson reluctantly campaigned upon, and then accepted under a dual-key arrangement, nuclear warheads for the Bomarc-B and the Genie nuclear air-to-air missiles for Canada's air defence interceptors. The warheads stored on Canadian soil remained under American ownership. In this way, the government could still claim that Canada was a non-nuclear state. Moreover, the decision was sold to the public as an alliance commitment rather than any belief in the strategic necessity or importance of nuclear weapons to the air defence of North America.

While no one said so, the Canadian preference was for missile defence in general, and subsequently ABM in particular, to go away. Even before the Vietnam War and the rising tide of a Canadian nationalism endowed with a strong streak of anti-Americanism made every aspect of Canada-US defence co-operation difficult, it was clear to all concerned that ABM would not be met with open arms in Canada. Unfortunately for the Pearson government, this preference was not shared south of the border. Instead, voices were growing in the United States for the rapid development and deployment of missile defences akin to uncontroversial air defences that had sprung up across the continent over the previous decade. If air defence was essential, so was missile defence. Furthermore, these voices became louder and louder, unbeknownst in any meaningful sense to the Canadian public and media, as intelligence estimates leaked information of a large-scale Soviet missile defence effort. For the United States, the earlier politically charged 1955 bomber and 1960 missile gap debates looked to be on the path to being replicated by an ABM gap debate.[16]

The United States identified missile defence as a requirement immediately following the first ballistic missile attack in history – the German V-2 rocket strikes in September 1944. Ironically, the effort to follow employed the same technology as the threat itself – ballistic missiles – despite initial US Air Force research on space-based interception under Project Bambi and Canada's artillery effort. The rockets or missiles for defence were fundamentally no different from those for offence (nor were they different from those used for civilian space research and exploration requirements). Offence required larger, heavier rockets to carry a comparatively larger, high-yield nuclear warhead necessary to destroy a city, especially given the inaccuracy of first-generation ICBMs.[17] By the 1970s, large ICBMs were still preferred, especially by the Soviet Union, even though accuracy had improved and technology had led to the miniaturization of nuclear warheads. Large ICBMs were also still needed to launch multiple warheads from a single missile each with a separate target, better known as multiple independently targetable re-entry vehicles (MIRU).

Defence, in contrast, needed to optimize a missile or interceptor for speed and rapid acceleration in order to intercept an incoming warhead either outside the atmosphere during the mid-course phase or inside the atmosphere during the terminal phase. With the incoming warhead travelling at about eight or more kilometres per second, and the distance from ground to an optimal intercept point at a range of hundreds of kilometres at the time, and depending upon the angle of attack, the interceptor needed at least comparable speeds with a flight time of less than a minute from launch to terminal intercept. Time was pressing also because of the complexity of the intercept process itself. With the total flight time of a Soviet ICBM from launch to detonation somewhere between thirty and thirty-five minutes, there was not a great deal of time to identify and characterize an attack, obtain release authority, program the interceptor, and then launch it. As well, identification and notification were relatively slower in the initial ABM days, having to rely upon the first generation of mechanical early warning and tracking radars and the state of communications technology. Initially, the ballistic missile early warning network was able to identify an attack only once the missiles, or more accurately the bus carrying the warheads, rose above the horizon within radar range. This shrank the amount of time available for an interception. It was not until 1970 with the deployment of the first generation of space-based infrared sensors in geosynchronous orbit, the Defense Support Program, that the United States acquired a capability to identify an ICBM launch in near real time. Even with the five or so minutes gained, time was still at a premium.

Before an ABM interceptor was launched, the attack had to be confirmed by assessors to avoid an accidental nuclear intercept (or otherwise apologize for destroying a satellite or a peaceful rocket in low earth orbit by mistake). This early warning assessor mission was assigned to NORAD and the Ballistic Missile Early Warning Center within the NORAD Cheyenne Mountain Operations Center in Colorado Springs. Once confirmed, release authority was given through the operational chain of command. In the case of nuclear retaliation, this was from the National Command Authority (the president) to Strategic Air Command in Omaha. In the case of Safeguard, this was to occur from the National Command Authority to US Continental Air Defense Command and

USAF phased array satellite tracking radar, Elgin Air Force Base, Florida.
NORAD Neg# 60-150, Courtesy Canadian War Museum

then to the planned operational Safeguard ABM command itself in North Dakota. During the process, the colocated tracking radar and associated support systems had to be notified where to look in space for the incoming warhead, find it, calculate its trajectory and optimal intercept point, program or cue the interceptor on target, and, finally once release authority was granted, fire the missile at a very small target.

For the US ABM system envisioned, this time-compressed, complicated process from early warning to interceptor launch was based upon 1960s computing technology – the speed and power of which are dwarfed by personal computers today. Given this technology and the much more complicated process and exacting demands for an effective defence, it is not surprising that offence considerations and investments took priority. All offence needed was a missile launching a warhead from one fixed point to another. However, the more complicated and technologically complex defence requirements did not alone determine the lag behind offence. Organizational interests and strategic thought also came into play.

NORAD Space Defense Operations Center, Cheyenne Mountain, Colorado Springs, Colorado. *NORAD Neg# 407-68, Courtesy Canadian War Museum*

As the Eisenhower administration confronted the rapidly approaching missile era, the question arose over which one of the military services would be responsible for the new missions that followed. Indeed, all three had engaged in ballistic missile research and development, and none wished to be left out, not least of all because success or failure had implications for future roles, missions, and budgets. All had also made a successful case earlier for a nuclear role, thereby giving each of them a rationale for missiles.

Early US decisions on missile defence roles reflected broader debates about the distribution of air roles and missions following the creation of the US Air Force in the National Security Act of 1947. Although the army lost responsibility for all air assets initially, it rebounded by gaining control of rotary wing (helicopter) platforms. For the air force, this decision was unproblematic. The organization never really wanted to waste too much of its time and effort supporting troops in the field. Its priority was the strategic or war-winning role of airpower. Combined with romantic notions of air-to-air combat and the love of speed, helicopters did not fit the air force's self-image. The same thinking drove the air force in the missile world – the strategic offensive ICBM mission over the defensive ABM one.

Besides, ground-based air defence had remained an army mission after 1947. Conceptually, missile defence was simply an extension of air defence, and thus assigning it to the army made perfect sense. The air force was unlikely to object as long as it was assigned the strategic offence mission. The first generation of ICBMs might be land based, but their purpose was a strategic air force role, not a traditional army one. Moreover, avoiding missile defence enabled the air force to concentrate its resources rather than dilute them between offence and defence. This did not mean, however, that the air force gave up its missile defence research and development effort entirely. It continued, but its focus was on outer space and anti-satellite systems developments, which also might have missile defence application. Similarly, adding missile defence to the army's air defence mission with the Eisenhower decision in 1957 did not mean significant army investments.[18] The state of technology could not be ignored. But for the army, traditional battlefield capabilities remained privileged. Soldiers, tanks, artillery (including short-range missiles), and then helicopters would always be more important than missile defence. Even within the air/missile defence world, tactical defences for forces deployed in the field would always be more important than the strategic defence of cities, never mind air force ICBM fields, as one would find out again in the 1990s.

In the end, organizational interests and preferences weighed in the lag of missile defence behind missile offence, even with the inherently larger technological demands of defence. Moreover, it is not surprising that, for technological and organizational reasons, the first missile defence program would be a high-level air defence system, Nike-Zeus. After its cancellation by McNamara, it was replaced by Nike-X, which formed the backbone of Sentinel and then Safeguard – the long-range Spartan interceptor for both ABM architectures. The Spartan interceptor had an estimated range of five hundred miles, employed a 5-megaton nuclear warhead, and was designed to destroy incoming warheads at the upper reaches of the atmosphere. Besides the blast, whose effect would be dependent upon the altitude (the vacuum of space limits the blast effect), the gamma, X-rays, and electromagnetic pulse emitted by the detonation might render the incoming warhead inoperative, assuming that it wasn't hardened. This effect had been demonstrated earlier in Operation Starfish Prime in 1962. A 1.4-megaton test not only caused a general communications blackout in the Pacific but disabled seven satellites in low earth orbit, including three flying into the field of trapped radiation.[19] The results of this and other nuclear tests paved the way for the 1963 Partial Nuclear-Test-Ban Treaty with the Soviet Union, banning nuclear tests in space, in the atmosphere, above ground, and under the sea, which in turn affected the testing envelope for ABM. Thirty Spartan interceptors, the first layer or line of defence, were eventually deployed at the Safeguard site. The second layer was the short-range Sprint missile. Produced by Martin Marietta, the Sprint missile had a range of roughly twenty-five miles and was armed with a one-kiloton warhead. It was designed to intercept missiles within the upper reaches of the atmosphere employing a combination blast and neutron effect. In total, seventy Sprints were deployed as part of Safeguard, bringing the system up to its ABM Treaty limit of one hundred interceptors, as set out in the 1972 treaty and 1974 protocol.

Both Spartan and Sprint went through extensive testing. Spartan's predecessors demonstrated successful intercepts when live tests were possible prior to the 1963 Partial Nuclear-Test-Ban Treaty. Afterward, neither Spartan nor Sprint could employ a live test, and the intercepts were only simulated. Nonetheless, test results warranted forward movement, and nuclear warheads covered a relatively wide area of space. Certainly, the defence would never be perfect, but no defence ever is. Yet given the stakes involved – thermonuclear war – any defence was better than none, or at least that was traditional thinking. However, new thinking began to view defence in an entirely different way. It was a danger to peace and security as provided by the superpower mutual suicide pact, MAD. This was the view of ABM opponents and of the Johnson administration and its defense secretary, McNamara.

The danger of missile defence, after September 1967 embodied in the Sentinel ABM program, emerged out of the new body of strategic thought seeking to come to grips with the implications of nuclear weapons on strategy and politics. During the first, or golden, age of deterrence thinking in the 1950s, the fundamentals of strategic nuclear deterrence were laid out. They included a new set of concepts that became central to strategic debate throughout the Cold War: first-strike pre-emptive attack forces (primarily ICBMs), second-strike retaliatory forces (primarily submarine-launched missiles and bombers), counterforce (military) and counter-value (cities) targeting, and the central ideas of strategic, crisis, and arms race stability. Stability, in any of its three guises, was a situation in which each superpower's ability to destroy its adversary, no matter what the adversary did, was assured. Thus was born the central concept and preferred condition for stability and peace: MAD. In this world, defence threatened assured destruction. However, this did not cover all defences. No one questioned air defences other than their utility in the emerging missile age.[20] Missile defences, however, increased the danger of war.

The initial US ABM effort began when the new thinking, especially regarding defence, was in its embryonic stage. It also was during a period of American nuclear superiority and invulnerability. Although analysts such as Albert

Wohlstetter raised concerns about the vulnerability of US Strategic Air Command bombers to a pre-emptive attack, the Soviet Union had only a limited long-range bomber capability.[21] Moreover, American nuclear strike forces forward deployed in Europe, the Middle East, and Asia, as well as its intercontinental long-range bombers, ensured the Soviet Union's destruction in the case of war. However, the emergence of ICBMs changed all this in ending American invulnerability and thus redrawing the strategic landscape.

The Kennedy administration inherited this new emerging strategic environment centred upon ballistic missiles. The 1960 presidential election campaign had raised the issue of a missile gap, with the Soviet Union in the lead. Upon entering office, the president and McNamara found the opposite to be true. The United States was far ahead of the Soviet Union in the development and deployment of strategic ballistic missiles – a lead that would not close for nearly a decade as the Soviet Union raced to keep up and the United States slowed down in part to let the Soviet Union catch up to create parity and stability.

During the early 1960s, the United States deployed its first generation of ICBMs – Titan – and submarine-launched ballistic missiles (SLBMs) – Polaris – to supplement its strategic bomber fleet and create the American strategic nuclear triad that would become the cornerstone of US strategic posture for the rest of the Cold War. At the same time, McNamara and the so-called whiz kids in the Pentagon grappled with developing a strategic doctrine for this new generation of weapons based upon the first wave of strategic deterrence thinking and practice of the Eisenhower years.[22] Eisenhower's New Look had been premised upon American nuclear superiority and invulnerability. Under the doctrine of massive retaliation, war would result in a full-scale strategic attack on the Soviet Union and its allies. Although the strategic bombing of cities had not met expectations during the Second World War, strategic bombing with nuclear weapons would. Nuclear weapons made the phrase "Bomb the Soviet Union back to the Stone Age" a likely reality. Eisenhower's secretary of state, John Foster Dulles, had qualified the conditions under which a massive strike would take place, yet critics immediately concentrated on the incredulous nature of the threat and called for more flexible options to meet with different scenarios.

Initially, McNamara proposed a doctrine based upon greater flexibility and military or counterforce targeting, which included military-industrial targets, in keeping with traditional thinking about the strategic or war-winning function of airpower. Originally announced in Athens, Greece, in May 1962 and reiterated in a public speech at Ann Arbor, Michigan, in June, the doctrine moved away from the targeting of cities as central to massive retaliation and existing US plans.[23] It became known as the "no cities posture," even though many military-industrial targets were located in or near major cities. The new doctrine also communicated US strategic thinking to the Soviet Union, as well as hinting at a tacit agreement to give cities sanctuary. Missile defence was not mentioned, but one could assume that the defence of military-industrial targets logically followed from a no cities doctrine.

However, almost immediately after the speech, McNamara began to back away from the no cities doctrine, gradually adopting one hinted at during the Eisenhower years: assured destruction based upon invulnerable retaliatory or second-strike capabilities. He also set about seeking to establish ground rules for the strategic nuclear relationship with the Soviet Union based upon the condition of MAD. As long as both sides could threaten the other with assured destruction (which McNamara also saw as a means to control the insatiable desires of the services for more and more strategic weapons for each leg of the triad to destroy an ever expanding target list), neither side would chance war.[24] Some academic thinking had been done on limited nuclear war, but the prospects of controlling nuclear escalation were very problematic, and the chance of gambling and being wrong was unthinkable.[25]

Part of the problem raised by no cities was the implicit treatment of nuclear weapons as just another weapon. Normalizing nuclear weapons made their use thinkable and thus undermined security based upon deterrence, even though it made the actual deterrent threat more credible. Defence against nuclear weapons also fit into this category. Defending cities in the abstract enabled one to think about prosecuting nuclear war. In fact, American invulnerability might be restored with a reasonably effective missile defence. A large-scale pre-emptive attack on the vulnerable Soviet land-based ICBM fleet and its command and control capacity would leave a thick missile defence system with a small retaliatory force to defeat. The problem, however, was that under these conditions the Soviet Union would be driven to a pre-emptive stance of its own: a use them or lose them situation. If this logic worked for the Soviet Union, and it too was engaged in a missile defence development program, it followed that the United States would also be driven to strike first in order to prevent the attack or limit its damage. In such a situation, which Thomas Schelling called the mutual fear of surprise attack, the danger of nuclear war, especially in a crisis, an abiding feature of the Cold War, would be high. Neither party might intend to go to war, but both might be driven to war inadvertently because of the new generation of weapon systems. In effect, defences provided an incentive for both parties to go first, whether one or both possessed missile defences.

Alongside this stability argument, ABM raised concerns among the European allies, despite the logic of the link to the US flexible response proposal. Since the failure of the alliance to meet the 1952 Lisbon Force goals, NATO had come to rely upon nuclear weapons as the cornerstone of European defence.[26] Unable to match Red Army conventional forces, the alliance adopted the strategy known as the shield and the sword – the shield being existing NATO conventional forces that would hold the line as well as possible from a Soviet attack, while the sword, primarily American nuclear forces, delivered a devastating nuclear strike against a full range of targets up to and including Soviet cities. This two-pronged strategy would deter the Communist leadership from attacking the West.

The credibility of this threat, consistent with the overall American strategy of massive retaliation, hinged upon American invulnerability at home. As long as US population centres could not be targeted by a Soviet first or second (retaliatory) strike, the threat to bomb the Soviet Union back to the Stone Age appeared credible. Once the Soviet Union developed a capacity to strike at the US homeland, this credibility evaporated. This began to appear with the development of a Soviet long-range bomber force. However, the United States and Canada had developed beforehand an integrated air defence network, culminating with the operational establishment of NORAD in 1957.[27] While this network was not perfect, it did provide an effective defence, especially with the addition of nuclear-tipped air-to-air and surface-to-air missiles and the number of intercept opportunities stemming from the large distances involved and layered defences.[28] North America might not be invulnerable, but its air defence network provided a significant hedge against a Soviet assured destruction capability, especially given the relatively small Soviet long-range bomber fleet.

The same could not be said about ballistic missiles. Without any defence, US cities were directly vulnerable to a Soviet strike, making any US extended deterrent threat for NATO Europe inherently incredible. As such, the European allies, and in effect Moscow (and, in its heart, Washington itself), were forced to wonder if the United States would sacrifice Washington, New York, Chicago, or Los Angeles for Bonn, Paris, Rome, or Brussels. If the answer was no, then NATO's deterrence posture was mirage and bluff. For some time, American strategic analysts had warned about the incredibility of the threat of massive retaliation, especially in treating the Eisenhower administration's pronouncements literally. Instead, analysts argued in favour of a strategy of graduated, measured responses. This was the forerunner of the idea of flexible response in which allied conventional capabilities would be strengthened, and, in the event of war, escalation to nuclear weapons use from the battlefield up to the strategic level tightly controlled, with the use of American strategic weapons at the top of the escalation ladder.

Although a declaratory policy that suggested the use of nuclear weapons in a meaningful, controlled manner might restore some measure of credibility, it raised two problems for the Europeans. First, increasing conventional forces suggested an attempt to avoid the threat of escalation and could be interpreted as a signal to Moscow that a conventional war in Europe was possible with the prospect of a nuclear exchange limited to Europe. In other words, flexible response meant that the United States and the Soviet Union might agree to their homelands as sanctuaries. Second, ABM in the equation reinforced the idea of sanctuary by providing a limited defence against an actual attack against the homeland. Flexible response and ABM for the Europeans meant the decoupling of US strategic forces and the American homeland from the NATO Europe deterrence equation. In response, the Europeans fought flexible response and opposed ABM, and, for the French, the strategic situation became one of the central rationales for its independent nuclear forces, known as the force de frappe. If there was some tacit sanctuary deal between Washington and Moscow (a European and Chinese suspicion of bilateral arms control negotiations), the French could deter an invasion of France and end sanctuary by striking at Soviet cities. In other words, the force de frappe ensured strategic coupling, whether the United States liked it or not.

Ironically, one might have expected full allied support for replicating air defence with missile defence, thereby restoring strategic credibility. The more the United States felt defended, the more credible its threat. Defence worked as much, if not more, to ensure US strategic forces remained coupled to Europe as it did to decouple these forces and thus US security. Regardless, decoupling remained a staple of the European critique of US missile defence throughout the Cold War and up through the 1990s. The Europeans believed that missile defence combined with flexible response served to delink America's security from Europe's and at least tacitly suggested to the Soviets a US willingness to fight a limited war in Europe while superpower homelands remained safe. This image was reinforced by the emerging contradiction in overall US strategy: the rejection of graduated response for assured destruction as the governing condition in the Soviet-US relationship. Flexible response, as the European manifestation of graduated response in the US strategic debate, did not fit well with assured destruction, at least at the declaratory level. In the end, it did not matter. The very year McNamara announced Sentinel, NATO adopted flexible response as its strategic concept – open to national interpretation to fit national needs.[29] For Europe, missile defence remained decoupling.

The deterrence, crisis stability, and decoupling arguments took somewhat of a backseat to a more pressing reason to eschew the deployment of an extensive or thick ABM system. It was not cost-effective. Defences could be easily overwhelmed by simply building more missiles, at a much lower cost than building defences, especially with the development of multiple warhead technology then under way. In one study commissioned by McNamara, a $40 billion ABM investment could be offset by a $10 billion offensive response even before the technology had been developed to place several warheads on a single ICBM.[30] Moreover, the prospects of massive offensive and defensive investments were daunting, especially with the ever growing costs of the war in Vietnam. As the Soviet Union's ICBM arsenal grew into the hundreds and thousands, the costs of a defence promised to become exorbitant.

With these arguments against the deployment of a US ABM system, there remained the Soviet Union's development of its own missile defence system. Politically, this was not to be ignored. Indeed, ABM proponents, especially in Congress, consistently pointed to these developments as a justification for US ABM. There was a certain strategic logic here. If the United States was able to defend a significant amount of its hard-kill ICBM fleet from attack, its capacity to defeat a Soviet missile defence system in a retaliatory strike was more assured. Defences might promote stability, depending upon what one defended, and this partially informed the Safeguard decision. There was also an element of simple political symmetry – if one did it, domestic pressures for the other to do so too would be high. Even though the Soviet Union deployed its own operational nuclear ABM system around Moscow in 1970, McNamara sought to convince the Soviet Union to give up on ABM, and the means was a relatively new diplomatic tool: arms control.

Arms control had emerged in the early 1960s as the offspring of nuclear deterrence.[31] Although it would become readily confused with disarmament by the 1980s, it was primarily conceived as a means to ensure a stable strategic nuclear relationship. The pursuit of stability through arms control sought to reconcile two competing propositions – that while the United States and the Soviet Union were political adversaries, they shared a common interest in avoiding nuclear war. It followed that war would be a likely result only unintentionally. Simply, either one party or both parties might be driven to fight because of a belief that the other was going to fight, and this belief might be acquired as a function of the countries' respective force structures or the technical characteristics of the weapons themselves – the type acquired and manner in which they were deployed, which might be misinterpreted by one or both as an intent to attack.

At a minimum, some means were necessary to ensure that neither side came to the wrong conclusion, especially in the midst of a high-stakes political crisis. This required communication between them in order to develop the ground rules for a stable nuclear relationship. Stability came to revolve around a guarantee that neither side would develop the capabilities to eliminate the other's ability to retaliate, and this guarantee, in the context of the adversarial political relationship, meant negotiations to confirm intentions and communicate beliefs.

It was not assumed that arms control necessarily meant the elimination of weapons. Stability might require one or both parties to build more weapons. In other areas, such as defence, stability demanded both sides to forgo or limit missile defence systems, or at least so thought the Western arms controllers, President Johnson and McNamara. For McNamara, communication also meant the education of the Soviet Union, as Soviet professional military journals seemed to indicate that the Red Army did not understand the real implications of nuclear weapons. Finally, if successful, arms control served to eliminate (or at least manage) domestic pressures for spending on missile defences by eliminating the argument that Soviet investments necessitated American investment (which worked both ways as well).[32]

The arms control dimension fed into the European side of the ABM equation. In the wake of the third Berlin crisis and the Cuban Missile Crisis, East-West tensions began to decline, symbolized in part by the first nominal arms control agreement between Moscow and Washington – the 1963 Partial Nuclear-Test-Ban Treaty. Thus began the era of peaceful coexistence, which would in turn usher in the period of détente in the 1970s, with arms control as one of its centrepieces.[33] In this environment, NATO in 1967 released *The Future Tasks of the Alliance,* also known as the Harmel Report, which offered the East improved relations through negotiations on the balance of forces in Europe.[34] This offer was followed by the establishment of the Mutual and Balanced Force Reduction Talks in January 1973. In the end, the actual impact of arms control on the strategic military relationship was marginal, largely codifying the strategic relationship between Moscow and Washington and reflecting existing strategic plans. Nonetheless, its political impact was substantial in generating an image of arms control as a path to disarmament and the end of the Cold War that Western decision makers could not ignore. Simply put, arms control promised an end to the arms race, and missile defence posed a threat to this promise. Although in North America many of the details of the missile defence – arms race relationship waited for the 1980s and the Strategic Defense Initiative (SDI), the linkage was clearly established in the West European imagination. Missile defence threatened to reverse declining tensions by igniting an offensive-defensive arms race between East and West and destroying the prospects for arms control. This would return Europe to the worst days of the Cold War, a situation more domestically sensitive for West European governments than the United States, as would become evident as détente collapsed in the late 1970s and the so-called second Cold War began in the early 1980s.

With ABM confronting the range of technological, organizational, strategic, cost, and alliance barriers, at least in the mind of the Johnson administration and its secretary of defense, the only issue was convincing the Soviet Union of the benefits of forgoing missile defences. This would be started with the first direct high-level meeting of the two superpowers since the 1961 Vienna meeting of Khrushchev and Kennedy. In June 1967, on the occasion of an emergency meeting of the United Nations Security Council, President Johnson met with Soviet Premier Alexi Kosygin at Glassboro, New Jersey. The primary purpose was to discuss the situation in the Middle East following the Six Day War. Johnson used the opportunity in private one-on-one discussions to seek progress on the Nuclear Non-Proliferation Treaty then under negotiation and to propose direct discussions on nuclear weapons, including an ABM ban. Whether simply a function of the lack of a negotiating mandate, clear instructions, or an understanding of the complexities of the anti-ABM arguments, Kosygin not only balked at the idea but also publicly stated that defences were good: "I believe that the defensive systems are not the cause of the arms race, but constitute a factor preventing the deaths of peoples."[35] Nonetheless, some initial progress was made, only to be derailed by the combination of the war in Vietnam and the Soviet invasion of Czechoslovakia in August 1968. It was left to Nixon and the Republicans to push the arms control agenda forward.

Such was the state of affairs when McNamara announced Sentinel. In perhaps one of the strangest announcements ever made by a secretary of defense, McNamara provided the aforementioned list of reasons against ABM, only to then announce the decision to proceed. The rationale for the thin system of fifteen ABM sites (of which roughly twelve would defend major cities and three ICBM fields) at a projected cost of $5 billion would be Communist China. Having detonated its first nuclear device in 1964, the Chinese ICBM capability was still in its infancy. In the midst of the Cultural Revolution, having fallen out with the Soviet Union and fully supporting the North Vietnamese (with reports of Chinese forces actively engaged in air defence and other functions), Mao's China reflected the very image of an irrational, undeterable actor. Moreover, the Chinese a decade earlier had labelled the United States a paper tiger in ignoring the American nuclear threat. A limited system might not only be effective against the Chinese but might also be deployed cost-effectively before China deployed its first generation of ICBMs, estimated to occur in the early 1970s.[36] It also signalled the Soviet Union that, in proceeding, the United States would begin laying the foundation for a larger system unless an agreement was reached limiting ABM – an implicit bargaining chip to push the Soviets to the table.

As this new strategic world emerged in the 1960s, the new minority Pearson government, like all the allies, had no choice but to come to grips with ABM, and for Canada, unique among all the NATO allies, its implications for Canada's continental defence relationship. On February 4, 1964, the minister of external affairs, Paul Martin Sr., was informed of McNamara's announcement of the development of Nike-Zeus in his testimony to the House Committee on Armed Services.[37] In the memorandum to the minister, however, the department did not take any formal position or speak directly to the implications for Canada. This decision on the part of department officials was not surprising, although officials were certainly well aware of American ABM developments. In McNamara's presentation, details about the purpose and nature of the new surface-to-air missile were sketchy. In the absence of these details, officials were unable to provide any informed judgement. Already in 1964, the situation that haunted every successive government on the missile defence file was at play – if not involved in some official way with the research and development effort, it was very difficult to acquire the detailed information necessary for the government to understand the implications of a decision one way or the other. It was another year before Canadian officials were provided with a formal briefing on ABM.

The memorandum occurred not long after the divisive nuclear weapons debate in Canada, which made everyone sensitive to waving another red flag in front of the government, especially when External Affairs and National Defence knew full well that Nike-Zeus would employ nuclear warheads. The employment of nuclear warheads to intercept incoming nuclear warheads was obvious to all concerned. Certainly, Canada's own missile defence research effort had concluded that intercepting incoming warheads using conventional means was technologically unfeasible. Given that this was a joint Canada-US effort, Canadian officials readily concluded that the only remaining option was a nuclear one. Moreover, close defence links between Canada and the United States, especially between the air forces via NORAD, meant that at least some Canadians were somewhat aware of the nature of the Nike-Zeus development program, even if it was the US Army. As well, officials were aware of the unforeseen consequences of the US test of a nuclear device in outer space in 1962, which had led to the 1963 Partial Nuclear-Test-Ban Treaty. Indeed, officials had to wonder why the United States would create a system that could damage vital military and civilian assets in space, and how such a missile defence system was to be developed, let alone deployed, if it could not be tested.

Despite the absence of detailed information on Nike-Zeus, Paul Martin Sr., the secretary of state for external affairs, raised the issue with Hellyer, the minister of national defence, pointing out that Canada faced two questions: what role might Canada want to play, and what role might the United States seek from Canada?[38] In a covering note to the letter, the under-secretary of state for external affairs (today known as the deputy minister) noted that McNamara's statement was not in conflict with the 1964 Defence White Paper, which had identified the US "anti-ICBM effort" as the most significant development for the future of North American air defence.[39] Moreover, the government, presaging the future, concluded: "So far as this present paper is concerned, the point is that there is no major questions of policy in this area which are ready for resolution at this time."[40]

Following the first formal technical US briefing on ABM at the Permanent Joint Board on Defence in the spring of 1965, External Affairs began to look in earnest at ABM. In July, the Defense Liaison Division warned that the United States was approaching a decision on whether to proceed with ABM and identified a series of seven questions that needed to be addressed. Did Canada have the capability technologically, fiscally, and industrially to participate; should possible participation be limited to the defence of North America; would the United States view Canadian participation positively, negatively, or indifferently; were there other roles, such as detection, and command and control, that Canada could play if active participation was either impracticable or undesirable; should Canada stand aside and limit its defence cooperation to air and civil defence; how did current intelligence estimates bear on these questions; and would an amended NORAD be a suitable venue for Canadian participation?[41]

At roughly the same time, the Division of Continental Policy commenced its first study on the Canadian position on NORAD renewal, with an emphasis on the ABM issue. This study emphasized four key considerations.[42] First, the extent to which the United States would welcome Canadian participation and whether the United States would actually deploy were unclear. Second, it was assumed that geography indicated that the system would be more effective if Canada participated. However, there was no evidence that the United States would look to deploy elements of the system on Canadian soil, and thus it was essential to gain information about US deployment studies. Third, a valuable Canadian contribution could be made from technical, industrial, and financial resources. Finally, the paper noted that engagements of incoming interceptors would probably occur over Canada, resulting in significant radioactive fallout.[43]

Although the paper neglected to examine broader international implications, these four points became central to the debate for the next two years, especially the issue of whether the United States wanted Canada to participate (in other words, seeking signals on whether to extend an invitation or not). Most importantly, this issue came to be interpreted differently by External Affairs and National Defence. External Affairs feared that the United States wanted and/or needed Canada to participate. For National Defence, the fear was the exact opposite. The reasons behind these fundamentally different views unfolded over the next few years. For the moment, the political climate in Canada had yet to begin to shift toward significant opposition to close relations with the United

States, not least of all because American escalation of the war in Vietnam had just begun, and the rise of the left in Canada as a significant political force was still in the future.[44]

At the same time, there were concerns about the alliance dimension. The idea of NATO and Europe as a political counterweight to American dominance in the bilateral relationship had begun to gain in prominence. It cut two ways when it came to ABM, in light of growing anti-Americanism. On the one hand, it could legitimize domestically bilateral defence initiatives as part of the wider multilateral alliance, as done with Bomarc and in the 1980s cruise missile testing. On the other hand, if Europe and the United States were in opposite camps, as they were on ABM, it made a Canadian decision one way or another very difficult.

For Ottawa, the ABM debate suggested the possibility of a major conflict between Europe and the United States, placing Ottawa in the potentially uncomfortable position of having to choose between Brussels and Washington. However, this conflict never materialized. European concerns were echoed within the Johnson administration, which would have been evident in the manner in which ABM was briefed to the allies, and especially by McNamara, who had become a convert to the idea of MAD as the key to a stable strategic relationship with Moscow and thus to American security.[45]

The ABM arguments were certainly not lost on Canadian officials. Nonetheless, as momentum for ABM deployment started to build south of the border, Canadian officials were understandably concerned about the implications for the government if Canada did not participate. Canada was not Europe, and the issues Canada faced were fundamentally different – Europe did not share a continent with the United States. At the fundamental level, alongside obvious domestic political concerns, if the United States proceeded with ABM and Canada failed to join in, it "would create political difficulty of explaining either why such measures were not necessary in Canada while they were necessary in the US, or why we had chosen to rely in toto on the US for such defence."[46] Pearson after Diefenbaker, like Mulroney after Trudeau, and Martin after Chrétien, had set repairing damaged relations with the United States as a priority. Canadian participation in the US ABM system might not be the panacea for improved relations. However, ABM was the only new major US continental defence initiative that was likely to affect the relationship one way or another.

As NORAD was seen as the natural venue for Canadian participation in the US ABM system, the discussion came to focus on the implications of not participating. The debate was similar to Canada not participating in the joint air defence of North America – specifically, non-participation undermined Canadian sovereignty and marginalized NORAD's and Canada's role in North

American defence activities. Indeed, External Affairs' continental defence policy paper reiterated the long-standing importance of NORAD to Canada, implying (but not stating) that the future of NORAD was at issue.[47] If NORAD was at stake, so was Canada's access to classified US defence information and planning that went beyond any specific joint program or project. If ABM was central to US defence plans for the continent, participation ensured Canada access to these plans (conversely, non-participation meant the possibility of no access). In other words, ABM participation meant more than just ABM. It might well be the linchpin to overall Canadian access, ensuring Canada had the need and the right to know about US defence developments and planning, and knowing about US defence plans for the continent was an obvious strategic interest. Furthermore, ABM participation was a possible symbol of Canada's credibility as a faithful ally, with all the intangible benefits that might follow.

The long-term strategic implications might be deductively obvious, but the immediate, practical implications of participation were not. Officials implicitly recognized the dilemma confronting Canada, which would be repeated over the next four decades. It was vital for Canada to access US missile defence thinking and plans in order to understand the specific implications for Canada and make an informed decision about participation. At the same time, obtaining this access meant not only a Canadian decision to participate in US ABM studies under way but also a willingness on the part of the United States to allow Canada to participate and thus gain access to highly classified and sensitive data long before a deployment decision was made. Canada might be the United States' closest ally and logical partner in the missile defence of North America, but to share such sensitive information carried significant political liabilities in the absence of a larger Canadian commitment to deployment itself, particularly when filtered through the lens of the debate in the United States.

In effect, Canadian involvement, if it became public, had potential domestic political ramifications on both sides of the border, or so Canadian officials felt. Canadian involvement might be used in the US domestic debate to legitimize ABM. In Canada, it could be interpreted that the government was fully committed to ABM rather than just interested in obtaining more information. Thus, the officials had to manage as best as possible the need to know while avoiding a commitment to proceed on a joint basis. Until the United States decided whether it wanted or needed Canadian participation, officials had to proceed only on the basis of the limited information provided by the United States in formal briefings or through unofficial contacts, reports, rumours, and speculation. Unfortunately, the Division of Continental Policy reported that the US ABM studies were too advanced for Canada to participate in.

In addition to the ABM-NORAD link, another component emerged: outer space. Although the ABM system was the responsibility of the US Army, all three of the services, as well as the Central Intelligence Agency and the highly secret National Reconnaissance Office, were engaged in developing space-based military capabilities for a wide range of purposes. The direct ABM linkage was space-based capabilities for early warning, ground-based systems for space surveillance, and ground- and space-based systems for space defence, better known as anti-satellite weapons (ASATs). Besides the early warning satellites, a space surveillance system evolved from the late 1950s consisting of numerous ground-based sensors designed to identify and track objects in space, and these included the ballistic missile early warning radars as adjuncts. Evolving into the Space Surveillance Network, space tracking was essential to ensure that warheads were not confused with other objects in space (and vice versa) during their mid-course phase, leading to a major assessment error by NORAD. As for anti-satellite research, Nike-Zeus was tested in such a role through Project Mudflap, prior to its cancellation in 1965 and replacement by Nike-X. The air force effort included Project Defender, which involved the aforementioned Starfish Prime test; Project Orion, an air-launched system; Project Saint, a co-orbital ASAT; and Project Blackeye, the first laser anti-satellite research work.[48] All three areas of military space development one way or another engaged Canada. Even before the Defense Support Program infrared satellites, the ballistic missile early warning mission had been assigned to NORAD. The space surveillance assets, originally labelled the Space Detection and Tracking System, or SPADATS, began to feed data to NORAD in 1960 and included the subsequent Canadian contribution of two ground-based Baker-Nunn cameras at Cold Lake, Alberta, and St. Margaret's Bay, New Brunswick.[49] As a matter of technology, ground-based anti-satellite systems employed the same technology as ABM. Intercepting a missile outside the atmosphere was little different from intercepting a satellite. In fact, it was easier because satellites flew in predictable orbits and were much larger targets than warheads. With little imagination, Canada's artillery missile defence research of the 1950s might have an anti-satellite role. Nonetheless, anti-satellites and ABM were destined for the time being to be kept separate conceptually and politically. A decade later, attention was directed to anti-satellite weapons by Prime Minister Trudeau, with little concern for their missile defence application. Reagan and SDI changed all of this.

Ironically, it was anti-satellite weapons, not ABM, that received initial political attention in Canada. Hellyer was asked in the House about the United States developing a weapon's system to counter potential threats from orbiting satellites. At a subsequent cabinet meeting on May 22, 1965, Hellyer reported on

NORAD Baker-Nunn camera for satellite detection and tracking, Cold Lake, Alberta.
NORAD Neg# 69-114, Courtesy Canadian War Museum

speculation in the press about new US weapons developments that were outside the NORAD framework.[50] He confirmed that the United States was developing two new nuclear weapons programs: anti-satellite and anti-missile capabilities. It was possible that neither would be allocated to NORAD, which he suggested Canada should agree to. He further believed that this should be the government's response to any queries on these weapon systems. Pearson, however, argued that no such statements should be made that might complicate Canada's NORAD policy. Recognizing that present NORAD arrangements were insufficient to deal with either the anti-satellite or anti-missile issues, flexibility was essential until further details were available.

Subsequently, External Affairs informed the Canadian Embassy in Washington and the Canadian delegation to NATO that, for the time being, Canadian policy was not to participate in the US development program – even though no invitation had been formally extended – but to continue to evaluate it.[51] On the same day, the Defence Liaison Division informed the under-secretary of state, H.B. Robinson, that the United States had two anti-satellite systems under development. According to a newspaper article, one system had been deployed on an experimental basis with US Continental Air Defense Command but could be transferred operationally to NORAD.[52] More importantly, officials speculated that the United States was consciously seeking new weapon systems that would not entail Canadian involvement and could be deployed entirely on US territory. This speculation was seen as a product of the Canadian position in negotiations on authorizing NORAD to use nuclear weapons as a function of their acquisition for the Bomarc-B surface-to-air missiles and the Genie air-to-air missiles.

Concerns about exclusion continued to be reflected in External Affairs. The Division of Defence Liaison made reference to a possible Canadian contribution from the Cold Lake Baker-Nunn camera and Tracker radar at Prince Albert established by the National Research Council and the DRB. Both facilities were designed to track satellites in orbit, and their data could be made available to US anti-satellite programs. On July 10, Defense Liaison raised the anti-satellite linkage to ABM in relation to the ongoing review of NORAD.[53] Senior officials pointed out that a US decision on ABM was near, and the issue of bringing anti-satellites under NORAD needed to be considered carefully. With regard to Canadian participation in the US ABM effort, the government needed answers to several issues: an assessment of Canada's technical, industrial, and fiscal capacity to participate; the advisability of Canada to participate in a North American-only defence; US views on Canadian participation (for, indifferent, against); potential secondary roles or the advantages and disadvantages of only a limited role in conventional air defence and civil defence; current intelligence estimates of direct relevance; and whether the NORAD agreement could provide an adequate basis for Canadian participation and influence.

One year later, the situation had evolved. Although officials remained concerned about whether the United States needed or wanted Canadian participation, they were much better informed on the status of the ABM development program.[54] A recent visit of defence officials from Canadian Forces Headquarters and the DRB at US invitation had been very valuable. In particular, the briefings appeared to indicate that Canadian participation would result in a more effective ABM system, though the nature of a Canadian role remained unspecified.

At the same time, the NORAD renewal issue had started to gain momentum. Initially, the US Permanent Joint Board on Defence (PJBD) section suggested that NORAD renewal be placed on the agenda at its next meeting in the fall of 1966.[55] However, US sensitivity to the potential political fallout in driving the agenda forward led to a change in approach. In September, External Affairs reported that the "USA has been very careful not to bring any pressure to open the discussion on this [NORAD renewal] subject. What they appear to desire is that both sections sponsor the item."[56] Accordingly, External Affairs recommended to the prime minister that the forthcoming PJBD agenda include aerospace and the role Canada desired to undertake after 1968.[57] The department added that the United States had not made a decision on whether to augment passive defence with an active ABM defence. Pearson subsequently approved the agenda but emphasized that the Canadian position in the PJBD must not prejudice any future options for Canada one way or another.[58]

At the fall PJBD meeting on October 14, the US chair pointed out that no decision had yet been taken on the ABM issue.[59] He reported that it was the administration's desire to consult closely with its allies. In what was clearly a US signal, the representative noted that ABM would likely have implications for Canadian territory and airspace. The chair expected that Canada would most likely wish to talk about the subject at a suitable time in the future – an invitation that the Canadian team accepted in large part because it would entail a full official briefing on ABM. Finally, a State Department representative provided a briefing on the international strategic and political implications of the deployment of ABM and requested Canadian views. These included the implications of the system for deterrence and the East-West relationship, China and Asia Pacific, nuclear proliferation and the non-aligned, and arms control.

In the January 24, 1967, budget speech, President Johnson announced that talks might begin with the Soviet Union on ABM limitations. Following this announcement, a secret memorandum was sent to the Prime Minister's Office suggesting that NORAD and ABM could be treated as separate issues.[60] Pearson replied that he was willing to go to cabinet on NORAD renewal without complete information on the implications of missile defence deployments. In so doing, he rejected the view that NORAD renewal could be separated entirely from ABM and felt that a commitment to renew might include a commitment to support or participate in ABM. Pearson also requested that the US paper on the global implications of missile defence be sent to him. Finally, the prime minister expressed concerns about the possible presence of US forces on Canadian soil and the monetary costs of an ABM system deployed under NORAD. In response, the under-secretary of state for external affairs informed the prime minister that the United States had promised a formal ABM briefing within a few weeks and that McNamara would be speaking directly to Hellyer on the subject.

By now, the Canadian strategy and policy dilemma had a circular quality to it. The goal was to use NORAD renewal as a means to leverage more information on ABM. This information would lead to a decision on two fronts. The first consisted of obtaining at least some indication of US preferences on ABM

– was Canada needed or wanted? The second was to decide whether Canada should participate, and the answer to this depended upon US preferences. Whether Canada was needed, or just wanted, by the United States would determine the political benefits and costs of Canadian participation, and these, of course, included an assessment about the future of NORAD itself. Thus, the circle was closed, with NORAD renewal intrinsically linked to ABM and the issue of Canadian participation, even though External Affairs had suggested they could be decoupled.

However, for the time being, Canadian discussions with the United States were to occur with as little, if any, disclosure of the Canadian position on either NORAD renewal or ABM participation. The day before the PJBD's February 1967 meeting, Pearson informed cabinet that the Canadian delegation would express "no views whatsoever on an extension of the NORAD agreement, either in duration or in responsibilities and activities ... Canadian officials had [sic] been warned to avoid saying anything one way or the other about Canada's willingness to consider the expansion of NORAD or the development of special programs such as might, for example, relate to anti-ballistic missiles."[61]

Putting this strategy into effect was key to the upcoming visit of the prime minister to Washington on March 8.[62] In the wake of President Johnson's announcement on upcoming ABM talks with the Soviet Union, Pearson could inquire about the status of these discussions. In so doing, Johnson's response could provide an opportunity for the prime minister to place Canada's policy dilemma on the table. Canada could not proceed with considering NORAD renewal until the government fully understood the implications of a deployed ABM system and from there decide upon the role Canada might take in the defence of North America.

During this period, External Affairs was in the process of drafting a memorandum to cabinet.[63] The draft outlined the various ABM deployment alternatives and motives facing the United States and concentrated upon their potential implications for Canadian territory. Not surprisingly, it focused on the technical question of whether existing air defence systems had any relevance for US ABM plans and the fundamental issue of whether the United States required or desired Canadian participation. Importantly, the draft posited that the use of Canadian airspace would be unavoidable, likely drawing upon the predicted deployment sites near the Canadian border and the type of interceptor missile that the United States would employ. In addition, the paper noted that some protection would be provided to Canadian cities; the NORAD air defence logic of moving engagements northward was also noted, even though this logic did not really apply.[64]

With regard to policy options, the draft effectively identified three: complete independence (no involvement); limited cooperation but no weapons; and full cooperation. Officials raised the question of the ownership of the nuclear warheads under full cooperation, implying that full cooperation meant interceptors on Canadian soil and the possibility of some alternative division of responsibilities in joint North American air defence. The draft further posited that the NORAD agreement could be separated from the ABM question but warned that the United States might still seek Canadian participation. Finally, the draft concluded that a Canadian decision on ABM participation was not pressing because a US decision on deployment and system architecture would not likely be made until 1968.

External Affairs' belief in, or perhaps more accurately preference for, separating NORAD and ABM also stemmed from disarmament and arms control considerations. Although a participant along with the United Kingdom and the United States in the Manhattan Project, which had developed atomic weapons, the King government, and its successors, had all rejected the nuclear option for Canada. Instead, these governments had become passionate advocates of nuclear disarmament, even though Canada was a member of a nuclear-equipped alliance and its forces acquired a nuclear role in the 1960s.[65] The tension between these two realities reached its apex in the bruising fight over the Bomarc-B acquisition program, which divided External Affairs and National Defence and their respective ministers, Howard Green and Douglas Harkness, and which would lead to the latter's resignation and the fall of the Diefenbaker government in 1963.

In the case of ABM and possible Canadian participation, the tension this time revolved around the negotiations of the Nuclear Non-Proliferation Treaty (NPT) and the Outer Space Treaty (OST), signed in 1968 and 1967 respectively. Both agreements were negotiated under the offices of the United Nations – the NPT through the Eighteen-Nation Committee on Disarmament, and the OST through the work of the Committee on the Peaceful Uses of Outer Space, established in 1959. Canada played a major role in both negotiations.[66] However, the driving force for both agreements was superpower strategic interests. In the case of the NPT, both shared an interest in maintaining their near-exclusive dominance over the possession of nuclear weapons. The Soviet leadership in particular greatly feared the possibility that West Germany would acquire nuclear weapons in the context of the US-sponsored Multilateral Force proposal.[67] As for the OST, both were concerned about the strategic implications of deploying nuclear weapons in space, and the US in particular sought to legalize its space-based intelligence assets, Project Corona, by having outer space defined as international in nature.[68]

External Affairs had also taken the position that an agreement prohibiting the deployment of ABM systems should become a primary objective of Canadian arms control and disarmament policy. Missile defence and Canadian participation flew in the face of Canadian policy on non-proliferation and potentially undermined Canada's role in drafting the NPT.[69] Missile defences prompted an expansion of nuclear arsenals, and Canada's participation might brand Canada as a nuclear hypocrite in a manner not unlike the exposé of Canada's nuclear role in air defence and NATO. ABM nuclear detonations in outer space also butted up against long-standing policy desires to go much further than the OST, such that ABM participation might again prejudice Canada's ability to take a leadership role in a more extensive follow-up treaty and again could make Canada appear to be hypocritical in calling for a ban on weapons in space but participating in an ABM system designed to explode nuclear weapons in space.

To this point, External Affairs had been unable to convince senior officials, including the prime minister, that the NORAD and ABM issues could be kept separate, as there had been no clear indication of either US views or the mechanism for separating the two. This all changed with Hellyer's meeting with McNamara in Washington on April 5 in the context of a meeting of NATO's Nuclear Planning Group, which had ABM on its agenda. Privately, McNamara informed Hellyer that not only could and should NORAD be kept separate from ABM but also a codicil to the NORAD Agreement could be negotiated.[70] Furthermore, the United States was not requesting Canadian participation. Instead, the secretary notified the minister that, if Canada did want to participate, it could be factored into US planning, which was interpreted as US willingness to sell Canada ABM interceptors as an add-on to the US system.

The ABM issue had been brought up in the House of Commons the day before Hellyer's meeting with McNamara. In response to two questions, Hellyer separated alliance considerations from a national decision on participation, reassuring the House that it would consider the ABM issue before any government decision was made. He also expressed the hope for a US-Soviet agreement not to deploy ABM. If the latter did not occur, then Canada would have to look at the "technical details carefully."[71] Three days later in reporting on his Washington visit, Hellyer referred only to the Nuclear Planning Group meetings, stating that discussions including ABM had occurred but that "no decisions have been taken in this respect."[72] Another three days later, after the release of the generic Planning Group communiqué, he stated that he had nothing further to add and again reiterated a commitment to a debate in the House.[73] The next day in a speech on amendments to the National Defence Act, Hellyer raised three points on the ABM issue. The first referred to the costs of the system, estimated at $20 billion but likely to double if a decision was made to proceed. Second, he noted McNamara's concerns about such expenditures and his differences with the Joint Chiefs of Staff, which was a strong ABM advocate. Finally, he delinked NORAD from the US ABM program by referring to it as a US Army project, while aerospace defence cooperation was an air force responsibility.[74] Except for three subsequent opposition references to the ABM issue, none of which raised Canadian strategic defence issues, missile defence disappeared from the House of Commons until the fall of 1968.

One could detect the hand of the civil service in the minister's response. Drawing from the Washington meeting, External Affairs concluded that a deployment decision remained in the distant future, likely after NORAD renewal, and the separation of NORAD from ABM was possible as a function of inter-service considerations in the United States. Above all else, the answer to the two burning questions was now available. The United States did not need Canada for its ABM system, and it was indifferent as to whether Canada became involved or not – a position the United States would subsequently adopt with all its NATO allies. Importantly, McNamara had answered the prime minister's concerns about separating NORAD renewal from the ABM issue.

With the issue of access to US thinking and planning having dominated the ABM issue, the separation of ABM from NORAD now clearly raised the issue of Canada's strategic interests in missile defence. It was now reasonably clear that a US system would provide only a limited defence of Canadian territory, which would not likely extend to all of Canada's major cities or other strategic Canadian targets. Even so, little immediate attention was paid, and NORAD renewal continued to dominate discussions in cabinet and the bureaucracy. In effect, it appeared that a decision not to participate had been the existing policy assumption all along, even though no such decision had ever been made. The ABM policy decision was strictly about the future of NORAD relative to the Canada-US defence relationship rather than about the possibility of defending Canada. This was reflected in the importance of the two questions regarding whether the United States needed or wanted Canada to participate and what may be assumed as a great sigh of relief after the April 5 McNamara-Hellyer meeting.

In May, further details were made available to Canadian officials after another meeting with McNamara that included Hellyer.[75] McNamara informed the Canadian delegation that US ABM plans did not include a requirement for installations on Canadian soil. In providing details on what would become the Sentinel light ABM program, he noted the likely construction of five perimeter acquisition radars close to the Canadian border, with the interceptors assigned to Detroit likely able to defend Windsor as well. Moreover, the longer-range

Spartan interceptor with a five-megaton warhead set to detonate between fifty and one hundred miles above the surface would cause little ground damage occurring beyond a twenty-five-mile radius. Overall, the system would be able to cover roughly 50 percent of the Canadian population at a cost of US$5 billion over five years, and the United States was prepared to discuss the exact siting of Sentinel components with Canada to maximize protection and minimize risk for the Canadian population without decreasing protection for the US population.[76]

On May 31, Paul Martin Sr., with the support of Hellyer, submitted a memorandum to cabinet on the ABM and NORAD renewal issues in advance of the PJBD meeting in Montreal. The next day, cabinet met to provide marching orders to the Canadian chair, with the expectation that US officials would seek to discuss NORAD and obtain a possible Canadian commitment to renew. In the discussions that followed, opposition to NORAD emerged from the president of the Privy Council, Walter Gordon, and the minister of justice, Pierre Trudeau. The former noted that, if the United States proceeded with ABM, there would be growing pressure to defend Canadian cities as a function of the relationship. Citing growing opinion in Canada against NORAD and the air defence of Canada, he felt Canadian NORAD participation "to be somewhat unreal and unsubstantial" and noted that the "forthcoming negotiations in relation to continental air defence might provide an opportunity for withdrawal."[77] Trudeau supported these views and endorsed Gordon's call for a thorough review of Canada's defence relationship with the United States, especially with respect to nuclear weapons – an implicit reference to Bomarc and Genie.

Martin and Hellyer rejected these views outright. Both argued that the current defence arrangement was limited to air defence. Hellyer noted that first there was no relationship between air defence and the ABM issue; second, the United States had no plans or need to deploy ABM components on Canadian soil; and finally, the question of Canadian participation was a Canadian decision only. He further added that the political, economic, and military costs of Canadian withdrawal were substantial. Martin Sr. also argued that, in light of the possibility of the ongoing Middle East crisis (the Six Day War) evolving into a global conflict, continental air defence remained essential.

Pearson concluded the meeting by noting that a full examination of the defence relationship was necessary, following the PJBD meeting. For the time being, it was vital that Canada not commit itself to any details other than a generic one to continue air defence cooperation. As such, cabinet agreed that the Canadian PJBD chair, Arnold Heeney, be instructed to tell US officials that Canada planned to continue air defence cooperation and was open to beginning negotiations on NORAD renewal. Without saying so bluntly, the government had begun the process of accepting McNamara's proposal to separate NORAD from ABM. In communicating this decision, the prime minister further requested that Heeney find out if the United States would entertain a new clause in the agreement specifying that the NORAD agreement was entirely independent from ABM. Such a clause would imply that a Canadian decision to renew NORAD did not entail a commitment to participate in ABM. All of this was agreed to at the subsequent PJBD meeting, where the US chair, speaking first, stated that it was US policy to renew NORAD in existing form and to keep it separate from ABM.[78] On July 14, Heeney informed the prime minister that the United States would agree to an ABM exclusion clause as part of the renewal.[79]

However, the results of the meeting did not include a Canadian commitment to renew NORAD per se. Following instructions, Heeney replied at the meeting that Canada sought to continue close cooperation with the United States but reserved its position on the extent and nature of such cooperation. This raised concerns regarding renewal within the American delegation. In preparations for the next meeting, the prime minister suggested that Heeney inform the United States that, even though it should not expect an automatic renewal, renewal was never really in doubt. The problem was the image and process of the ABM and NORAD issues being linked together in the context of growing anti-American sentiment because of the Vietnam War.[80]

This did not mean that Canadian policy on the ABM issue was formally decided. Separating ABM from NORAD did not mean that the government had formally decided on its position on ABM itself, though the very nature of the debate suggested that the government would say no if it had to. Knowing that an operational ABM system was still years away, the issue became whether Canada should close the door entirely. Two options emerged. The first was to announce that ABM and NORAD were separate, and the independent issue of ABM would be dealt with in the future when necessary. Arguably, this position would manage concerns that might emerge if the public wondered why American cities were being protected but not Canadian ones. The second was to go further and indicate that the government under the current circumstances could not envision deploying ABM in Canada. For External Affairs, the preference was the former, as it would provide more flexibility.

On September 5, 1967, External Affairs and National Defence submitted a follow-on memorandum to cabinet on NORAD renewal to obtain "authority to negotiate with the United States ad referendum to cabinet the renewal of the NORAD Agreement and approval of the terms of reference for the negotiations [sic]."[81] In making what had become the traditional arguments supporting NORAD, the document reiterated the decoupling of NORAD renewal from the ABM issue: "Canada should not become committed through these arrangements to participate in a possible active United States ballistic missile defence."[82] On September 12, cabinet met to authorize Canada's formal position on NORAD renewal for the forthcoming negotiations at the PJBD meetings on September 18-22.[83]

Drawing from the September 5 memorandum and the report of the Cabinet Committee on External Affairs and National Defence, which had met the day before, cabinet agreed with the recommendations calling for a five-year renewal, a one-year notice of withdrawal, and an ABM exclusion clause. In the discussions leading to these conclusions, Hellyer provided a clear exposition on the state of US ABM thinking a week before McNamara formally announced Sentinel. The United States would proceed with a light ABM system designed primarily to defend US ICBMs located mostly in the Mid-West and against the growing threat from China, which had made unexpected significant technological advances. Domestic pressure in the United States made it necessary to develop an effective defence against the estimated Chinese threat of twenty-five ICBMs by the early 1970s. The US ABM system would not require Canadian territory, though the system would provide limited defence for some Canadian cities. As well, the United States was planning to seek an arms control agreement with the Soviet Union that would limit ABM deployments. Finally, Hellyer emphasized that US plans for an ABM system did not signal a US abandonment of deterrence as the cornerstone of its strategy to deter Soviet aggression, as US scientists had concluded that no effective defence could be created against a massive Soviet attack.

Pearson, in fully endorsing the renewal of NORAD, stated the desirability of Canada not to participate in missile defence and support for any effort to ensure that the United States and Soviet Union avoid a competitive defence-driven arms race, implying that Canadian participation would be at odds with Canada's nuclear, arms control, and disarmament policy. Echoing future policy considerations, he noted some empathy with US concerns about China but, like Hellyer, attributed these concerns more to domestic politics than to an immediate strategic nuclear threat from China. He concluded by stating: "As the adoption of this form of defence by either or both of the great powers would not assist in any way in deterring an attack by one against the other, it would be in the interests of peace to convince both sides that the system would not work and would have devastating economic consequences."[84]

In some ways, the decision on NORAD renewal and the ABM exclusion clause was the easy part. What would be of greater difficulty was how the government would handle the more complicated issues raised by McNamara's announcement. Hellyer, for one, preferred an ambiguous, qualified response indicating that the issue was under review, so as not to pre-empt Parliament. Pearson preferred a statement before Parliament opened on the 25th but after the conclusion of the PJBD negotiations. The cabinet tentatively approved this statement to be made prior to the opening of Parliament: "Based on information presently available, Canada does not contemplate active participation in an anti-ballistic missile defence system."[85] In other words, Canada would say no but not necessarily no to the US Sentinel ABM proposal.

On September 18 – the day the PJBD met and McNamara announced Sentinel – cabinet returned to the question of a public statement.[86] As discussed at the previous meeting, the draft noted that, on the basis of available information, Canada did not contemplate active participation in the US ABM program. It added an expression of regret that the Soviet Union had not agreed to US proposals for offensive and defensive arms negotiations. The cabinet discussions also addressed the economic dimensions of the ABM question. The finance minister, Mitchell Sharp, made it clear that the financial situation facing the government effectively foreclosed participation as well. Bud Drury, the minister of industry, expressed his concern that the announcement should not prejudice the possibility of Canadian firms winning contracts in the US program, particularly by exploiting developments that had occurred in the space sector. Hellyer made clear the US position that there would be no allied involvement in ABM for security and balance of payment reasons, and as a result there would be no contracts for Canadian firms, regardless of the government's position, unless sites were located in Canada for construction work. He also raised the concern that, if "active" was included in the statement, some people might come to think there was some form of hidden involvement – a concern that would haunt National Defence officials two decades later, following the Mulroney decision.

Four days after the McNamara announcement and three days after Pearson moved Hellyer to Transport, replacing him with Léo Cadieux, who had been associate minister of defence, the prime minister issued a public statement on Sentinel at a press conference in Ottawa: "We have no intention *at this time* of taking part in any such ABM system ... Naturally, we are keeping the matter under careful review. We do not wish to commit the government to any particular course of action in the future as to what might be the best solution to the security problem that Canada will face."[87] Three days later in the House, Pearson was a little less equivocal in voicing the closest thing to an actual no: "We are not involved in any way, nor do I think we shall be involved in the particular project ... Nor do I have any information to indicate that the US government will be asking us for facilities of any kind on our soil."[88] In effect, the government said no in the absence of a US invitation from a US administration skeptical of its own proposed ABM system. Publicly and formally, however, the government had not actually said no. Lingering concerns about the potential implications if the United States did proceed – for both the relationship and the possibility of a domestic demand for defence – dictated ambiguity. In the end, there might be an ABM architecture, but there was no deployment date per se, no invitation, and thus nothing for Canada to decide formally.

Even though the ABM question had been managed in the cabinet discussions of September 1967, opening the door for the quick and painless renewal of NORAD, some uneasiness remained, as reflected in the final two cabinet meetings on NORAD, held on February 8 and 27. The explicit purpose of these meetings was for External Affairs and National Defence to obtain approval to proceed with NORAD renewal. Approval in September had been premised on a forthcoming US *aide-mémoire* at the PJBD meeting that had been delayed because of the US defence review, which included the ABM program, followed by McNamara's statement on US air defence on January 28, 1968. Regardless, at these cabinet meetings, officials not only clearly expressed how and why Canada could have its cake and eat it too but also provided a harbinger of what would become a significant shift in the next government's attitude toward the US relationship.

At the first meeting, the new minister of national defence, Léo Cadieux, recommended renewal on the grounds that it would not commit Canada to any new types of defences (essentially ABM).[89] He further informed his colleagues that the United States was developing new options and new detection capabilities to support a defence against a missile attack from China, without the need for facilities in Canada and by employing an airborne detection system.[90] He added that the US ABM system could provide some degree of protection to Canadian cities in the event of a ballistic missile attack. He then suggested that a full defence briefing be provided to cabinet, at which point the issue was placed on hold.[91]

At the second of these two meetings, Martin began the discussion by seeking cabinet authorization to begin formal NORAD renewal negotiations given that the agreement was set to expire on May 12.[92] He reiterated the three parameters of the government's negotiating mandate: NORAD would be renewed in its present form; the agreement would be for five years rather than ten; and the negotiations would not touch on the issue of ABM and Canadian participation in a future operational system. He stated that US officials had agreed to the last two conditions. He then raised the possibility of deferring the decision and passing NORAD renewal on to the new government after the Liberal Party leadership convention scheduled for April 3. However, he warned that doing so would affect the willingness of the United States to provide economic assistance to Canada and would damage Canada-US relations in other areas.

Cadieux reinforced these arguments by reporting that the United States was deeply concerned that Canada's contribution to North American defence had been reduced to the bare minimum. NORAD was in Canada's national interest, and, echoing long-standing arguments, NORAD granted Canada with an air defence at much lower cost than otherwise possible and provided direct access to US defence plans. Most importantly, these two key benefits were the product of one factor: Canada's geostrategic location increased US warning of an air attack by several hours. This was held to be so essential to the United States that Canada could use it to its advantage. Cadieux emphasized: "So vital was this factor in US minds that Canada could even use it to negotiate agreement on a level of impact for the US ABM defence so as to provide maximum protection for Canadian cities."[93]

However, these arguments did not go unchallenged. One dissenting voice raised the implications of the relationship for Canada's relations with other countries.[94] In addition, the political implications of NORAD were not a one-way street that just favoured Canada. NORAD provided the United States with an avenue to pressure Canada that might not otherwise exist. Finally, fears about a slippery slope to participation in ABM were expressed, even though it would be outside NORAD. Specifically, "there were grounds for suspecting that Canada would be under additional pressure because of NORAD membership."[95]

Cadieux replied that the United States did not believe that Canadian participation was necessary, there were no incentives to move the system into Canada, and the United States was willing to negotiate the level of impact on Canada as a means to obtain access to Canadian airspace for early warning purposes. Supporting Cadieux, Hellyer, now in Transport, stated that there was no need to fear being dragged into the ABM warning system. Drawing from earlier discussions with the United States when he was minister of national defence, he noted that the United States would have to pay the entire costs if the system was physically extended into Canada. Besides, the United States was not interested in moving the system into Canada. The only crucial issue was to seek an agreement to prohibit low-altitude nuclear intercepts over Canadian cities, as these would be nearly as harmful as a direct attack.

Thus ended the 1968 NORAD renewal debate. Pearson concluded the discussion by injecting one key factor – cost – that would become a fundamental consideration in future missile defence discussions. He accepted the assurances of the three proponents that NORAD renewal would not commit the government to engage in a new, expensive, and far-reaching defence program with the United States. On that note, cabinet agreed to renew NORAD, followed by an order-in-council on May 12 and an exchange of notes with the United States. But the ABM debate was far from over and in some senses only about to begin. Beneath the surface of the NORAD renewal debate was a distinction between ABM as a weapons system consisting of nuclear-tipped interceptors and ABM as a defence system consisting of the operational planning and employment process – a distinction first raised by External Affairs in 1965. The latter in particular included early warning for, and command and control of, the proposed ABM system. In other words, the meaning of participation and the actual interpretation of the new ABM exclusion clause were still open for consideration.

This distinction was important politically, economically, and strategically. Politically, Canada could participate without actually participating and thus keep the nuclear side of the equation at arm's length. Economically, Canada could avoid the costs associated with the ABM system itself. Strategically, the nation could obtain access and input into the operational defence plans for ABM, thereby potentially optimizing the defence of Canadian cities. Above all else, the newest major North American defence initiative would come under the binational command, thereby reinforcing a North American approach to defence rather than a national one – an issue of some concern with the perceived declining value of air defence in a world of ballistic missiles and the growing significance of outer space as a domain that negated territorial significance. In other words, the future for NORAD didn't look very good if it was outside the primary defence developments of the future.

Politically, the interpretation of participation was picked up early on by a former defence minister from the Diefenbaker years, Douglas Harkness. The exchange with Paul Martin Sr. in testimony before the Standing Committee on External Affairs, which followed defence minister Cadieux's statement that the expectation was that ABM would be under a different US command than NORAD, revealed the actual state of ABM thinking in the government, even though it was not to be exploited politically after the fact:[96]

Harkness: "NORAD headquarters would still be, we will say, the command centre and the chief warning centre for missile attack."

Martin Sr.: "But there is no commitment on our part in the event of renewal [NORAD] to participate in an anti-ballistic missile defence."

Harkness: "No, but in view of the fact that the deputy commander of NORAD is a Canadian and might very well be in command at NORAD when a missile attack took place, there really is participation, whether we like it or not."

Martin Sr.: "I mean there is no commitment by Canada of resources."
Harkness: "This is a different thing. The two concepts have to be separated – any commitment on our part as far as providing hardware is concerned, and our participation, on the other hand, from a command and warning point of view."
Martin Sr.: "Yes."[97]

With the resulting confusion on the government's exact position on the ABM issue, Cadieux returned to the formal line: "There was no approach to ask Canada either to participate or not participate ... This is the stage at which we are now. So there is no known negotiations, no propositions; it is just that we are facing possible technological developments. The Americans have indicated that they plan to introduce an ABM system on their territory. They have not requested us to do anything about it."[98] Following this testimony, the issue was dropped entirely by the committee until the next year.

National Defence officials had long been concerned about missile defence operational planning issues and how to ensure access to critical information. One approach was to seek to optimize ABM to ensure an adequate level of protection for Canadian population centres, even if External Affairs officials were correct in believing that the Chinese were not purposely targeting Canada.[99] To meet this goal, Canada would need to obtain access and input into US ABM deployment and employment planning. McNamara might have promised consultations and Canadian input, but he had made it clear that the allies would not have access to highly sensitive secret information. For National Defence, the means to get access was through command and control. If it was assigned to NORAD, it would require involvement in operational planning. Just as NORAD provided a binational solution to continental air defence planning, it could provide the same solution for ABM.[100] In so doing, Canada would have both a need and a right to be involved in planning for the operational use of Sentinel, and this then extended deeper down into access to the highly classified characteristics and capabilities of the ABM system. As this did not include any direct involvement in the ABM sites, a US Army responsibility, this would not necessarily violate the NORAD missile defence exclusion clause. Thus in some senses was born the question of the meaning and nature of participation and the possibility of participating without participation – the equivalent of Mackenzie King's credo "not necessarily conscription, but conscription if necessary."

The problem, however, for National Defence was that US officials were interpreting the ABM exclusion clause as a no to Canadian participation in ABM across the board, and this included command and control. Certainly, Canadian officials might suggest to their US counterparts that no actual policy decision on ABM had been made, and certainly on the surface they were correct. But with all the angst surrounding ABM in the NORAD renewal process, evident in PJBD discussions and Hellyer's meetings with McNamara, there was little doubt in the United States' mind that Canada was out. Foreshadowing the future, the ball was in the Canadian court to reopen the ABM door, and in this proponents of Canadian participation had one key ally: the US commander-in-chief of NORAD.

Shortly before the September cabinet decision and the Sentinel announcement, the NORAD commander, General Reeves, met with senior Canadian officials, including the ministers of external affairs and national defence. In preparation, External Affairs officials informed the minister that Canada would have to take the initiative. The United States would have to be formally told that Canada's view of the ABM exclusion clause "does not mean that Canada is indifferent to the arrangements of C^2 [command and control] nor that use of NORAD for this purpose [C^2] is necessarily excluded."[101] At the meeting, Reeves made it clear that he did not see the ABM exclusion clause as prohibiting a NORAD command and control role for ABM and emphasized the need for bilateral discussions on the issue.[102]

The roots of this possibility partially drew from the outer space and space defence issues, which had in many ways been more central to NORAD-US Air Force concerns than ABM had, partially because ABM was under army jurisdiction, whereas space and space defence were under that of the air force. For Canada, the space issue had first emerged in the early 1960s regarding NORAD's terms of reference. Despite the air force's space-based anti-satellite/missile defence research programs, the existing terms of reference were restricted to air defence even though NORAD's early warning mission extended into missiles and space with the operational deployment of the Ballistic Missile Early Warning System and Space Detection and Tracking System providing information to NORAD. Both the Canadian chairman of the chiefs of staff and the minister of national defence informed External Affairs that they agreed with the NORAD commander's request that NORAD's mission include responsibility for space defence under the umbrella term "aerospace."[103] External Affairs agreed but expressed two concerns. First, that if this new term of reference became public, it would embarrass the ongoing work of the United Nations' Committee on the Peaceful Uses of Outer Space.[104] Second, External Affairs wondered if a new term of reference was possible under the provisions of the existing agreement or whether the agreement would need to be formally amended.[105] On October 15, 1963, the under-secretary of state for external affairs informed the minister that any change to NORAD's terms of reference required a formal amendment and suggested the issue be referred to the Cabinet Defence Committee. Before the issue moved any further, the United States interjected that it would be premature to undertake diplomatic negotiations prior to internal US deliberations, thereby stalling any formal discussions on the terms of reference and the issue of amending the agreement.[106]

The issue remained dormant until the following year, just after the beginning of US ABM briefings to the PJBD. At the Military Cooperation Committee meeting in late 1966, US military officials raised changing NORAD's terms of reference from air to aerospace defence, citing the role of the Space Detection and Tracking System in NORAD's Integrated Tactical Warning/Attack Assessment mission.[107] At this time, a possible future missile defence mission for NORAD was raised and tied to the need for new terms of reference. At the January 1967 meeting, the Canadian chief of defence staff's support for new terms was communicated to the US military, but no discussions on ABM and NORAD appear to have taken place. At the October meeting, reflecting the understanding already reached between Canada and the United States on excluding ABM from NORAD, Canadian officials stated that Canadian policy was to renew NORAD for a period of five years without any concern for, or reference to, ABM/Sentinel.

At the Military Cooperation Committee's next meeting, soon after the renewal of the NORAD Agreement in May 1968, two interpretations of the ABM exclusion clause appeared. The Americans interpreted the clause to mean that the ABM system in any sense could not be part of NORAD. The Canadians disagreed. The exclusion of ABM command and control, not specifically mentioned in the clause, was not implicit in the exchange of notes governing the renewal. It was necessary to examine the issue of command and control of the proposed US ABM system. If US military authorities sought to include it in NORAD's mission, then Canada would like to consider the possibility. Canadian military officials also noted their responsibility to be involved in ABM operational doctrine planning owing to the possible direct or indirect implications of an ABM intercept for Canadian territory and citizens. For the moment, Canada had no policy on this because of a lack of information on US plans. Once the ABM Sentinel program moved further forward, Canadian military officials expected the United States to present ideas about how planning could occur on a bilateral basis. In the subsequent September meeting, Canadian military officials again reiterated their responsibility in being involved in ABM planning.

External Affairs was sensitive to the practical problem of separating ABM from air defence, especially in tracking a range of public statements in the United States suggesting such an interrelationship.[108] In the fall of 1968, officials were looking for answers to three key questions after the discussion of NORAD's terms of reference at an earlier PJBD meeting: the technical feasibility of separating air and space defence, the possibility of separating the two in the same complex in Cheyenne Mountain, and the implications of a separate ABM command and control system on NORAD effectiveness.[109] At a subsequent PJBD meeting in Colorado Springs following NORAD renewal, General Reeves again argued that space and air defence could not be separated and that command and control for both, including ABM, should be assigned to NORAD.[110] He further noted that, as a result of the ABM exclusion clause, all Canadians had been excluded from ABM planning. Finally, he suggested that such an arrangement could be based solely upon a change in NORAD's terms of reference from air to aerospace defence, which had been agreed by the binational Military Co-operation Committee and reported directly to the US chairman of the Joint Chiefs of Staff and the Canadian chief of defence staff, with no need to change the agreement. As the committee was tasked with managing the technical-functional implementation requirements of the binational arrangement, adding ABM command and control was just such a requirement, one that extended into the early warning function for the ABM system.

The exclusion clause hampered Canadian officials in obtaining information from the United States on whether Sentinel would be programmed to protect Canada as well as on the blast effects of ABM nuclear intercepts over Canada. To pry open the door, External Affairs officials suggested that the relative silence of US State Department representatives at the Colorado Springs PJBD meeting could be interpreted as an implied invitation.[111] They also suggested that Canada agree to adding space to the NORAD mission; this would bring ABM interceptors under NORAD without requiring Canada to commit to host interceptors on Canadian soil. And it would ensure Canadian input into the ABM planning and employment process, thereby maximizing coverage for, and limiting damage to, Canadian cities. National Defence concurred, adding that, if ABM command and control went to US Continental Air Defense Command, the erosion of the NORAD concept of joint command would begin – an argument NORAD and National Defence had regularly made during the NORAD renewal debate.

The Opposition also picked up on the question of the exact meaning of the exclusion clause and its implication for Canadian participation. On October 21, Ed Schreyer (NDP) raised in the House of Commons the question of differences between the United States and Canada on the interpretation of NORAD renewal as it concerned missile defence. Cadieux sidestepped the question by noting Canadian interest in technical developments of Sentinel relative to North American defence.[112] A month later, in the context of the government's announced review of foreign and defence policy, Schreyer raised the issue twice, which led the new prime minister, Pierre Trudeau, to confirm that the US ABM system would not fall under NORAD and that there were no difficulties between the two parties on interpreting the agreement as it related to ABM.[113]

Although NORAD and the US military in the Military Cooperation Committee might agree that the ABM exclusion clause did not answer the question of ABM command and control, neither necessarily spoke for the Pentagon, State Department, or White House. Besides, in November 1968, the Republicans won the presidency, and the new administration brought with it a somewhat new perspective on ABM. The ABM-reluctant Democrats – led by their chief doubter McNamara – had been replaced by the Republican Party, which had supported a thick missile defence, under the leadership of Richard Nixon, who had initially gained some notoriety as a junior senator during the McCarthy anti-Communist witch hunts in the 1950s, served as Eisenhower's vice-president, and been narrowly defeated by Kennedy in the 1960 presidential election. However, whatever fears there may have been in Ottawa (and elsewhere) of a harder right-wing foreign and defence policy – which might lead to an expanded ABM effort far beyond Sentinel – were quickly dissipated. Indeed, one of Nixon's first major policy decisions was to announce the replacement of Sentinel with Safeguard. An ABM thin defence was to be even smaller, and eventually reduced to only two initial interceptor sites, which would be subsequently codified in the 1972 ABM Treaty and then further reduced to one with the 1974 Protocol.[114]

McNamara's thin defence, Sentinel, was to consist of fifteen sites deployed around US cities. Nixon's Safeguard reduced the number of sites to twelve and refocused the effort somewhat to the defence of US ICBM fields.[115] The initial deployment was to take place at Malmstrom Air Force Base in Montana and Grand Forks Air Force Base in North Dakota – the home of two ICBM fields. Nixon's announcement on March 14, however, set off a major debate in Congress, which brought the issue back to the forefront in Canada. Even though Safeguard passed the House by one vote on August 6, and the Senate by a margin of two votes the next day, the future of Safeguard remained uncertain, making the Canadian government further reluctant to commit one way or the other.

While proceeding with Safeguard, the Nixon administration also moved forward on the arms control front. However, the Canadian government's opportunities to comment were constrained by its NATO commitments not to say anything that might compromise the US negotiating position.[116] To this end, External Affairs in a prepared statement for the prime minister spoke of the Strategic Arms Limitation Talks as a virtue that did not affect stability or entail unacceptable risks for Canada.[117] Not surprisingly, External Affairs did not raise ABM or NORAD, not least of all because of the new prime minister's well-known opposition to ABM; the department had already been taken to task for its

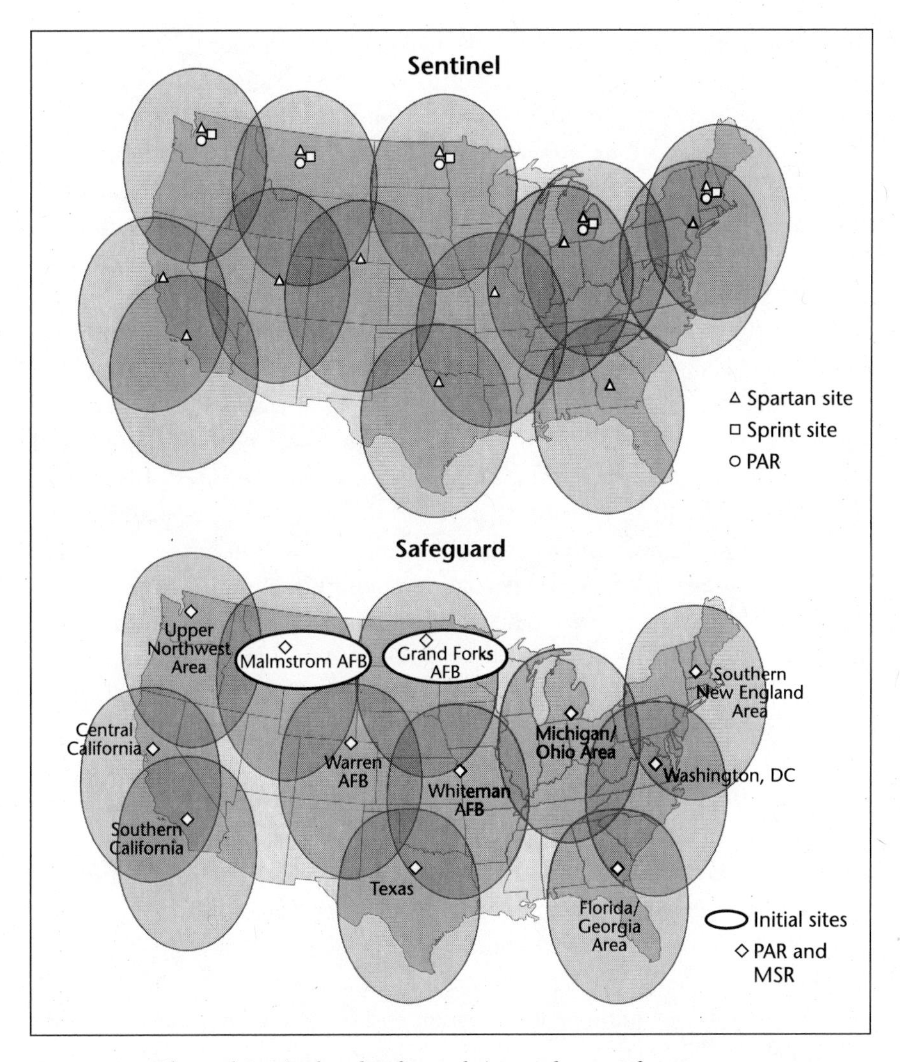

FIGURE 1.1 Planned Sentinel and Safeguard sites and areas of coverage
Source: ABM: An Evaluation of the Decision to Deploy an Antiballistic Missile System, ed. Abram Chayes and Jerome B. Wiesner (New York: Signet Broadside, 1969), 43, 45. Originally published in *Scientific American.*

foreign-policy review document that had legitimized the status quo, much to Trudeau's dislike. Indeed, much to the regret of External Affairs, Trudeau would not let the ABM issue go. On February 2 and May 8, Trudeau publicly criticized US ABM plans. In both cases, External Affairs was deeply concerned that his criticism would enmesh Canada in an internal US debate and be used as ammunition by the anti-ABM side, and it implicitly worried about the impact on the government's relationship with the Nixon administration.

On May 5, 1969, Cadieux met with the new secretary of defense, Melvin Laird, in Washington.[118] At the meeting, Laird stated that Safeguard was moving forward on the assumption that Canada did not wish to participate, based upon the ABM exclusion clause and other public statements. He did, however, raise the NORAD command and control issue, thereby opening the door to movement. But Cadieux, while reporting that the government had not taken a formal position on the ABM issue, provided no direct answer to the command and control question. In the end, Laird, presaging the next four decades of the missile defence question, concluded by expressing an openness to have Canada engaged, though in a manner that conveyed US doubts that Canada would do so.[119]

Ten days later, an interdepartmental paper on Nixon's Safeguard proposal was forwarded to the prime minister; the secretary of state for external affairs, now Mitchell Sharp; and the minister of national defence.[120] The paper identified and addressed five implications that Safeguard would have for Canada. First, Safeguard would not upset deterrence stability between the United States and the Soviet Union. Second, Safeguard, and the Soviet Union's ABM system defending Moscow, Galosh, did constitute a further development in the arms race. Third, Safeguard would not likely be detrimental to arms control. Fourth, the risks to Canada from the possible detonation of an ABM warhead were relatively small, especially when compared to the destruction from a nuclear attack. Finally, it was possible that Safeguard could protect Canadian cities, though this had not been broached with the United States.

In turning to the Canada-US defence relationship, External Affairs and National Defence agreed that the 1968 NORAD exclusion clause was not a Canadian decision on missile defence per se but was only to avoid an automatic Canadian commitment to the developing US system. Numerous questions about the ABM system remained, including early warning and command and control, which might include NORAD. However, the United States had made no decisions, and officials (likely led by National Defence) warned that even the extension of NORAD's terms of reference to include aerospace defence might be insufficient to obtain any level of Canadian participation in US ABM planning, and Canadian access to intelligence information on aerospace activities might still be threatened.[121]

Shortly after the interdepartmental paper was forwarded, Sharp laid out the ABM issue in a letter directly to Trudeau.[122] In assuming that Safeguard might not proceed, he argued that there was no need for Canada to take a final position. In the subsequent memorandum to cabinet submitted in July 1969, which included the interdepartmental paper as an annex, preliminary discussions with US officials related to the command and control question indicated that the

United States was unwilling to re-examine NORAD's terms of reference at that time. Alongside a reference to unspecified political concerns, it was recommended that there was no need "at this time to re-examine the Canadian position as stated in the caveat to the NORAD Agreement."[123]

Foreshadowing the position taken by National Defence thirty years later on Clinton's National Missile Defense program, the memorandum raised numerous technical and operational questions that the United States had not yet addressed and reiterated doubts about whether Safeguard would actually proceed. As such, any decision would certainly be premature. In reality, any attempt by National Defence to push the issue forward would most likely lead to a negative decision. The unspecified other political concerns related to growing Canadian nationalist sentiment that targeted in many ways the United States as a threat to Canadian independence and sovereignty. This sentiment was also strongly opposed to the US war in Vietnam, as demonstrations in the streets of Canada mimicked similar demonstrations in the United States. Any decision to participate in ABM was certain to ignite significant domestic opposition. Moreover, the new Liberal majority government under Trudeau reflected this nationalist sentiment. At best, the Liberal Party might be divided if the issue was forced. At worst, it would be united against Canadian participation. Besides, it was highly unlikely that any recommendation to signal the United States that Canada was open to participating via NORAD in command and control of Safeguard or any other US ABM program would receive the support of the prime minister. Pearson and his cabinet had been concerned about the implications of ABM for the future of NORAD and the Canada-US defence relationship, which had led to a cautious approach. Trudeau was on record as an opponent of ABM and had questioned the NORAD arrangement.

Trudeau had always been on the left wing of the Liberal Party, and his views on strategic defence questions were in some ways closer to the NDP (though he did not necessarily share the left's position on neutrality). He was more in line with McNamara in concerns about ABM – implicitly at least accepting the logic of MAD reflected in some degree by the nationalist thrust of the 1971 White Paper and the reduction of the Canadian military presence in Europe. Elements of this were also at play in the arguments Trudeau subsequently laid out in the House. The initial occasion was the issue of the implications for Canada of the proposed US interceptor site to defend Detroit.[124] This site would also defend (or from the perspective of the NDP threaten) Windsor. In a reverse of previous concerns during the Pearson government about the amount of protection Canada might obtain from Sentinel, Trudeau turned the question upside down. Particular Canadian interests in obtaining more protection were secondary to global peace and security interests. Trudeau stated: "I do not want, in other words, to negotiate with the United States for a little more protection for Canada, if I feel that that protection is such that it might entail more danger for the whole world."[125]

However, neither Trudeau nor anyone else in the government ever answered the question on the amount of defence Sentinel or Safeguard might provide for Canada. Instead, the government increasingly became more nuanced in its responses to ABM questions in following the advice from External Affairs that blunt criticism of US plans served little value and might indeed be dangerous. Mitchell Sharp, the external affairs minister, in a memorandum to the prime minister, argued: "The ABM issue is relatively quiescent in the U.S.A. and to volunteer a statement on ABM (for example, a statement on motions) might stir up political controversy in Congress and perhaps irritate the United States administration needlessly without any real prospect of influencing its recent decision to deploy the first limited phase of Safeguard."[126] In the end, the government would only reiterate that NORAD was separate from Safeguard and had no role in US decisions but would continue to track the issue from the outside looking in.

From there, the ABM discussion began to wane. The final thrust occurred in July 1970, when cabinet was informed of the proposal to limit ABM defences to national capitals in the ongoing Strategic Arms Limitation Talks (SALT) and with it the likelihood that the United States might abandon Safeguard.[127] In the memorandum to the prime minister for the meeting, National Defence noted that both the United States and the Soviet Union were developing missile defences and that the three primary purposes of Safeguard were to defend ICBMs and bombers, protect against an accidental Soviet launch, and counter the limited Chinese threat.[128] Defence also argued that it was unlikely that the United States would seek Canadian cooperation, such that missile defence would not likely become an area of Canada-US defence cooperation. However, Defence cautioned that, if SALT failed, the United States might proceed with a thicker and/or a mid-course-phase ABM system. If so, the United States might seek Canadian territory and cooperation.[129]

During this period and in the wake of Nixon's Safeguard announcement and as part of Trudeau's review of Canadian foreign and defence policy, the Standing Committee on External Affairs and National Defence also turned its attention to North American defence centred primarily upon NORAD issues – a function it would regularly perform in the lead up to future NORAD renewals. In so doing, the committee turned brief attention to the ABM question. In its initial hearings, committee members raised the issue of the relationship between ABM and NORAD with Lieutenant General Sharp, the deputy commander of

NORAD. Consistent with the relative vagueness of Canadian policy, and the implicit assumption that the 1968 NORAD exclusion clause did not mean a Canadian no to participation, Sharp sidestepped the question of NORAD assuming an operational command role, placing the onus on the United States: "Since the US has not decided to go ahead with it [Safeguard], they cannot possibly have made a decision like that."[130] He suggested that it would be improper for him to comment on the current American debate on Safeguard but did offer his own view that NORAD should acquire operational control, as the air defence and missile defence missions were functionally identical and both relied upon NORAD early warning data.[131] He did, however, note that NORAD was not necessary for operational command.[132]

Sharp would be followed a week later by Dr. George Lindsey, chief of the Operational Research and Analysis Establishment (ORAE) beneath the Defence Research Board (DRB). Lindsey had had a long distinguished career with the defence research establishment dating back to the development of radar in the Second World War and had served as Canada's representative on the High Level Group supporting the NATO Nuclear Planning Group. Lindsey provided the committee with a three-part study on strategic considerations entitled *Strategic Weapons Systems, Stability, and Possible Contributions by Canada.*[133]

In many ways, Lindsey's brief and testimony served as an educational primer for the committee on the complexity of strategic deterrence issues, especially when it came to missile defence. In particular, he clarified (implicitly at least) some of the ambiguity of Canadian policy and the likely parameters of Canadian participation. First, Canadian territory could be valuable for the location of future warning and tracking radars to provide in-flight missile, bus, and warhead tracking; obtain a more accurate determination of trajectories and targets; and provide dispersed radar coverage to protect against possible nuclear blackout.[134] Lindsey noted, however, that no request had been made by the United States in this regard. Second, little would be gained from forward-deploying interceptors on Canadian soil, because Safeguard was a terminal system and had to be deployed close to the targets it was designed to defend. If, however, the United States moved to a mid-course intercept system, the situation would be different.

Finally, on the question of the amount of defence Safeguard might provide to Canadian cities, Lindsey argued unequivocally that it would be very little. At best, Toronto and Montreal might be at the periphery of Safeguard's defensive umbrella, depending upon a variety of considerations. On the related issue of intercepts over Canada, Lindsey recognized that the Spartan missile would certainly detonate over Canada, but whether that would occur in Canadian airspace or international outer space was for lawyers to argue over. The coverage provided by the short-range Sprint would depend upon how close it was deployed to the border. In this regard, Lindsey noted that the use of Spartan or Sprint would occur only after a hostile act by the Soviet Union in launching ballistic missiles against North America. Pre-empting part of the questioning by the committee, Lindsey emphasized that the damage from an intercept above Canada would be much less "than suffered by an ICBM burst, saving lives, resources, and radiation effects."[135]

Lindsey was followed by Dr. Tom Sherman, a Canadian political scientist with the Hudson Institute in the United States. His brief provided a detailed overview of the various positions and options within the United States on ABM in general and Safeguard in particular.[136] Although seen by members of the committee as pro-ABM in many ways, especially when asked about the amount of funding the institute received from the US Department of Defense, Sherman raised two interesting points. First, in response to the question of whether more installations on Canadian soil would make Canada more of a target, he replied that Canada was part of the Soviet Union's integrated North American target list, and ABM would make no difference.[137] Second, he clearly noted that, whatever Canada did or did not do or say about ABM and Safeguard, it would have no effect on the outcome of the US debate.[138]

In its questioning, the committee covered the range of international strategic and bilateral policy questions on ABM, with an underlying theme being the question of what the United States might or might not want from Canada if Safeguard proceeded. What was missed entirely was any consideration of the utility of a limited missile defence shield for Canada. The issues of costs, effectiveness, deployment options, and access to the American technology (as promised in 1967 by McNamara) were not raised by any members of the committee. Of course, it is not surprising that neither Sharp nor Lindsey broached these issues, given the potential political implications in the absence of high-level approval. Even though neither Trudeau nor Pearson had ever bluntly said so, the idea of a missile defence for Canada remained beyond the pale.

In the end, the Standing Committee on External Affairs and National Defence recommended to the House nothing of value on ABM/Safeguard. The report sidestepped the NORAD question except in noting the logic, but not the necessity, of the operational command link between air and missile defence. It recommended no change in the absence of ICBM early warning and detection installations on Canadian soil. And finally, and most indicative of Canadian policy throughout most of the life of the missile defence issue, it stated: "In view of the uncertainty about whether the US will proceed with ABM and the extent to which Canada might be asked to participate, if at all, the committee is unable to make any recommendations concerning Canadian involvement in such a system."[139] Echoing Canadian policy thirty years later, there was as yet no US decision and no invitation and thus nothing to decide.

By the summer of 1970, the ABM issue was drawing to a close without the government having ever publicly said no to participation. No was a default position, and by then, with the United States having dropped the question of the revision of NORAD's terms of reference, US officials had obviously come to recognize that Canada was out. In a final brief to the PJBD on the state of Safeguard on June 23, 1970, the US representative went even further by stating that there was no significant value in even fielding ballistic missile early warning radars in sites in northern Canada. Satellites were more effective. NORAD would subsequently acquire the early warning mission for Safeguard, relying upon the US early warning system, and command and control fell to the US-only Continental Air Defense Command. The early warning mission would never, however, receive any public scrutiny. The 1971 Defence White Paper would simply reiterate these points, including the absence of a US invitation, while applauding the tentative ABM agreement between the United States and the Soviet Union, which had been announced on May 20.[140] All would be quiet on the missile defence front until March 23, 1983.

Act 2

The Strategic Defense Initiative: Much Ado about Very Little (1972-85)

ON MARCH 23, 1983, President Ronald Reagan took the opportunity of a national television address to surprise the American public – especially the US nuclear weapons freeze movement – and shock members of his own administration – especially the Joint Chiefs of Staff – by proposing to launch a massive research and development effort to defend not just the United States but also the entire world from the spectre of a ballistic missile thermonuclear war.[1] Thus began the most controversial defence program in American history – the Strategic Defense Initiative (SDI), popularly known as Star Wars. Roughly a year after the announcement, the Strategic Defense Initiative Organization (SDIO) was created to coordinate the effort. A year later on March 26, 1985, Secretary of Defense Caspar Weinberger issued an open invitation for all the NATO allies to participate. Six months later, on September 7, Prime Minister Mulroney officially declined the offer but agreed to allow Canadian companies to participate in the SDI research effort – a decision that had been hinted at in various circles at least six months earlier and was subsequently formally proposed by the senior US defence officials in private discussions with Arthur Kroeger, a senior Canadian civil servant whom the prime minister had tasked with the SDI question.

In 1967, McNamara had made ABM safe for Canada. In 1985, senior US officials led by Under-Secretary of Defense Fred Iklé made SDI safe by proposing the very policy announced by the prime minister in September and viewed as entirely unproblematic by the president. Unbeknownst to both, the decision would come to have a significant negative effect on the Canada-US defence relationship in the years to follow and play a major role in Canadian assessments of President George H. Bush's 1991 plan to develop and deploy a two-layered missile defence system based upon the SDI research effort – Global Protection against Limited Strikes (GPALS). Above all else, SDI became the touchstone in every future Canadian debate about missile defence, and the arguments directed against SDI by Canadian critics would return on each occasion to haunt the discussions, even after the Cold War had become ancient history. ABM may have resonated unconsciously in the SDI era and beyond, but Star Wars would never be forgotten.

Even before the signing of the ABM Treaty in 1972, missile defence had fallen completely off the Canadian political agenda. There were no missile defence references in the public domain or the House of Commons until shortly after Reagan's speech, and this amounted to only a single question. Trudeau would call for an anti-satellite weapons ban at the United Nations Second Special Session on Disarmament on June 18, 1982, but he made no linkage to the commonality of missile defences and anti-satellite weapons (ASAT) systems. There was also little public reaction to the removal of the 1968 ABM exclusion clause with the 1981 NORAD renewal.[2] Canadians, including government officials, cognizant of the issue felt comfortable under the ABM Treaty safety blanket in which the United States and the Soviet Union agreed "not to transfer to other States, and not to deploy outside its national territory, ABM systems or their components limited by this Treaty."[3] Besides, there were no indications that the United States was even thinking about resurrecting ABM, following the cancellation of Safeguard in 1975.

Reagan's election on a platform of restoring US military strength reinforced concentration on the modernization of US strategic nuclear forces, which had begun with the Carter administration. At the time of the SDI speech, the dominant strategic issue facing the United States concerned the basing mode for its next generation of ICBM, the MX Peacekeeper. Internationally, attention was focused upon NATO's controversial Intermediate-range Nuclear Forces (INF) modernization program. In Canada, the most prominent issue was the air-launched cruise missile testing agreement with the United States, which would allow the United States to launch test missiles from B-52 bombers in the Far North and track their progress down the Mackenzie Valley Corridor to termination at the Cold Lake test range. For all intents and purposes, missile defence was non-existent until Reagan's speech, and it would take another year and a half before it exploded onto the Canadian scene, in late 1984.

Following the end of the Safeguard debate in 1970, the early warning mission for the brief life of Safeguard was assigned to NORAD. On the surface at least, this new mission required some form of Canadian consent or acquiescence. On June 1, 1970, cabinet discussed NORAD's early warning mission but did not appear to have considered the possibility of NORAD undertaking the early warning mission for Safeguard.[4] Nor did the discussions indicate a formal US request. Instead, cabinet dealt with three specific US proposals: the refuelling of US strategic bombers over Canadian territory, the dispersal of USAF (United States Air Force) air interceptors onto Canadian bases, and the establishment of an alternative NORAD command centre at North Bay. Three years later, cabinet ignored Safeguard and the early warning mission in the process of considering the two-year renewal of NORAD in 1973, though the 1975 renewal acknowledged NORAD's space surveillance and aerospace warning of attack missions.[5]

The early warning mission for Safeguard fit easily into existing NORAD early warning mission protocols, such that no formal change to its terms of reference was necessarily needed, and if no change was needed then cabinet had no role to play. US Continental Air Defense Command was already one of the recipients of this information. With it assuming command of Safeguard, providing this information from this US command to the missile defence command site was a US decision, not Canada's. Moreover, as Safeguard employed nuclear weapons and was a terminal-point defence, a release decision required authorization from the US National Command Authority, and the time available for this decision-making process was little different from the decision to release US strategic nuclear retaliatory forces under Strategic Air Command – another recipient of NORAD early warning information. In other words, just as Canada through NORAD's early warning mission kept the US strategic nuclear deterrent at arm's length, so too could the same arrangement easily function for Safeguard. Assigning Safeguard early warning to NORAD was also consistent with Canadian defence policy. The Trudeau government had established the principle of Canadian cooperation to maintain a stable nuclear deterrent relationship between the United States and the Soviet Union. With early warning vital to ensure that the Soviet Union could not pre-empt US land-based retaliatory forces, Safeguard, as a point defence for these forces, was clearly consistent with this principle, whether this was truly understood or not by cabinet and the prime minister.

At best, then, the issue was a technical one and at most the responsibility of the Canada-US Military Cooperation Committee. Besides, why raise a politically contentious issue when there was no need to do so? Missile defence was off the agenda, and it was best to keep it that way. One thing was clear, however: a precedent was established. The early warning mission for missile defence was outside the ambit of the ABM Treaty and its Article 9 prohibitions on third-party involvement, even though the treaty did have formal restrictions on the location of early warning radars.[6]

Missile defence beyond Safeguard, however, had not entirely disappeared. As early as 1974, National Defence was aware of research on the possible use of lasers in a missile defence role from a seminar at the University of Toronto given by Dr. Merher of the Hudson Institute.[7] National Defence officials were also aware of pressure in the United States to respond to Soviet missile defence research.[8] Nonetheless, as far as Canadians were concerned, the ABM Treaty and the US decision to cancel Safeguard in 1975 had put missile defence to rest. There were no major dedicated US missile defence research programs. Defence research programs that did have some public profile and possess missile defence applicability were nicely embedded in the ASAT envelope.

As ballistic missile research begat missile defence research, so satellite research begat ASAT efforts. Both the Soviet Union and the United States began developing and testing ASATs in the early 1960s. The Soviet program evolved into a co-orbital ASAT system, with testing having begun in 1968.[9] It consisted of a killer satellite launched into the same orbital plane as its intended target. After three or fewer orbits, the killer would catch up to its target and detonate at close range, sending a mass of metal or pellets outward to collide with the target satellite. The kinetic energy generated by the collision would be sufficient to disable or destroy the target. Of course, this type of ASAT would have no missile defence utility because of the time involved from ASAT launch to target interception.

The United States took a different approach, while also conducting research on a co-orbital interceptor. The US effort was in part informed by 1960s intelligence reports of a Soviet fractional orbital bombardment system, in which nuclear warheads would be launched on a southern trajectory, reaching partial orbit, before tracking from the south to US targets – effectively negating the US ballistic missile early warning network and planned ABM systems. To deal with this possibility, alongside generic missile defence requirements, US intercept efforts concentrated upon direct ascent weapons, which included the testing of the Nike-Zeus, the predecessor of Nike X and the Spartan interceptor, in an ASAT role. Such efforts effectively blurred the missile defence–ASAT functions. The capacity to launch a direct-ascent warhead to intercept a satellite in low earth orbit was little different from that required to intercept a ballistic missile warhead during its mid-course phase in outer space. Both missile defence and ASATs subsequently benefited from the US Army's Homing Overlay Experiment between 1983 and 1984 – the forerunner of today's kinetic-kill missile defence warheads.[10] The United States also developed and deployed a high-altitude F-15 air-launched kinetic-kill interceptor in the 1980s to strike at satellites in low earth orbit.[11] And both the United States and the Soviet Union proceeded with research on new exotic ASAT technologies with missile defence application – particle beams, lasers, non-nuclear electromagnetic pulse, and electromagnetic rail guns.[12] In the US case, SDIO would absorb all these research efforts.

It was only after SDI emerged onto the public agenda in late 1984 that critics made the ASAT–missile defence linkage in seeing Star Wars almost exclusively as weapons in space. Closely related, opponents also pointed directly to the 1981 deletion of the ABM exclusion clause from the NORAD agreement as part of the grand conspiracy. Either the Trudeau government had been duped by defence bureaucrats into dropping the ABM exclusion clause or had willingly gone along with the decision to reopen the door into missile defence. In reality, the process leading to the elimination of the clause began with informal discussions and got onto the agenda by mutual agreement, as was the case with the first official ABM briefing in 1965 and the first formal discussion of NORAD renewal in the PJBD in 1966. Both Canadian and American officials readily agreed that the clause had become wholly redundant after the signing of the ABM Treaty, and no one even contemplated the possible death of the treaty. A 1979 US State Department draft text did not contain the clause. National Defence officials felt that it was an unnecessary irritant in the relationship, with the protection provided by the ABM Treaty. In a subsequent memorandum to Mulroney on SDI, officials bluntly stated that the idea had been a Canadian initiative, motivated in part by the view "that the exclusion clause might be taken to imply that the USA had planned to deploy an ABM system in violation of the [ABM] Treaty."[13] As well, it was thought that dropping the clause might help Canada's relationship with the new administration, especially given the deep ideological divide between Trudeau and Reagan.

Regardless, the missile defence issue itself was simply off the political and strategic agenda at the time of the renewal, making the decision to drop the exclusion clause a cost-free one at least for the time being. Instead, the ASAT question dominated, and, as had been the case with ABM, this naturally led to the question of NORAD's and thus Canada's role. Following the same logic as ABM of participating without participation, National Defence felt that ASATs posed few problems for NORAD and Canadian personnel. Although the United States preferred to operate its ASAT capability from the Space Defense Operations Center colocated with NORAD, command and control would be a US-only function under the US commander-in-chief Air Defence rather than under the commander-in-chief NORAD. Even so, it did raise the question of Canadian NORAD personnel in the Space Defense Operations Center and the auxiliary Space Computation Center because of the early warning mission.[14] Regardless, early warning was not ASAT, and the ASAT mission was USAF only. Canadian defence officials were relatively unconcerned: "Since ASAT launch and control, wherever and by whomever executed, would depend on NORAD data inputs, then so long as Canadian/NORAD personnel [were] not involved in the actual control sequences, conditions of the NORAD agreement would be satisfied."[15] Moreover, National Defence felt that there was no definite Canadian position on ASAT and space-borne air defence until Trudeau's formal call for an ASAT ban in 1982. Previously, there had been no formal public government statements since the 1967 Outer Space Treaty, and nothing bearing on space defence except dormant Canadian opposition to the weaponization of space, which implicitly referred to weapons deployed on orbit in space.[16] The USAF ASAT effort was not weapons on orbit.

Following Trudeau's call for an ASAT ban, the Cabinet Planning and Priorities Committee agreed to table a formal proposal in the Conference on Disarmament.[17] While External Affairs agreed the proposal should go ahead, Alan Gottlieb, the Canadian ambassador to Washington, voiced strong objections. In direct conversation with the prime minister, Gottlieb suggested that the proposal be stopped cold because of US sensitivities and broader concerns related to the current international political, strategic, and arms control environments. It would contradict the allied pledge of solidarity in the midst of the INF modernization debate.

Reagan's landslide 1980 victory had emphasized rebuilding American military strength after a decade of neglect. He called for a reinvigorated set of policies and capabilities to challenge the Soviet Union. For Reagan, détente was a one-way street. It had served only to embolden the Soviet leadership, as witnessed by the Soviet Union's penetration into Africa employing Cuban proxy troops, growing involvement in Latin America, and invasion of Afghanistan. Central to policies designed to challenge the Soviet Union worldwide would be the rebuilding of the US Armed Forces after the post-Vietnam era of the hollow army. Central to rebuilding would be the modernization of US strategic nuclear forces – a program that had actually begun in the Carter years, which Reagan took full credit for.[18] Strategic modernization, in turn, was driven not only by Reagan's goal to restore US military strength, alongside the inexorable life cycles of weapon systems and technological developments, but also by concerns about an unfavourable and dangerous strategic balance in favour of the Soviet Union.

In the 1960s, McNamara's decision to put a ceiling on US strategic forces through the enunciation of the criteria for assured destruction – the capacity to destroy 50 percent of the Soviet industrial plant and kill 20 to 25 percent of the population by each leg of the US strategic triad of ICBMs, SLBMs, and bombers – had been partially motivated by an expectation that this would constrain the Soviet Union's strategic force levels. Recognizing that the Soviet Union would not accept numerical inferiority forever, the limit on US forces would provide a ceiling for Soviet forces. Once parity was achieved, a foundation would be laid to institutionalize parity – arms control – and perhaps in the distant future move toward a reduction in the levels of strategic forces. Parity would create the conditions of strategic (numerical) and thus political equality for negotiations. The negotiations, which would begin in earnest with Nixon in 1969, led to the SALT basket of agreements – the Agreement on Measures to Reduce the Risk of Outbreak of Nuclear War (1971), the Strategic Arms Limitation Talks treaty (SALT I), and the ABM Treaty (1972). The SALT process and agreements also served on the strategic-political level as the centrepiece of the

Nixon-Kissinger strategy of co-opting the Soviet Union by recognizing it as a legitimate equal and thus a status quo power. Internationally, and especially among the European allies, the SALT process and agreements were the symbolic cornerstones of détente.

Too many observers would mistake détente with the end of the Cold War and SALT as a harbinger for disarmament rather than a strategic adjustment to the US strategy of containment and the codification of existing US and Soviet strategic plans and force levels. Regardless, SALT I and its successors, the 1975 Vladivostok Accords and the 1978 SALT II agreement, served only to cap the number of strategic launchers (ICBMs, SLBMs, and bombers), while allowing warhead numbers to grow. This growth was the direct product of the development of multiple warhead technology – multiple independently targetable re-entry vehicles, or MIRVs, which had begun to replace single-warhead missiles by the time of SALT I. The backbone of the US ICBM fleet, the Minuteman, deployed three MIRVs on each missile. The Soviet Union deployed much heavier ICBMs, such as the SS-18, which possessed much greater throw weight than the Minuteman. Each SS-18 could carry roughly twelve warheads.[19]

Throughout the 1970s, the United States continued to possess more strategic warheads than the Soviet Union, though the Soviet Union had more launchers (aggregate missiles and bombers). Thus, it wasn't the overall number of Soviet warheads that raised concerns among US strategic analysts per se. It was the overwhelming concentration of warheads on heavy ICBMS (see Table 2.1). This concentration, in contrast to the more balanced US distribution in which SLBMs made up a significant proportion, raised concerns in the United States about the vulnerability of its ICBM deterrent. In strategic deterrence thinking, ICBMs were seen primarily as first-strike weapons. With a large concentration of its strategic forces invested in heavy ICBMs, the Soviet Union appeared to be seeking the capacity to eliminate the US ICBM fleet in a pre-emptive first strike. Such a capacity would threaten the stability of the strategic relationship, with significant political implications, as voiced most directly by a bipartisan group of expert US strategic analysts in the Committee on the Present Danger.

The Committee on the Present Danger was established in 1976 and led by Paul Nitze, a long-standing senior analyst on strategic issues who had worked for several administrations dating back to Truman, and by Eugene Rostow, who had served as the under-secretary of state for political affairs in the Johnson administration.[20] As a result of Soviet ICBM deployments and the codification of launcher limits in SALT I and Vladivostok, the committee argued that US strategic forces were vulnerable to a Soviet first strike. They estimated that the Soviet Union could eliminate 90 percent of the Minuteman ICBM force with only 50 percent of its ICBMs. After such an attack, the president would face two

Table 2.1

Strategic arms limitation agreements and US-Soviet Union strategic nuclear forces

		United States				Soviet Union			
		SALT I		Actual		SALT I		Actual	
		Launchers	Warheads	Launchers	Warheads	Launchers	Warheads	Launchers	Warheads
1972	ICBMs	1,054	–	1,054	1,726	1,409	–	1,530	1,600
	SLBMs	656	–	656	5,107	9,50	–	494	481
	Bomber	–[a]	–	455	7,360	–	–	140	596
	Total	1,710	–	2,165	12,363	2,359	–	2,164	2,815
		Vladivostock		Actual		Vladivostock		Actual	
		Launchers	Warheads	Launchers	Warheads	Launchers	Warheads	Launchers	Warheads
1975	ICBMs	1,320[b]	–	1,054	2,251	1320[b]	–	1,618	2,277
	SLBMs	–	–	656	6,586	–	–	784	869
	Bomber	–	–	432	6,911	–	–	135	596
	Total	2,400[c]	–	2,142	15,748	2400[c]	–	2,537	3,743
		SALT II		Actual		SALT II		Actual	
		Launchers	Warheads	Launchers	Warheads	Launchers	Warheads	Launchers	Warheads
1980	ICBMs	1,128[d]	–	1,054	2,251	1,128[d]	–	1,398	5,630
	SLBMs	–	–	656	5,914	–	–	1,003	1,636
	Bomber	–	–	316	6,239	–	–	156	596
	Total	2,250[c]	–	2,026	14,404	2,359	–	2,557	7,862

a A dash refers to no amount specified.

b Number of multiple warhead (MIRVed) launchers permitted.

c Total number of strategic nuclear delivery vehicles (launchers) allowed.

d Includes 820 multiple warhead ICBMs and 308 heavy multiple warhead ICBMs allowed.

Sources: International Institute for Strategic Studies, *The Military Balance* (London: International Institute for Strategic Studies), 1971-72, 1975-76, 1979-80; warhead totals from Natural Resources Defense Council, *Archive of Nuclear Data*, www.nrdc.org.

untenable decisions. He could either order immediate retaliation with the surviving ICBMs, the full SLBM fleet, and strategic bombers against Soviet cities or surrender. If the United States retaliated, the Soviet Union would undoubtedly strike back at American cities, which had been spared from the first strike given the location of the US ICBM and strategic bomber fleets in the rural western parts of the United States. In such a situation, the president would have little choice but to surrender in order to spare the bulk of the American population.

Their point, lost on many critics, was not the esoteric question of nuclear war. Rather, it was the effect of this strategic situation on the management of the political relationship between Moscow and Washington. The Soviet Union with strategic superiority could become more politically aggressive in its global dealings with the United States, whereas the United States would become more reluctant to stand firm. Indeed, Soviet and US behaviour in the late 1970s could be pointed to as evidence of this. Even worse from a stability perspective, during a crisis, US vulnerability to a first strike would create incentives for the Soviet Union to go first and as a result drive the United States to go first – the "mutual fear of surprise attack" condition, as argued by Schelling a decade earlier.[21]

Alongside these strategic concerns, US nuclear deterrence strategy had been under close scrutiny for some time. McNamara's rejection of a no cities strategy had left the United States with few options in the case of war. Critics argued that a strategy of assured destruction relying upon the threat of a massive retaliatory strike in a condition of MAD was no strategy at all, and in the absence of a strategy to fight a nuclear war, if deterrence failed, it implicitly lacked credibility. In other words, the credibility of any threat was dependent upon making one's opponent believe that the threat would be carried out. If one's posture at the declaratory level was only based upon the notion of mutual suicide, why would the opponent believe it? If, however, one's declaratory posture was premised on a clear strategy of prosecuting the threat (nuclear war) to victory, the opponent was much more likely to see the threat as a credible one. This, in effect, appeared as the Soviet posture through its deployment practices and military writings in treating nuclear weapons as simply larger artillery shells. The Soviet leadership might speak publicly of nuclear war as unthinkable, but Soviet actions spoke to a belief that fighting and winning a nuclear war were possible.

Of course, key to this debate on nuclear deterrence credibility was the distinction between declaratory and operational doctrine. The former, as used above, related to the manner in which a nation's leadership spoke or communicated its beliefs about the strategic nuclear question. The latter concerned the actual top-secret operational plans to employ strategic nuclear weapons, embedded within the US Single Integrated Operational Plan. It was not necessarily the case that the two were identical and operational plans followed directly from declaratory statements. Indeed, the complexity of the strategic nuclear question in light of the large number and type of weapon systems and target lists made operational planning difficult. Declaratory strategy as designed to communicate strategic beliefs by its nature had to be relatively simple.

All administrations had operationally grappled with the problem of what to do if deterrence failed. In the first instance after McNamara, it led Nixon's secretary of defense, James Schlesinger, in National Security Defense Memorandum 242 to enunciate the doctrine of lesser nuclear options as an alternative to a spasmodic counter-value (cities) retaliatory strike.[22] From there, the Carter administration took the next step and enunciated in 1980 a somewhat new declaratory strategy in Presidential Directive-59: "If deterrence fails initially, we must be capable of fighting successfully so that the adversary would not achieve his war aims and would suffer costs that are unacceptable or in any event greater than his gains from having initiated an attack ... These requirements form the broad outline of our evolving counter-vailing strategy."[23]

The countervailing strategy inherited by the Reagan administration, which it never truly deviated from, was firmly embedded in deterrence thinking and requirements. However, as Reagan would take credit for the rebuilding of US military strength and the modernization of US strategic forces, so Reagan would get credit for the countervailing strategy. Even before SDI, this strategy was represented as a fundamental shift in US strategy from deterrence to warfighting. In proposing a missile defence umbrella over the United States, SDI was seen as consistent with this shift in US strategic thinking. By negating the Soviet strategic deterrent, the countervailing strategy and SDI made the world safe for war again, or so academic critics and the peace movement believed.

Naturally, there was a symbiotic relationship between concerns about US strategic vulnerability and the countervailing strategy, which informed the three options available for strategic modernization. Of these, one option was resurrecting missile defence to provide protection for the vulnerable US Minuteman fleet. Even though this option spoke to a strategic stabilizing function for missile defence, it meant a return to Safeguard, which had been killed by Congress. The prospect of obtaining funding for another missile defence effort was very low. Moreover, there had been no major technological leaps that might provide an alternative to the use of nuclear weapons for interception. The complexity of the task had grown as a function of MIRV technology. Multiple warheads meant the capacity to employ very sophisticated decoys to confuse any missile defence.

The British, for example, had made a significant effort through their Chevaline program to ensure that their Polaris SLBMs would be able to penetrate the Soviet Galosh missile defence system around Moscow.[24]

With missile defence an unlikely option, the US debate turned to two other possibilities, which were somewhat intertwined. One was the deployment of a new generation of ICBM with capabilities similar to Soviet ICBMs – the ten-warhead MX Peacekeeper. The other was to move beyond MX to the next generation of ICBMs – the smaller mobile Midgetman ICBM. The direct linkage between them was the basing mode, and it was this issue that received the greatest attention. The attractiveness of Midgetman was directly linked to the idea of an underground rail system in the open spaces of the western US. Midgetman missiles would be moved around the system in a shell game and thus be difficult for the Soviets to target for a pre-emptive attack. MX was too large for such a basing mode relative to the costs entailed. There was, however, the alternative idea of using the above-ground public rail system to shuttle MX (and Midgetman) in a crisis situation – a public relations nightmare. The most prominent of the MX basing schemes was the somewhat bizarre dense-pack idea – a large number of missiles deployed in silos close together, where the detonation of the first Soviet warhead would destroy follow-on warheads.

Unable to get agreement on any immediate alternative, the decision following the SDI announcement was to deploy a small number of MX missiles in existing Minuteman silos for the time being. Regardless, it was this strategic modernization debate that had been of primary concern within the Pentagon when Reagan proposed SDI. Being in direct contradiction to ICBM modernization, SDI made the basing debate somewhat moot. Why invest in new ICBMs and complicated basing modes if SDI would make them all obsolete? It was for this reason that senior military officials and Pentagon civilians were shocked, to say the least, with Reagan's surprise announcement.

With these developments, strategic arms control negotiations played a significant role in images of a fundamental shift in US strategic thinking. President Carter's initial strategic arms control proposal for deep cuts in strategic forces had caught the Soviet leadership by surprise. They rejected it outright. Instead, negotiations proceeded on a follow-up to SALT I and Vladivostok, which led to SALT II. However, with the political climate in the United States deeply opposed to any further agreements, especially after the Soviet invasion of Afghanistan, Carter had no choice but to remove SALT II from consideration by the US Senate.[25] Among the critics of SALT II was Ronald Reagan. But Reagan and the Republican right wing's opposition to SALT II extended much further. Reagan and part of his administration viewed arms control as a one-way street in favour of the Soviet Union and opposed any and all arms control negotiations. Reflecting this view, Reagan initially appointed Richard Perle, an ardent opponent of arms control, director of the US Arms Control and Disarmament Agency. Arms control for this administration was apparently dead.

Of course, Reagan's team, including Paul Nitze and Richard Burt, knew that such a position was politically untenable at home and abroad. As such, Reagan resurrected Carter's proposal for deep cuts, renaming the process Strategic Arms Reduction Talks (START). However, to opponents and critics, the proposal was seen as simply propaganda designed to ensure outright rejection by the Soviet Union – reminiscent of the disarmament proposals by both parties during the 1950s. Whatever expectations the Reagan administration might have had of seizing the political high ground were for naught. For many experts and the peace movement (especially in Europe), responsibility for the end of détente and the outbreak of the Second Cold War was the Reagan administration's fault in general and the president's doing in particular.

Alongside these developments in the United States, Reagan's SDI speech took place in the midst of the publicly contentious NATO decision to modernize its theatre-level nuclear forces. In December 1979, alliance leaders agreed to the modernization of NATO's intermediate-range nuclear forces, INF. Consistent with the 1967 Harmel Report linking NATO military decisions to arms control negotiations, this decision consisted of two tracks. One track was to seek a negotiated return to an INF balance upset by the Soviet deployment of a new generation of intermediate-range ballistic missiles (IRBMs), the SS-20. If this failed, the other would see the deployment of NATO INF – 108 Pershing II IRBMs and 454 ground-launched cruise missiles (GLCMs). In reality, there was little expectation that the Soviet leadership would agree, and the first track was as much a public relations endeavour as anything else. The need to modernize NATO's aging INF capability existed independently of the Soviet SS-20 deployment. Furthermore, the modernization decision was driven largely by European leaders, who nonetheless didn't want to take too much public credit for the decision, fearing significant domestic political fallout. Central to the decision, however, was the agreement that the allies would maintain solidarity so as to avoid a repeat of the neutron bomb fiasco.[26]

After growing public demonstrations on the streets of European capitals to the initial dual-track decision, Reagan on November 18, 1981, announced the global double zero option.[27] The United States and NATO would forgo the deployment of Pershing II and ground-launched cruise missiles in return for the elimination of all Soviet SS-4, SS-5, and SS-20 missiles. Not surprisingly, the Soviets rejected it outright and responded by calling for a freeze on all new deployments and negotiations to reduce existing arsenals. At the same time, the Soviets threatened to walk away from all arms control negotiations in Geneva if NATO proceeded with INF deployment. The leadership gambled that, with significant public opposition to INF deployment in Europe, NATO would have no choice but to give way or face a potential split among the allies. With no success at the negotiating table in Geneva, on November 22, 1983, the first Pershing II missiles arrived in Europe. The next day, the Soviet Union walked away from all arms control negotiations.

The Soviet gamble, which ultimately failed, was driven by negative European public images of Reagan. For the Europeans, détente had always been more politically important than for the United States, and the core symbol of détente was arms control. The end of arms control signalled the end of détente, and the end of détente meant a return to the crisis-filled era of the first decade of the Cold War. To many in Europe, and indeed North America, détente was mistakenly seen as the end of the Cold War, and Reagan would take full responsibility in their minds for the outbreak of the Second Cold War. Reagan was seen as a dangerous, ill-informed, and not-too-bright cowboy or gunfighter – not dissimilar to the image of former president George W. Bush today. Reagan's hard-line anti-Communist rhetoric, labelling of the Soviet Union as the evil empire, and on-air radio announcement that nuclear strikes against the Soviet Union were about to begin simply reinforced the worst fears of critics and the peace movement. Despite Carter's earlier actions, worsening political relations with Moscow were laid at Reagan's feet. If Reagan's election signalled the beginning of the death of détente, the new strategic warfighting posture, US strategic modernization, the INF deployment, and the end of arms control negotiations were the nails in its coffin.[28] SDI was one more nail. It simply reinforced fears, especially within Europe, not only of a new arms race but also of a drift to war – a perception that would also emerge within the Soviet leadership of the United States contemplating a pre-emptive strike.

In Canada, officials in External Affairs and National Defence closely monitored these strategic developments, as did the growing and publicly vocal peace movement. The movement, however, never truly understood the complicated nuances of American strategic discussions. Instead, it, like its European counterparts, was obsessed by the image of Reagan as a nuclear cowboy. The collapse of détente and the beginning of the Second Cold War were entirely attributed to Reagan and his coterie of hawks within the administration, which would include Fred Iklé as the deputy under-secretary for defense for policy and one of the administration's point men on SDI. This environment would inform Trudeau's call for an ASAT ban and, more importantly, his 1983 peace initiative, which starkly contradicted the alliance solidarity pledge on NATO's INF modernization program. It was also informed from the fallout of Trudeau's decision to allow the testing of the US air-launched cruise missile (ALCM) over northwestern Canada.

The Canadian decision to agree to the US ALCM testing request in July 1983 took place under the umbrella of the Canada-US Test and Evaluation Agreement signed on February 10, 1983.[29] The agreement allowed both parties to obtain cost-free access to the other's test and evaluation facilities upon request and availability. For National Defence, this held the promise of major cost savings, as US facilities far outnumbered Canada's, covered the full gambit of military technologies, and in many cases were more technologically advanced than Canadian facilities. Its terms, as such, were highly favourable to Canada. It was also widely understood that the United States saw the agreement as the means to access Canadian territory for testing its new ALCM to rejuvenate the US B-52 and B-1b strategic bomber fleet in an environment that approximated its flight path to targets in the Soviet Union. The US bomber fleet equipped with nuclear gravity bombs had become extremely vulnerable to the extensive Soviet air defence system. ALCMs enabled bombers to stand hundreds of kilometres away from their targets and release nuclear-tipped missiles, which would fly close to the ground beneath Soviet air defence radars, and employed a terrain-mapping radar to guide them to target.

The decision to allow ALCM testing unleashed a firestorm of public opposition, leading to the largest public demonstrations the country had seen during the Cold War.[30] For the peace movement, the government had exposed its hypocrisy in directly contributing to the nuclear arms race. Trudeau partially justified the decision as part of Canada's NATO alliance commitment. To some extent, he was correct. As US strategic forces were coupled to the fate of NATO Europe within a complicated vertical escalation ladder, the US strategic bomber fleet was one element of this coupling, and supporting the rejuvenation of the fleet did have an element of an alliance commitment. However, in reality, bilateral Canada-US defence considerations dominated the Canadian decision.

What further complicated matters at the time of the SDI announcement were discussions between Canada and the United States on the modernization of the North American air defence system. The combination of the age of the early warning system deployed in Canada, a new generation of Soviet bombers (Backfire and Blackjack), and Soviet ALCM developments had led US military officials to examine air defence modernization requirements in the late 1970s.[31] In 1979, the Joint Canada-US Air Defence Study identified significant gaps in the air defence of North America. In January 1981, the United States released its Air Defense Master Plan calling for new air defence capabilities. Two months later, President Reagan and Prime Minister Trudeau released the US-Canada

Joint Policy Statement on Air Defense. From there, discussions moved forward on what would become known as the North Warning System (NWS), replacing the old Distant Early Warning (DEW) network, along with the elimination of the Pinetree and Mid-Canada radar lines. This would be made official with the signing of a joint modernization program at the Mulroney-Reagan Shamrock Summit in March 1985. With these modernization discussions, air defence would become entwined in the internal SDI debate as one Canadian option and publicly as part of the conspiracy to get Canada into missile defence through the back door.

Two other factors also came into play that would affect the SDI debate. In December 1982, the United States initiated the Strategic Defense Architecture (SDA) 2000 study to examine potential defence requirements against bombers, cruise missiles, and ballistic missiles under the lead of the commander-in-chief NORAD. He invited Canadian NORAD, Air Command, and National Defence officials in Ottawa to participate in Phase I, which was completed in early 1985. From there, an invitation was extended to participate in Phase II. Not surprisingly, this invitation would run afoul of SDI, even though the study was not initially focused upon missile defence. As Canadian military planners would find out in the late 1980s, Mulroney's SDI decision would be interpreted as a blunt no to SDI, and a no to SDI meant in US military eyes a no to further involvement in SDA 2000.

The other factor was the NORAD space dimension. By the early 1980s, US military officials had begun to recognize the growing significance of satellites for a range of functions beyond the strategic nuclear world – the forerunners of the force enhancement role satellites would first play in the 1991 Gulf War.[32] Arguments began to gain currency in the United States to establish an independent command for military space. In 1982, USAF Space Command was established with its commander also in command of NORAD and US Continental Air Defense Command.[33] For the USAF, it was seen as the forerunner of a unified command led by the USAF for all military space issues, and Reagan's SDI speech provided the catalyst for the Joint Chiefs of Staff subsequent agreement to establish US Space Command.[34] In the lead up to its formal establishment in 1985, the US Continental Air Defense Command was eliminated, and discussions took place with Canadian officials over the possible relationship of the new command with NORAD. In the end, Canada agreed with the US decision to have the commander of Space Command triple-hatted with NORAD and the new US Element NORAD, even though it was a US-only decision outside the bounds of the NORAD Agreement. The early warning and space surveillance assets supporting the NORAD mission simply moved to Space Command, as did the air defence mission of the former Air Defense Command. Nothing had really changed, except all this occurred in the midst of SDI, fuelling suspicions of a conspiracy behind closed doors.

Space Command, air defence modernization, ALCM testing, a strategic war-fighting doctrine, INF, ASATs, and the removal of the ABM exclusion clause could all be wrapped into a compelling conspiratorial picture of Canada being captured by dangerous exotic US defence initiatives or willingly looking for a back door into missile defence and around the ABM Treaty. This scenario was reinforced by Reagan's public image and the election of a Conservative government under Brian Mulroney seeking to restore good relations with Washington. SDI simply confirmed the worst, helped along by two growing misperceptions. First, SDI became portrayed as an imminent deployment rather than simply a research program. This served to drive the public debate in Canada to the margins of the real issues facing the Canadian government. Second, SDI was seen as exclusively weapons in space, despite the reality that space was only one of many elements of the multifaceted research program, and it was almost impossible to predict which technologies would prove fruitful and which would obtain monetary support from Congress. These two combined to create strange bedfellows with the anti-nuclear, anti-US peace movement ostensibly aligned with the expert MAD community in opposing SDI.

In this climate, it was hoped by many within the government that SDI would just go away, as ABM had. For some time, this appeared possible given the lag between Reagan's speech and putting the research program in motion. It was possible to hope that SDI would fade, having been met with significant opposition in the United States, including concerns expressed by the Joint Chiefs of Staff and the expert academic community. This, however, was not to be. The initiative moved forward, and Canada had no choice but to confront it one way or another. Thus, in early 1984, some direct but private attention began to be paid to SDI. In February, External Affairs proposed a meeting with National Defence on SDI.[35] The ostensible purpose was to evaluate US Deputy Secretary of Defense for Policy Fred Iklé's suggestion that Secretary of State George Schultz brief cabinet on SDI during his official visit on April 1-2. The briefing never took place, which is not surprising on several counts. National Defence officials argued that it was premature, especially given the embryonic status of the SDI effort in the United States. These officials also feared a negative response from cabinet. In this context, it butted directly up against Trudeau's peace initiative, which he had announced in October 1983. If driven, the government might come out early and publicly to condemn SDI, which might affect other elements of the relationship already somewhat difficult on ideological grounds. Why risk further damage to the relationship on an issue whose future remained unclear and ill defined? The government was also facing growing opposition to the

Canada-US cruise missile testing agreement, and to have the public suddenly become aware of SDI discussions would add fuel to the fire and feed conspiracy thinking.[36]

The lack of attention paid to SDI reflected the vagueness of its scope. The first formal SDI briefing to NATO occurred on August 22, 1984. Following the ABM pattern, SDI was presented as a modest research project. This, of course, reflected the status of the SDI effort at that time, as well as concerns within various circles in the United States, including the Pentagon, about SDI and its costs. But little detail was provided. It wasn't until the end of October, after the election of the Mulroney government, that Canada received its first formal briefing on SDI from Iklé and a US team, including officials from the SDIO, to the minister of national defence, Robert Coates, and senior Defence officials.[37] At the meeting, Coates not only expressed support for the US initiative in arguing that eliminating the immoral threat of nuclear weapons would alleviate Canadian concerns but also inquired what Canada could do to indicate its support. He offered to help SDIO officials identify possible areas of Canadian technological expertise that could be of value. Iklé replied that the SDIO program was being structured and managed to ensure the inclusion of Canadian expertise and noted possible areas of Canadian participation as ballistic missiles; microelectronics; and command, control, and communications. Nonetheless, before the meeting broke up, a senior External Affairs official emphasized to Iklé that the minister's view was not the official position of the Canadian government.

On November 19, Coates expressed the government's interest in SDI for the first time in the House. He described SDI as an opportunity to change the strategic relationship between the Soviet Union and the United States from "assured destruction to assured survival." He added that SDI held out the promise "to remove the nuclear sword of Damocles."[38] Somewhat surprisingly, there was no immediate response from the ranks of the Opposition until a month later, on December 18, when Lloyd Axworthy requested clarification of the government's position on Star Wars.[39] Two days later, Jean Chrétien, framing Star Wars as weapons in space, inquired why the government was not following the previous government's policy of opposition to SDI, especially given British opposition. The prime minister responded that Canada was an adherent to the 1967 Outer Space Treaty and added, "we have not been approached in regard to what is not an undertaking, but a research project ... We have not been asked to consider anything, but we have ventured the view that the Canadian government will press incessantly for the elimination of all instruments that damage the cause of peace."[40] The next day, Axworthy again sought clarification of the government's position, calling for a renunciation of Star Wars technology. In response, the prime minister reiterated that the US SDI research program had not been officially brought to the government's attention.[41]

Thus, by the end of the first foray into Canadian policy on SDI, two initial parameters for the subsequent debate were in place. The Opposition sought to frame the debate around the weaponization of space question in trying to get the government to expose its position on SDI one way or another. The prime minister in particular looked to avoid exposing the government by emphasizing the absence of conditions for a Canadian policy decision. In a refrain that would re-emerge more than a decade later, there was no formal approach to the Canadian government by the United States. Besides, SDI was only a research program. This was echoed by Clark, the minister of external affairs, a month later in the House as part of the government's first formal policy statement: "To date, the full extent of the programme has not been explored and it would therefore be premature to draw definitive conclusions about it."[42]

Immediately following these exchanges, two significant events occurred. On December 22, Margaret Thatcher, the prime minister of the United Kingdom and a long-time close friend and supporter of the president, met with Reagan at Camp David. Thatcher used the occasion to outline her views on SDI, which would serve as a compass point for all the allies in seeing her criteria as an American commitment. She identified four points: the Soviet Union is engaged in missile defence research, and it is right and prudent for the United States to do so as well, there is a clear distinction between research and a decision to develop and deploy, any decision to develop and deploy should be subject to negotiations, and the end result should strengthen deterrence and stability. Two weeks later, on January 6, US Secretary of State Schultz met with Soviet foreign minister Gromyko in Geneva and after two days of meetings announced that new arms control negotiations would begin. Following these two events, Clark was to prepare a government statement on SDI to be presented to the House on January 21. On January 11, Clark received a written briefing on SDI, with input from the ad hoc interdepartmental working group.[43]

From the onset of discussions, divisions existed between the two departments, as they would throughout the entire life of the missile defence debate in Canada. Not surprisingly, National Defence, while conscious of the broader strategic questions, remained fixated upon the binational relationship. Canada-US defence relations in NORAD were at issue, including Canadian access to US defence thinking, plans, and technology. The risk of losing this access was simply too great, even without knowing exactly what type of access might be obtained or maintained. Nonetheless, they initially favoured a yes, regardless of where it might lead. External Affairs officials remained deeply skeptical of SDI, not knowing where it might lead, especially in terms of those broader strategic and political implications for East-West relations. This division extended to the ministerial level as reported in the press at the time.[44]

However, External Affairs was not united either. A pronounced internal division existed that would plague the ability of the department to develop a single corporate perspective. On one side stood the international security/arms control faction and on the other the Canada-US relations group. Initially, they divided on the likely economic and technological benefits to be accrued from possible participation, the issue that had initially captured the minds of Coates and others within the new government. Subsequently, the internal division reflected the relative weight to be placed on international security and alliance considerations versus the explicit foreign policy strategy of the Mulroney government – close relations with the United States predicated upon close personal ties between the prime minister and the president – Brian and Ronnie in their direct correspondence.

The January 11 brief to Clark returned to the issues that had first been played out during the initial ABM debate.[45] First, the Canadian situation was distinct from that of Europe, because Canada would be defended. Second, SDI's implications for NORAD directly concerned the next phase of NORAD modernization and the relationship between missile defence and its early warning and surveillance mission. Third, it was unclear whether Canadian territory and airspace for a possible mid-course or terminal phase of any SDI system, à la Safeguard, would be required, as no operational architecture yet existed. Finally, Canadian opposition or refusal could eliminate Canadian involvement in advanced research and industrial benefits linked to SDI and NORAD modernization with direct economic implications.

Interestingly, the initial brief failed to include arms control considerations. This failure was a function of several considerations. As a research program, SDI was consistent with US ABM Treaty obligations. There was no need to consider arms control, unless one suspected that the United States was planning to violate the treaty. Like the defunct ABM exclusion clause, expressing this concern suggested that the United States would act illegally, and such a public suggestion would simply feed conspiratorial thinking within the Opposition. In addition, the prime minister's priority was to improve relations with the United States through close ties with the president. To suggest that the president might violate the nation's ABM commitment was for all intents and purposes a public insult to the president. Nonetheless, arms control could not be entirely ignored. Clark had been co-author of the Progressive Conservative Party's pro-arms control policy paper prior to the 1984 election. The Opposition too had directly raised the issue in the context of space weaponization in December, and the memory of public demonstrations on cruise missile testing remained fresh. Something had to be said about arms control, and this omission was rectified four days later.[46] An additional paragraph was to be added to the formal statement characterizing the nature of the ABM Treaty issue, the threat to the future of arms control if the treaty was violated, and the importance of monitoring this element of the SDI question.

As Clark prepared a draft statement for the House, the day after the Schultz-Gromyko meeting in Geneva, Robert Fowler, the assistant secretary of the foreign and defence policy secretariat of the Privy Council Office (PCO), forwarded a memorandum on SDI directly to the prime minister, which was returned with no comment a week later.[47] The memorandum provided an overview of SDI, including European concerns, and emphasized the need for a formal government statement on SDI, especially with the return of the House on January 21. Fowler identified four options for the prime minister: publicly oppose SDI, which would appease "'peace groups' and many informed people" but would produce "anger in Washington and particularly in the White House"; publicly support SDI, which would alienate the above groups, could be popular, especially among those attracted by possible technological benefits, but would be unexpected in Washington and in some cases unwelcome by forces opposed to SDI within the administration and elsewhere; take no public position whatsoever, which would be difficult and please no one; and, finally, publicly support research, but not testing, development, and deployment, which might be seen by some as going too far.[48]

In recommending the final option, Fowler emphasized that the position would be consistent with the US effort and international arms control considerations while not intruding on the US-Soviet negotiating process. He suggested that the government consider going beyond Thatcher's public position by calling for a future ban on weapons in space, while remaining consistent with British policy. Fowler warned, however, that the public believed that SDI was a deployment program such that the government's position "would have to be presented very precisely."[49] He concluded by recommending that this position be communicated to the Policy and Planning Committee meeting of cabinet on January 16 as basic instructions for all ministers to follow.

After the committee meeting, where this approach was formally adopted, save for the idea of calling for a ban on weapons in space, Clark set out the government's position in a statement to the House on the 21st. Echoing earlier statements, he stressed that SDI as a research program did not violate either the Outer Space Treaty or the ABM Treaty. He noted that if, however, the United States proceeded to the development and deployment of SDI as "currently constituted," without defining what this exactly meant, the ABM Treaty would be violated.[50] In response, John Turner, Axworthy, Pauline Jewitt (NDP), and

Svend Robinson (NDP) spoke directly to government duplicity. In particular, Jewett's remarks directly implied that the US Department of Defense had duped Canada on the reality of SDI because it was much more than just a research project. Robinson raised the issue of differences within the United States on Star Wars. Clark responded that it was of no use to create division, especially at a time of delicate negotiations (referring to the resumption of talks in Geneva between the United States and the Soviet Union). For his part, Brian Tobin reiterated the Liberal Party's opposition to Star Wars. In this instance, Clark replied by hinting that the best policy for Canada was to pursue private quiet diplomacy with regard to the strategic relationship between the two superpowers.[51]

Thus, by the end of January, two other threads were added to the debate. The Opposition played the nationalist card by implying that the government was being subservient to the United States. The government, courtesy of the external affairs minister, had fallen back on the traditional Canadian foreign policy refrains of allied solidarity and quiet diplomacy as the best means to influence US policy. Over the next month, the Liberals and the NDP pushed their arguments a little further by implicitly linking SDI to other issues on the Canada-US agenda and emphasizing mythical US deployment plans. Thus, for example, in the debate on the Canadian Investment Act, which had implications for relations with the United States, Axworthy and Chrétien both spoke of the government kowtowing to the United States on SDI.[52] Derek Blackburn, the NDP defence critic, identified Pentagon plans to assign Star Wars and ASAT weapons to NORAD.[53]

Most interesting during these debates in the House was the sudden departure of the external affairs minister from the official government line. On February 13, Chrétien rose in the House and referred to numerous articles in the American press that alluded to Washington establishing nuclear installations in Canada that were to be part of Star Wars.[54] Clark responded by going further than the twenty-first statement by stating the government's opposition to Star Wars.[55] Likely, the minister was referring to the deployment of SDI in the context of the ABM Treaty and Canada's long-standing commitment to arms control and disarmament and the sanctity of a rule-based international order.[56] Nonetheless, it was the first time that anyone on the government side of the House had bluntly departed from the policy mantra of SDI as a research program and, in effect, had condemned SDI carte blanche. Regardless, it was a public signal of a division within the government on SDI – one that reflected the contentious nature of SDI within External Affairs, between External Affairs and National Defence, and within the Progressive Conservative Party itself.

Throughout this initial phase of the SDI debate, work on a joint External Affairs–National Defence memorandum to cabinet on SDI had begun. On February 5, a memorandum to the prime minister from the PCO's Foreign and Defence Policy Secretariat included as an annex a joint External Affairs, National Defence, and Privy Council paper that identified all the various policy questions facing the government on SDI.[57] It also noted ongoing work on the upcoming memorandum to cabinet, which in the end would never see the light of day.

In the debate on the memorandum, National Defence officials staked out a strong position on Canadian participation. Coates had clearly indicated a pro-SDI position in his meeting with Iklé in November 1984 and subsequent statement in the House. For National Defence, SDI could not be readily separated from the bilateral defence relationship as a whole range, including North American air defence modernization and NORAD renewal in 1986. Although air defence modernization requirements provided some degree of protection for NORAD, Defence believed that participation in SDI was vital to ensure complete access to US thinking and planning about North American defence. Moreover, a Canadian decision would have major implications for Canada's ability to keep abreast of future US technology developments.

Not surprisingly, External Affairs took a much more cautious approach, reflecting its advice to Clark in the lead up to the first formal government statement in January. In the end, External Affairs triumphed, only to see the agreed-upon memorandum scuttled in Clark's office with the intervention of Douglas Roche, the ambassador for disarmament. The position and appointment had been Clark's doing. Roche had long been a vocal opponent of SDI, and, strangely enough, his overall views on defence and security were closer to those of the Liberals and New Democrats than the government. In one of his first acts as ambassador, Roche had unequivocally condemned SDI, employing the standard left-wing disarmament movement's destabilizing arms race critique at a conference in Toronto in October 1984. Roche and Clark had also been responsible for the centre-left Progressive Conservative arms control policy posture adopted by the party before the 1984 election, which Mulroney had privately deep-sixed in his first meeting with Reagan.

The draft memorandum only recommended that the government avoid a complete refusal or acceptance in response to any future US invitation, in order to keep the door open to more detailed information as SDI unfolded – the compromise position of the two departments. Roche, however, saw it as the first step toward participation. Only full condemnation of SDI would suffice. Regardless of whether Roche was instrumental in Clark's decision or simply reinforced Clark's own doubts about SDI, the memorandum was killed and with it the closest the two departments would ever come to an agreed memorandum throughout the life of SDI.

After Clark's condemnation of SDI in the House, the exchange on SDI returned to form. In mid-March, Axworthy asked whether the prime minister had communicated to the new leader of the Soviet Union, Mikhail Gorbachev, that the next version of North American defence included Star Wars and whether the prime minister had asked Reagan to put Star Wars on the negotiating table. Mulroney responded that the topic would be on the superpower bilateral agenda. In response to the March 19 non-confidence motion by Ed Broadbent, the leader of the NDP, Clark quoted from the Shamrock Summit final communiqué, which in turn echoed Thatcher: "In this regard, we agree that steps beyond research would, in view of the ABM Treaty, be matters for discussion and negotiation."[58]

Broadbent had directly tied Star Wars to the North Warning System (NWS) modernization program, which Paul Nitze, the special advisor to the president and secretary of state on arms control, had linked during a press conference on the occasion of his visit to Ottawa on March 6.[59] Even though the future new air defence NWS radars had no capacity for tracking missiles, Nitze suggested that the relationship between NWS and SDI "remains to be seen ... [SDI] is a research program that hasn't yet resulted in the development of specific systems."[60] In some sense, Nitze was correct. If SDI proceeded to actual deployment, one of the possible Soviet responses would be to develop more offensive air-breathing systems to sidestep SDI, requiring further new air defence systems – the logic behind the subsequent option from National Defence of a Canadian focus on air defence in response to an SDI invitation. This might save NORAD, while keeping SDI at arm's length, as US strategic nuclear weapons were. However, this argument was much too sophisticated for a press conference with any expectation of being reported fully and accurately. As well, Nitze's linkage raised suspicions of not only an existing invitation but also one that had already been accepted. External Affairs immediately recognized the damage that had been done. Thus, External Affairs officials immediately contacted US Embassy officials seeking public clarification or retraction, which followed. Nonetheless, the damage had been done, and the linkage of air defence modernization and SDI had now been confirmed. SDI was everywhere in the imaginations of its opponents.

External Affairs had for some time expected some type of invitation from the United States to participate in SDI research. Officials had advised Clark of this in the first formal briefings on SDI in January. Moreover, officials had begun to wonder if an invitation had already taken place and they had been left in the dark. Throughout January and February, External Affairs asked allies through its representatives at NATO whether any had received formal invitations to participate.[61] The allies confirmed that no invitation had been extended. In early February, the invitation rumour received a shot in the arm with reports that the allies had been invited through Defense Secretary Weinberger's public address in London on February 11, which had actually been delivered by Perle.[62] He spoke directly to the importance of allied support to SDI and hinted that allied participation would be welcomed if it remained within ABM Treaty limits on technology transfer. However, it was not a formal invitation. Indeed, when Canadian officials contacted SDIO the next day, they were informed that no specific invitation had been issued.

Canadian concerns, especially within External Affairs, then turned to the possibility of a direct invitation during the forthcoming Mulroney-Reagan summit in Quebec City. These concerns were elevated by Gottlieb's assessment of the US administration's commitment to SDI in early March.[63] From idealism (Reagan) to loyalty (Schultz) to concerns about strategic stability and MAD (Iklé), SDI appeared like a religious movement in which no doubts were permitted. More disconcerting, Gottlieb reported that the administration would likely view allied responses to SDI, invited or not, as a loyalty test.

The question of what the two governments would say to each other – and their electorates – during the Shamrock Summit in two weeks' time, including the final communiqué, emerged during Nitze's March 6 visit to Ottawa.[64] Clark, with a visceral need to demonstrate differences between Canada and the United States, emphasized the political danger of a post-summit image of Canada being too close to the United States. Clark's wish to have the government avoid any questions on SDI, as Martin would similarly wish in November 2004 during President Bush's visit, was already under conflicting pressure from the wish for a close personal relationship between the prime minister and the president on the one hand and the vocal criticism of the peace movement and the Opposition on the other.[65]

In this context, Canadian and American officials began negotiating the text of the summit communiqué. Canadian officials sought to downplay, if not eliminate, any reference to the controversial missile defence research program.[66] If SDI had to be mentioned, Canadian officials preferred that it be embedded in a direct reference to the ABM Treaty and the requirement that testing and deployment be a matter of negotiation – a position consistent with existing policy based upon the Thatcher criteria. By now, however, US officials had begun to depart from a verbal commitment to the criteria, especially with regard to the issue of testing and deployment requiring negotiation.[67] They sought a very strong joint declaration in support of SDI, without any reference to testing and deployment. This fundamental difference between the two countries continued in discussions right up to the final minutes, acquiring almost a crisis-like quality within External Affairs.

In the end, Canada got its reference to the ABM Treaty and testing and deployment, using the Thatcher language of testing being a matter for discussions and negotiations. Little did the officials know that, within the year, the interpretation of testing would be challenged by the Reagan administration. More immediate, within a week, External Affairs (and the government) faced the very invitation they had feared.

On March 26, Ottawa became aware that Weinberger had sent, or was about to send, letters to all the NATO defence ministers inviting their nations to participate in SDI research and requesting that they respond within sixty days. With that, officials immediately sought to contact the Canadian delegation to the NATO Nuclear Planning Group (NPG) meeting in Luxembourg to find out if the defence minister, Erik Nielsen, had received a copy. Unfortunately, the information and the decision to contact the minister were too late. The die had been cast, apparently by the initiative of Weinberger without the full knowledge of the rest of the administration.

The issue of extending an invitation to allies had been discussed extensively within the administration. No decision had been reached, even though the text of a letter of invitation had been agreed to, which, strangely, was not taken by Weinberger to Luxembourg.[68] Instead, Weinberger surprised allied delegations with a verbal invitation in the morning meeting of defence ministers. In response, the British defence minister, Michael Heseltine, requested a formal written text. This text was drafted on the spot and sent to the State Department, which wasn't planning to release the text publicly. However, at the very moment in the afternoon that the written text was provided to the delegations, Perle held a press conference in which he delivered the invitation basically as it had been written.

Roughly around the time that Weinberger provided the formal written invitation, the presence of an invitation had been reported in the press. Iklé, the day before in Vancouver, stated that, although no formal invitation had been issued, "we have already made clear in many ways to our allies that we want their cooperation in terms of technical research ... It's broader and more complex than some thing you can handle in a single note."[69] After question period on the 26th, Clark, when asked about the invitation in a scrum outside the House of Commons, flatly denied that any had been extended. Sadly for Clark, the press was better informed, whereby one of its members, in possession of the press release, confronted the minister.[70] Clark's public embarrassment subsequently fuelled a range of speculation about the internal politics of the government; the relationship between the prime minister, Nielsen, and Clark; and cooperation and coordination between External Affairs and National Defence. Perhaps, at the end of the day, the failure of the NATO delegation to inform Ottawa was simply an unintended consequence of bureaucratic procedure – an expectation within the Canadian NATO NPG delegation that Weinberger would not have released the text without having it approved by Washington, provided directly beforehand either to national capitals, their embassies in Washington, or the allied delegations in Brussels.

The day after the invitation, Gordon Smith, the deputy minister for political affairs in external affairs, forwarded an explanation to the PCO on the failure. He confirmed the Weinberger invitation, explaining that Nielsen had simply put the letter in his pocket without the knowledge of anyone else. As far as the failure to communicate with Ottawa, Smith attributed this to Nielsen not wanting any "reporting from Luxembourg," the lack of secure communications, and that "no one thought it of sufficient importance to report urgently by telephone."[71]

External Affairs may have been in the dark in the lead up to Luxembourg, but National Defence and the defence minister were not. The Monday before the Luxembourg meeting, Nielsen was briefed on the possibility that the United States might invite Canada to participate in SDI research at the NPG summit. This warning, however, was not passed on to External Affairs or Clark, ostensibly because defence sources could not confirm its likelihood. Nonetheless, it is surprising that no one at External Affairs became aware of the possibility, especially the External Affairs members of the informal SDI ad hoc working group. Regardless, senior officials concluded that Weinberger had a calculated strategy of how to approach or trap Canada into SDI.

Of course, speculation that an invitation had been forwarded months earlier had circulated among the Opposition and the press for quite some time. Senior officials in the lead up to the Shamrock Summit lived in trepidation of a formal invitation. Of course, External Affairs may simply have assumed that the United States would not extend an invitation publicly unless it was confident of a positive response. Clark had made it clear to Nitze that the government did not want to answer questions on SDI. After the bruising political battles on INF, one would also have to wonder why the United States would entertain another such battle if it was unsure how the allies would respond. Washington was as well aware as Ottawa of significant differences of opinion among the NATO allies on SDI. Contemplating an invitation was one thing, but the timing and method appeared strange, to say the least.

In addition, the prime minister's lukewarm views on SDI had been directly communicated to senior members of the administration the week before the Shamrock Summit. On March 13, on the occasion of the funeral of Soviet

Communist Party general secretary Konstantin Chernenko in Moscow, Mulroney met privately with Vice-President Bush and Secretary of State Schultz. Among the topics discussed was SDI. Mulroney communicated Canada's position that SDI was problematic for Canada, but the government would place no obstacles in the way of Canadian firms. In effect, Mulroney suggested that the best manner of approaching SDI for the time being was to avoid a public invitation, while permitting Canadian engagement within the long-standing parameters of the Canada-US defence industrial- and technology-sharing relationship. In other words, cooperation did not need anything formal, as SDI was simply a research program. In effect, Mulroney communicated months in advance of the policy formally announced on September 7. However, in failing to tell anyone else, he allowed a tortuous process to continue.

Mulroney's private direct communications with Bush and Schultz were consistent with his preference for personal diplomacy, evident, for example, in his approach to Canada-US relations and in the Meech Lake process. This preference was clearly understood within External Affairs as the basis of his foreign policy in general. And it was consistent with the growing imperial power of the Prime Minister's Office that had begun with Trudeau. As for communicating his discussions or decision with his cabinet colleagues and senior officials, this appeared unnecessary assuming that Bush and Schultz provided reassurances that no formal public invitation was forthcoming. As previous memoranda to the prime minister noted, even a relatively benign policy of allowing Canadian firms to participate would spark a negative response. Besides, Canadian firms already had privileged access to the US defence market through the Defence Production Sharing Agreements (DPSAs), in spite of existing irritants. Moreover, there were no guarantees that the prime minister's private communication of the decision would necessarily remain private. If it leaked, the government would become very vulnerable to a range of charges, including duplicity. In other words, with no invitation, there was no need for the prime minister to make a decision on participation, and if there was no need for a decision then no decision should have been made and communicated.

From this perspective, Mulroney was as shocked as his colleagues upon hearing of the Weinberger invitation. Had the invitation been officially approved by the administration, the prime minister would have had advance knowledge from the president or vice-president – Ronnie or George. Of course, the invitation may have been actually addressed to America's other allies, who didn't possess privileged access to the US defence market. If so, Canada became unintentionally caught in the crossfire. As far as the SDI opposition was concerned, however, the invitation to the allies was simply cover for Canada. Regardless, the time had come for the government "to fish or cut bait," and time appeared to be at a premium.[72]

In placing a sixty-day deadline for a response, Weinberger explicitly framed allied participation within the constraints of US treaty obligations by limiting it to technology research outside the ABM Treaty component level.[73] Exactly where those boundaries were drawn, however, was open to speculation and could be decided only by the United States. In addition, Weinberger requested from the allies a response to two questions: whether the country was interested in participating, and what areas of a nation's research effort would be most relevant to SDI. As with the earlier ABM question, the Canadian dilemma resurfaced, and this time it directly applied to all the allies.

Any decision required clear information from the United States on the nature and limits of available research opportunities. No ally could identify the areas of research to which it might contribute. In this regard, only the United States as party to the ABM Treaty could interpret what areas of research would be compliant with Agreed Statement G: "The Parties understand that Article 9 of the treaty includes the obligation of the United States and the USSR not to provide to other States the technical descriptions or blue prints specially worked out for the construction of ABM systems and their components limited by the Treaty."[74] Obtaining this information also hinged upon US willingness to provide highly classified details, and this willingness turned on a positive answer to the first question. However, to answer yes to an interest in participating implied an acceptance of the invitation as a whole and left little room to manoeuvre after the fact. The United States would not expect a no afterward, with the repercussions very high for the bilateral relationship, especially if Gottlieb was right about SDI being a loyalty test. Besides, it was unlikely that a yes to question one could be kept private, and once it became public it would be interpreted as a commitment to everything, including deployment. Once this ball started to roll, especially after the Liberals and NDP connected the dots, the government would be portrayed as willing to violate the sacred ABM Treaty and conspiring to place weapons in space.

With this dilemma front and centre, National Defence and External Affairs immediately took opposite sides. Upon his return from Luxembourg, the minister of national defence argued that the lack of discussion with the United States on SDI had two effects. First, it limited the amount of information that the government could use to make a decision – information essential to a sound decision. Second, it created an unfavourable impression of Canadian reluctance, if not indifference. Nielsen requested authorization from the prime minister to engage in direct discussions with the United States and recommended that he write a letter to Weinberger requesting meetings with US officials to obtain more information upon which to move forward.[75]

Clark and External Affairs, however, preferred to put any discussions with the United States on hold until a decision on participation had been made.[76] Seeking more information through discussions might create a US expectation of a formal commitment. If this occurred and the result was negative, the implications could not be ignored. In this context, senior officials sought to provide clearer background for the minister. They informed Clark that the invitation was a political one that the United States had not thought fully through.[77] It was highly unlikely that the United States would be willing to transfer highly classified technology, alongside US contracts, to non-US firms. Nonetheless, Canada was well placed to benefit ahead of its allies from whatever technology transfer and contracts did become available because of the long-standing defence industrial and technology transfer relationship with the United States.[78] In addition, External Affairs officials noted that a foundation for participation had been somewhat laid at the Shamrock Summit (despite the battle over the final communiqué).

PCO officials were also concerned about public perceptions surrounding any consultations and discussions with the United States.[79] They advanced that, with a divided Canadian public, the government should not overestimate, as a function of its popularity, public support for space initiatives and increased employment opportunities that might result from SDI. Public sensitivity to Canada-US defence and sovereignty questions remained high. Any possible meeting of the deputy prime minister, who was also the defence minister tasked with the SDI question, with senior US officials on Star Wars would not remain private and would likely spark a major public debate. At a minimum, the government needed to make a clear public statement beforehand that the meetings were only to consult with the United States prior to a Canadian decision and reference possible employment and economic benefits.

In proposing direct discussions with the United States, Nielsen also raised fears about the implications of saying no on Canada's future economic and technological competitiveness.[80] Early on in the debate within External Affairs, some officials believed that participation would lead to significant economic and technological benefits for Canada. Others, however, felt that not much would be gained by pointing to the limited payoffs to date from the Defence Production Sharing Agreements. In a memorandum to Minister of Industry Tom Siddon on April 1, External Affairs officials outlined the benefits and costs of Canadian participation. On the benefits side of the equation, the possibility of substantial gains in high-technology research and technology transfer were noted, especially in the areas of lasers, particle beam accelerators, computers, sensors and radars, micro-electronics, and advanced materials.[81] Echoing the view that would dominate the French perspective, SDI represented a research project whose scale and consequences with regard to US technological leadership had not been seen since the Apollo landing on the moon in the 1960s.[82]

However, going along for the technological ride would require a careful coordinated selection of Canadian industrial and research engagements, and these would have to take into account the interests of other possible allied participants, especially the British and the Germans, who were seen as likely to jump on board early on. In this sense, the need for a quick decision was pressing in order to avoid losing opportunities to other allied companies. At the same time, External Affairs raised questions about economic implications for Canada if SDI were to be abandoned by the next US administration and the pressure SDI participation could place on Canadian defence-spending choices especially if or when the program moved to deployment.

A month later, the Department of Regional and Industrial Expansion (DRIE) provided an evaluation of the potential economic benefits of SDI.[83] The Electronics and Aerospace Branch argued that participation would be consistent with Canada's long-standing defence industrial niche strategy to develop an internationally competitive industry capable of producing specialized goods for export markets rather than one designed to meet the operational needs of the Canadian military. SDI would fit directly into this strategy, especially if Canada engaged early on in the research effort and before other allies acted.

However, DRIE did not oversell the benefits of SDI. Immediate payoffs in terms of employment would be limited to a relatively small number of engineers and scientists, though second-order effects concerning employment and technological competitiveness directly related to SDI might be important. Third-order effects associated with general technology diffusion or spinoffs into other economic areas should also not be overestimated, as Canada would receive many of these anyway because of the close economic relationship. Other areas of greatest significance (micro-electronics and computing) were not ones of great Canadian strength. In other words, the ability of Canada to participate and thus obtain significant benefits was limited by the nature of Canada's industrial and technological capacity relative to SDI. In addition, the relatively small size of Canadian capacity limited the extent to which Canada could participate.

Overall, DRIE concluded that the most significant factor in Canadian participation would be in the field of access to US defence technology and improved terms of defence trade with the United States, warning that Canadian discussions with the United States on defence trade could well collapse if Canada did not participate – a position also advocated by National Defence. As for options, passive participation (no government resources) would produce little payoffs, with only a public optics affect. Significant benefits in the medium to long term would require active participation; "significant and focused economic benefits are unlikely to result unless Canada is willing to put money on the table," warned DRIE.[84]

Alongside internal considerations, the defence industrial and technological dimension received an ironic boost from the Canadian Chamber of Commerce Science Council.[85] The council supported Canadian participation in SDI research on grounds that SDI wouldn't work. Drawing from discussions between Canadian and US scientists, the council held that SDI would neither be feasible nor be built. US scientific support was largely for the money. Thus, there were no problems for Canada politically or strategically, as Canada could gain important technological benefits without any consequences. On March 29, the Aerospace Industries Association of Canada also gave its endorsement to SDI participation.

The long-awaited joint External Affairs and National Defence memorandum to cabinet was by now expected for the April 23 meeting of the Policy and Priorities Committee.[86] In the meantime, the pressure abated somewhat when it appeared that the United States would not hold the allies to Weinberger's sixty-day deadline for a response, though fear remained that the administration might change its mind again or at least seek a statement of support at the Bonn Economic Summit on May 2. Also somewhat reassuring were the conditions Nitze placed upon SDI deployment in the Alistair Buchan Memorial Lecture on March 28 in London:

> They [SDI] must produce defensive systems that are reasonably survivable; if not, the defenses could themselves be tempting targets for a first strike. This would decrease rather than enhance stability. New defensive systems must also be cost-effective at the margin – that is, it must be cheaper to add additional defensive capability than it is for the other side to add the offensive capability necessary to overcome the defense. If this criterion is not met, the defensive systems could encourage a proliferation of countermeasures and additional offensive weapons to overcome deployed defenses, instead of a redirection of effort from offense to defense.[87]

Not only did he raise doubts about the likelihood that SDI would ever be deployed on strategic grounds, but Nitze also, as Thatcher before him, clearly separated participation in the SDI research program (which did not violate the ABM Treaty) from the future issue of deployment, whenever the technology actually became available. If Canadian and other scientists were to be believed, that was a long way off. Thus, the situation appeared to provide the government with the means to accept formally the US invitation without committing to the deployment of SDI (which would likely be a question for a future government anyway). Moreover, not only could the government rely upon the Thatcher and Nitze criteria for support, and Article 9 of the ABM Treaty for protection, but, more importantly, such a position was consistent with government policy most recently enunciated by Nielsen on March 26 echoing Thatcher's rationale – the prudence of research relative to Soviet activities in the same area.

Besides, the decision had to be placed in the context of the significance attached in Washington to SDI and the potential impact of saying no on the overall strategy of the government to improve Canada's relationship with Washington and its international standing on the basis of a close personal relationship between Brian and Ronnie. As one senior official wrote, "such a decision [no] would fly in the face of efforts the government has been attempting vis-à-vis the U.S."[88] Moreover, the United States was not ignorant of the Canadian internal situation on SDI. Nonetheless, government pronouncements to date would not lead Washington to expect a negative decision, such that a no could have even more profound implications on the relationship – a situation that would result in February 2005.

Problematically, officials in External Affairs were nearly paralyzed by the Canadian dilemma, with no clear guidance from above. Their preference was to delay a decision on the grounds that more information would be needed from the United States before a decision could be made. Yet they were reluctant to engage the United States in discussions for fear of misinterpretation. Officials feared that a research commitment now would dictate a deployment commitment later. At the same time, officials were sensitive to the costs of indecision in terms of Canadian engagement with SDI and other future programs as well as public attitudes toward SDI. The former reflected the NORAD air defence modernization program, of which SDI might or might not be a component. The latter concerned the fallout from the previous government's decision to allow the United States to test its ALCM over Canada.

Officials recognized that Canada's final decision would be significantly affected, if not trumped, by any consensus that might emerge out of the Bonn summit.[89] The original invitation, of course, had been issued to all the NATO allies, and even though the implications for Canada were distinct from those of the other allies, in large part because of geography, Canada would be hard-pressed to ignore a NATO consensus. Moreover, NATO offered an opportunity to manage potential domestic political fallout of participating in Star Wars, with the same rationale that had been employed to sell cruise missile testing. Justifying such participation as part of Canada's multilateral alliance commitment would also play upon deeply held beliefs among the attentive and general public about the fundamentals of Canadian foreign policy, including the notion that the United States, not Canada, was responsible for keeping North American defence largely separate from European (NATO) defence. Finally, embedding a Canadian yes within an alliance yes would deflect arguments of Canadian subservience.

Unfortunately, the alliance had not and would not produce a consensus on SDI participation. Initially, the allies as a whole responded favourably to the NPG communiqué on March 26 on the grounds of alliance solidarity, the prudence of research in light of Soviet efforts, and the legality of possible allied participation under the ABM Treaty permitting research. However, regular tracking by External Affairs indicated that the alliance divided not only on SDI participation but also on interpretations of the motives and implications of SDI. In early April, External Affairs reported that the United Kingdom and West Germany were most likely to agree to participate, and if so Italy would follow.[90] The Netherlands was likely to come on board, whereas Belgium and Norway were difficult to predict. Denmark and Greece were out, and Portugal, Spain, and Turkey lacked the technological capacity to participate regardless. Outside Europe, Japan was viewed as a likely participant, while Australia had already said it would not participate. Finally, and arguably most importantly, it was unclear for some time how France would respond.

French president Mitterrand had been critical of SDI, not least of all because the Soviet development of a similar system would undercut the independent French nuclear deterrent. Moreover, the French had significant doubts that the United States would provide foreign firms real access to SDI contracts. Nonetheless, the French saw SDI largely in terms of its broader implications for the technological gap between the United States and Europe. As a massive research investment program with significant dual use or civilian payoffs, SDI would likely widen this gap. The French, while strategically opposed to SDI for reasons similar to ABM, recognized that Europe could not simply condemn SDI and stand aside. Instead, the French believed that a coordinated European position was needed – one that would include greater European technological investment through the European Community. To this end, the French initially looked to the possibility of French companies participating in SDI research under two conditions: real access to US technology and no diversion of investment from European space programs.[91] In the end, however, the French would say no.

The initial French position hinted at a middle-of-the-road option beyond the yes or no to the Weinberger invitation – one in which governments would stand aside but their national companies would be allowed to participate. This alternative had been hinted at in the February PCO memorandum to the prime minister, suggested by Mulroney to Bush and Schultz in March, and finally directly communicated to the government on April 16 as the means to finesse the issue.[92] However, an idea was one thing, the implications of such policy were another, especially if one accepted the Canadian ambassador's view that SDI was a loyalty test. What was needed was a signal from the United States of the acceptability of a middle-of-the-road policy, and that process had begun almost immediately after the Weinberger announcement.

From the onset there had been concern, if not opposition, within the Reagan administration to issuing a formal invitation to allies on two specific grounds. First, Washington recognized the political problems the invitation held for allied governments, especially with public attitudes toward Reagan and the legacy of the bruising INF fight (and, in Canada, the ALCM issue). The United States was sensitive to the situation in Canada, as communicated by the president directly to the prime minister: "I understand that SDI remains a sensitive issue in Canada and I know that it has come under repeated scrutiny in the House of Commons and in the press. Thus, I appreciate both your steadfast endorsement of the prudence of our research program and your thoughtful consideration of possible Canadian participation."[93]

Second, there was no consensus within Washington on technology access and transfer issues. These concerns became evident fairly quickly when administration officials began to back away from Weinberger's sixty-day deadline for a response. On April 10, US officials informed Ottawa that there was no deadline for a Canadian answer and that the United States was willing to provide additional technical information on SDI. This would be reinforced several days later when Weinberger informed Nielsen that the deadline was not an ultimatum.[94]

With all these considerations at play, the prime minister decided to take the issue out of the hands of his ministers and the departments and assign it to an outsider. On April 18, Arthur Kroeger, a senior civil servant then without a formal position in government, was asked to examine Canadian participation in SDI and report directly back to the prime minister. His terms of reference were to investigate the SDI research program, examine the economic and financial dimensions for Canadian scientific and technological interests, and evaluate the strategic and arms control issues, especially those concerning the ABM Treaty.

Even though Mulroney took SDI out of the hands of External Affairs and National Defence, the two key departments still had direct influence on the report. Kroeger's task force was drawn directly from the ad hoc interdepartmental SDI working group. Its task was not just to provide initial briefings and expert opinion to Kroeger but also to influence as best as possible Kroeger's final recommendations to the prime minister and keep senior officials informed (which Kroeger clearly understood). In the end, however, the report was Kroeger's and directly for the prime minister, leaving External Affairs and

Kroeger Task Force on SDI

Chair
Arthur Kroeger

Members
Tim Garrard, director, Electronics and Aerospace Branch, Department of Regional and Industrial Expansion (DRIE)
Rick Clayton, policy analyst, Space Policy and Plans, Ministry of State for Science and Technology (MOSST)
Dr. Derek Schofield, chief, Research and Development, Department of National Defence (DND)
Jon Anderson, assistant deputy minister, Policy, DND
Arthur Mathewson, chief, Policy Planning, DND
W.M. Joe Beckett, director, Nuclear and Arms Control Policy, DND
Col. Bill Weston, SPA-3, DND
David Karsgaard, director, Defence Relations, Department of External Affairs (DEA)
William Reid, Defence Programs Bureau, DEA
John Bryson, Defence Relations, DEA
Robert Fowler, assistant secretary to the cabinet, Foreign and Defence Policy, Privy Council Office

National Defence to speculate on its final contents, having been party only to the numerous drafts. Cabinet did not see the report either and was limited to a fifteen-or-so-minute briefing provided by Kroeger in August.[95]

As far as External Affairs and National Defence officials on the task force were concerned, there was agreement that no was not an option. Both departments communicated to Kroeger the importance of a clear and simple communications strategy, regardless of the decision. The reality of SDI as simply a research program within the limits of the ABM Treaty had been completely lost in the public debate. Instead, SDI had become mostly understood as a deployment program under the rubric of Star Wars. Regardless of the ultimate decision, government and departmental officials were deeply concerned about the manner in which any decision was communicated to the public. A public no could not entail an outright condemnation of SDI for fear of US reaction. A public yes had to clearly demarcate research participation from deployment. Anything in between had to manage the possibility of both potential conclusions.

Despite these two areas of agreement, External Affairs and National Defence continued to disagree on timing. Whereas National Defence looked to move quickly, External Affairs was more cautious, in large part because of concerns that moving too far forward could create significant problems in the future when or if SDI moved from research to deployment. The more Canada indicated a positive position on SDI participation, the more difficult and potentially costly it would be for Canada to break with the United States over deployment – in effect, in for a penny, in for a pound. Conversely, the more cautious Canada became without saying no, the less access Canada would have to US defence developments, technology, and knowledge about SDI itself.

This is not to suggest that there was now unanimity within External Affairs. After years of Trudeau's foreign policy during the heyday of arms control and détente, and the first four years of the so-called Second Cold War, it would be surprising if there were agreement among officials that the issue of Canadian participation was simply one of timing and caution about the future. In fact, one expected (and some in the Mulroney government feared) a bureaucracy dominated by a centre-left perspective on foreign and defence policy that would undermine new policy initiatives. This did not prove to be. Nonetheless, voices of concern did exist, and these were raised in internal discussions and found their way to Kroeger. In one such case, an official argued that Canada should not only decline the US invitation as tactfully as possible but work behind the scenes with allies to lower expectations about SDI in public and insert SDI into the Geneva negotiating process.[96] In so doing, he presented the panoply of critiques of SDI that encompassed its destabilizing nature; the likely death of the ABM Treaty; few, if any, economic and technological benefits; dubious technologies; little impact on NORAD; and its negative public impact that could undermine other important objectives regarding closer relations with the United States.[97]

In the conclusion of the primary internal response to this critique, a senior official emphasized that relations with the United States were the prime minister's top priority, and Reagan had made SDI one of his top priorities. In so doing, the official concluded: "These ideas [critique] would fit very well into another time and another government, but I suggest they are rather foreign to the current PM's thinking."[98] Insightful about Canada's unique position, the official argued that Helmut Kohl, the West German chancellor, and Thatcher understood the importance of the relationship with the US president, despite their reservations about SDI, partially because they could not automatically rely upon the United States as Canada could in defence. Finally, public opposition should not be overemphasized. The government should not necessarily be a slave to public opinion, for if it had been Canada would have never agreed to cruise missile testing.[99]

The internal divisions within External Affairs and the lack of clear direction from the minister affected the nature and quality of the department's involvement in the process leading to the final decision. External Affairs was seen as directionless, lacking a "corporate view."[100] Instead, External Affairs officials on the Kroeger task force provided a panoply of views ranging from opposition from the disarmament perspective to support from the Canada-US side to confusion from defence relations to silence from the international trade side.

Echoing National Defence's preference for quick movement forward, Nielsen warned Kroeger that to concentrate on broader strategic-related issues would cause months or years of delay and that it was better to focus strictly on the research invitation.[101] Until late May, three basic options dominated government thinking: yes, no, and a qualified yes employing the Thatcher criteria with or without direct government engagement. In late May, the military representative on the task force, Col. Bill Weston, communicated to Kroeger a fourth option focused on air defence modernization. In earlier discussions with the United States on the modernization of the North Warning System (NWS), Canadian officials suggested that the countries jointly pursue a space-based air defence surveillance system. Such a system was in Canada's long-term strategic interest in recognizing that space-based assets would eventually obviate the importance of ground-based systems, such as the NWS – a concern identified originally during the ABM debate and present in every subsequent debate after SDI. The research and technology involved in developing such a system would have significant overlap with SDI research. As such, the government could move forward with a proposal to engage SDI from an air defence modernization perspective on the basis of the difficulty of separating missile defence and air defence research in the areas of space-based surveillance, reconnaissance, and target identification and tracking. Weston warned that the United States would go it alone if necessary, and unless Canada participated in some manner in SDI research the United States would tighten access restrictions – if not close off access entirely – on Canada on North American defence.

In communicating this option to Kroeger, Weston argued that the minimalist option of private company involvement had met with little public opposition or concern.[102] But this option would likely not suffice in obtaining major economic and technological value for Canada. To obtain such value, Canada would have to go further and provide government funding and undertake work within government research facilities. However, if the government had reservations in going this far, it could follow through with the minimalist option and propose a bilateral cooperative research program on future surveillance technologies against bombers and cruise missiles. US military officials recognized such research as being vital to deal with a Soviet expansion of its air-breathing strategic forces as one response to SDI but that such research was important even if SDI did not materialize. As well, such a proposal was consistent with the historical nature of cooperation in the defence of North America. It did not affect the ABM Treaty question whatsoever. It was also directly in Canada's strategic interests, ensured a voice and access for Canada in the future of North American defence, and kept SDI at a distance. In other words, Canada could participate without participating because the research on future air defence simultaneously had independent importance and value for missile defence.

Certainly, the option made a great deal of sense, especially as a means to inform and guide Canadian engagement and thus answer Weinberger's second question. However, it needed to be communicated very carefully, and the message was somewhat complicated for a public, which thanks to the peace movement and the Opposition increasingly confused SDI as research with SDI as an operational weapons-in-space program. Moreover, as the option required Canadian investment, it was in effect a yes to government participation, and any attempt to differentiate air defence research from SDI research would smack of duplicity or at least be readily seen that way by the Opposition and critics of SDI. If the government was to be in, it might as well bluntly say so. The specifics of Canadian research engagement could be left to private discussions with the United States.

Of course, the key for this and any of the other options was predicting the US response. Although not a unanimous view, senior US defence officials had already made the SDI invitation safe in their discussions with Kroeger, just as McNamara had made ABM safe for Pearson. Less than a week after his appointment, Kroeger travelled to Washington to meet with senior officials and non-governmental experts. His first meetings were with Iklé, the SDI point man, and Richard Burt, the assistant secretary of state for politico-military affairs. At this crucial meeting, he received two conflicting messages. Iklé suggested that the United States did not need a formal answer from governments. Rather, Weinberger's invitation had simply been a signal that the SDI research process was open to the bidding and participation of qualified allied companies. Government involvement would become an issue only if or when a system moved to deployment as part of alliance strategic concerns.

Burt, however, was more cautious in noting that there were differences of opinion within Defense and the administration on whether a government-to-government agreement or memorandum of understanding (MOU) would be required. Certainly, a formal MOU would provide more benefits to Canada and, perhaps most importantly, enable Canada to influence the SDI decision-making process and program direction. But Burt recognized the political sensitivity of such a formal commitment for Canada.

The next day, April 26, Kroeger met with Leslie Gelb, who had been Burt's predecessor in the Carter administration. Gelb argued that there was no need for a refusal, as the program would go on regardless. In contrast to External Affairs' fears, conditional support within the parameters set by Thatcher would act as future constraint on a SDI deployment decision. Moreover, with almost unanimous expert opposition to SDI, public support would eventually dissipate within a couple of years. As far as experts were concerned, John Steinbrunner of the Brookings Institution, a long-time arms control advocate and staunch opponent of missile defences in general and SDI in particular, told Kroeger that a Canadian refusal would bolster anti-SDI forces in the United States.

Kroeger's final meeting was with Lt. Gen. James Abrahamson, the director of SDIO. Abrahamson told Kroeger that the issue of technology transfers would not require a new agreement but could occur under the existing agreements governing the protection of classified information between Canada and the United States. In addition, he reassured Kroeger that no nuclear information of substance, or blueprints, would be transferred that would ostensibly run afoul of the ABM Treaty.[103] In the end, he went further than Iklé and Burt by noting that neither an MOU nor an exchange of notes was necessary – simply a letter in which Canada would identify areas of interest and request US assistance in obtaining more information.

On May 7, Kroeger provided an initial report on his work to the prime minister. He emphasized that the United States had backed away from seeking formal allied agreements: "No formal 'signing up' for the program will be required, nor will the US be seeking financial contributions from participating countries."[104] Instead, existing arrangements primarily concerning the treatment of classified material sufficed. Kroeger added, however, that the government would still have to make a formal decision and communicate it to the public and Parliament. As for his forthcoming report, he recommended that it remain private and classified to protect the government relative to its final decision.

Kroeger's initial views were reinforced by External Affairs. In a memorandum to Kroeger, External Affairs reported that Iklé had clearly indicated that it would be enough for governments to signal they did not object to their companies bidding and engaging in SDI research, largely because of a recognition of the political sensitivity of going any further forward.[105] As such, External Affairs now believed that, perhaps, the allied response to SDI was not a loyalty test. More importantly, the United States had signalled, as clearly as one could expect, the type of positive response that it would deem acceptable – the very response announced by Mulroney on September 7. If there were any doubts, these were cleared up by the US ambassador to NATO and Iklé on June 25 and 29 respectively. Reflecting Ambassador Abshire's statement to the NATO

Council on the 25th, Iklé reiterated that the Canadian government did not have to reply to the invitation but only decide whether it would allow or block company involvement.[106]

Just before submitting his report, Fowler forwarded his views to Kroeger. First of all, he argued that Canada could not say no: it would contradict the prime minister's policy of improving relations with the United States "at a time and in an international economic environment in which good relations are likely to be vital to Canadian prosperity."[107] It would also contradict government policy of offering full support to the United States during ongoing arms control negotiations in Geneva, as Clark stated in the House on January 21. Second, he rejected full, unqualified support because of significant public opposition in light of public preferences for multilateral solutions to international issues. In this regard, US unilateral proclivities on SDI were problematic. Outside expert opinion also believed that SDI was technologically unachievable and, if deployed, destabilizing.

Instead, Fowler argued for a "shade of grey" in Canada's response to the invitation. In so doing, he noted that a companies-only decision would contradict repeated government statements that SDI research was prudent. Thus, government policy should reiterate the prudence of research; permit company engagement, especially since Canadian companies were already engaged and would continue to act according to their interests; and announce that the government would engage in areas vital to future Canadian defence needs. Although Fowler was sensitive to the issue of the militarization and weaponization of space, he felt that space-based military assets were already of significance for the Canadian Forces, the Outer Space Treaty banned only nuclear weapons in space, and, above all else, failure to be engaged would "surrender the protection of our territory to the Americans." Finally, failure would undermine the future of Canada-US defence cooperation in North America, while eliminating any possibility of Canadian influence on future US plans that might affect existing and future arms control. Presaging the Liberal rationale for entering into discussions with the United States on missile defence in 2003, the "camel in the tent" was the preferable policy solution.[108]

On June 6, Kroeger completed his report, which was formally sent to the prime minister on June 26. Not surprisingly, it contained four options: decline participation, limit participation to the private sector, complete government support, or support concentrating on relevant air defence research.[109] In covering the wide range of issues involved, Kroeger concluded that the second option, limiting participation to the private sector, was optimal.[110] In so doing, he suggested that Canada's reply had already been given. The government should simply reiterate its January 21 statement of SDI research as prudent, adding only that no work would be undertaken at government facilities and no government money allocated for SDI research.[111] With the exception of the no money recommendation, which had never been central to the internal debate on SDI, and after months of private and public debate, no significant change in government policy was to occur.

Shortly after the report was submitted to the prime minister and two months before the formal public announcement, the government established a joint parliamentary committee to review Canadian foreign policy, initially tasking it to examine two issues: SDI and free trade.[112] It began by undertaking extensive cross-country consultations on SDI, similar to the private Liberal Party initiative led by Jean Chrétien, which naturally concluded its efforts by reporting widespread public opposition to SDI. Whether a function of the process itself and the type of witnesses heard from, the chair, Tom Hockin, in the lead up to final deliberations, suggested that the committee could be employed as a means for Canada to say no. Linking numerous issues, including free trade, to the government's decision on SDI, a no on SDI was "a unique opportunity to prove its [government] bona fides as a defender of our national interest while also appearing 'internationalist.'"[113] In so doing, future contentious issues (free trade) would be easier to manage publicly. Further, Hockin argued that a committee no would help pro-SDI forces in cabinet accept a no, communicate the situation in Canada to the United States, and demonstrate an effective role for Parliament. The clerk of the PCO in supporting Kroeger's recommendations responded to Hockin's idea that such an act would "needlessly annoy the United States which is backtracking at a furious pace on SDI ... In essence, the issue [SDI] seems to be dying – let it do so in peace."[114] In the end, the Conservative-dominated committee would report little of value and recommend nothing. The Progressive Conservatives might be divided on SDI, but the prime minister was in charge. As far as public opinion was concerned, during the final stage of the government's decision-making process, significant support existed that would be long forgotten in future debates. In a May Gallup poll, 53 percent of Canadians supported participation, with only 40 percent opposed.[115] Two months later, a July 2 CROP poll found 49 percent of the Canadian public favoured Canadian participation, with only 36 percent opposed. In a follow-up poll on August 10, support for Canadian participation rose to 57 percent, with 65 percent in favour if participation translated into jobs. Most importantly, those opposed to SDI were more likely to be opposed to the government regardless. SDI might not be a vote winner, but nor was it likely to be a vote loser. Besides, an election was years away.

In the end, it would be misleading to suggest that the ultimate decision was made in Washington. Certainly, as in the ABM case, the government, or perhaps more accurately the prime minister, waited until the United States made up its own mind in terms of the actual choices available. Officials long recognized that no was simply not an option, even before the US clarification of Weinberger's unauthorized invitation. Indeed, just before the formal public announcement of the decision of September 7, External Affairs expressed concerns to Clark that the decision had not gone far enough.[116] The administration, especially Reagan and Schultz, would neither expect nor be prepared for such a decision from its closest ally and friend. It would likely have a negative impact on the prime minister's strategy of close personal relations and be a symbolic blow to the United States with regard to the coming Reagan-Gorbachev Summit. The decision might also affect administration support for free trade, an acid rain deal, and defence trade. External Affairs believed too that the decision would affect defence relations directly, noting similar significant concerns from National Defence. It recommended that no decision be taken on government participation pending more information and a greater alliance agreement and that the government should simply announce that Canadian companies could participate.

This last interjection came to naught. Just before the public announcement on September 7, Mulroney phoned Reagan with the news. Reagan's response was positive in viewing the decision as support of SDI. He also believed that there would be no problems as a result.[117] Close personal relations with the president emerged unscathed from the SDI decision. Indeed, within two weeks, any fears of a negative linkage to other areas of concern evaporated with Reagan's request for fast-track legislation for free trade negotiations with Canada – a policy priority of the Mulroney government. This was followed by agreement on the acid rain issue. Free trade, along with the 1987 White Paper's proposal to acquire nuclear-powered submarines, would push SDI off the agenda. With a majority in the House, the final report of the joint committee served to put missile defence to sleep. However, all was not fine for the bureaucracy. SDI would linger, especially with the new broad interpretation of the ABM Treaty in the fall of 1985, the issue of defining the line between company and government involvement, and, most importantly, the negative reaction of US-North American defence planners, which took the decision in a narrow literal fashion as a no. These three issues the bureaucracy confronted out of public sight, awaiting the next resurgence of missile defence – the Gulf War and George Bush Sr.'s Global Protection against Limited Strikes (GPALS) program.

Act 3

Global Protection against Limited Strikes: Too Close for Comfort (1986-92)

On the evening of January 28, 1991, twelve days after the launching of Desert Storm, President George H. Bush in the State of the Union Address announced that SDI would be translated into a limited defence system, somewhat reminiscent of Sentinel and Safeguard, and that the United States would proceed quickly to full development and deployment. The system, known as Global Protection against Limited Strikes (GPALS), was to entail two layers: ground-based and space-based interceptors. The announcement followed immediately the start of the Scud-Patriot missile duel in the Gulf War and received two major shots in the arm over the next eighteen months. In the fall, Congress passed the first National Missile Defense (NMD) Act, mandating the deployment by 1996 of one hundred ground-based interceptors as allowed by the ABM Treaty and calling for renegotiation of the treaty to permit multiple ground-based sites and interceptor numbers of between 750 and 1,200. The next year at a summit with Russian president Yeltsin on June 17, the two presidents called for a global protection system (GPS) to be developed through a joint Russian-American program and deployed following either an agreed legal reinterpretation of the treaty or an entirely new treaty.

Roughly a year or so later, following the defeat of Bush in the 1992 election, GPALS disappeared, and the Strategic Defense Initiative Organization (SDIO) was renamed the Ballistic Missile Defense Organization (BMDO). In October 1995, the Ballistic Missile Defence Act replaced the NMD Act in recognition that a limited defence system for the United States would not be possible by 1996. The next year, President Clinton's secretary of defense, William Perry, announced the NMD development and deployment program called 3 plus 3 – three years to develop and three years to deploy operationally, presuming that the technology was available and a deployment decision was made.

As far as Canada was concerned, GPALS didn't just pass into oblivion; it never really existed except for a few experts, and, of course, the bureaucracy. Fears inside government reached near-fever pitch that George would extend an official public invitation to his close friend Brian for Canada to join in. Regardless, many in Canadian officialdom saw the passing of GPALS as evidence that missile defence would never, ever see the light of day. Technological barriers were too high, and his successor's 3 plus 3 program was perceived by the Republican mob as Clinton guarding his political flank against charges of being weak on defence and national security issues. In reality, however, GPALS indicated the inevitability of missile defence, and Canada's failure to respond generated a growing drift toward US unilateralism in continental defence – the very behaviour that Canadian decision makers had long hoped to avoid.

Despite lingering public attention following the September 7 SDI announcement, by 1987 SDI had simply disappeared from the public agenda, even though the large US research program continued. As for GPALS, it was as if it never existed. The GPALS announcement passed with virtually no comment.[1] The Scud-Patriot missile duel in the Gulf War found no public comment on its possible implications for the SDI program and other, strategic (North American) applications.[2] Even the 1991 NMD Act passed in silence, as did the 1992 GPS announcement.[3] The Canadian academic community of relevance, overwhelmingly obsessed with the new world of UN peace operations, said nothing. In the House, GPALS was worthy of only two passing questions, in February 1992. For all intents and purposes, one could easily believe that nothing of significance concerning missile defence had occurred since the fall of 1985. It was as if the Mulroney decision had killed SDI – unless one was a senior civil servant in External Affairs, National Defence, or NORAD. For them, SDI had not disappeared, and GPALS raised the spectre of another invitation to precede the actual deployment of an operative missile defence capability for the defence of North America.

GPALS could not be ignored, even if the Mulroney government, increasingly tired and under siege, wished it to go away. The question of Canadian participation loomed once again, and this time the close personal relationship between Brian and Ronnie had been supplanted by an even closer relationship with George. A new policy was needed, not least of all because the 1985 decision had actually been a non-decision: the government had only said no to official participation in SDI research with the clear blessing of the Reagan administration. It had avoided an actual decision on participation in a future missile defence system, because no decision was needed. Moreover, a new policy was needed as a function of movement on the SDI-testing question south of the border, the impact of the 1985 decision on the defence industrial and technology front, and growing problems of Canadian access through NORAD.

The ABM Treaty was Canada's safety blanket. It provided an automatic ideal answer for Canadian governments wishing to avoid any potential problems stemming from US missile defence developments. Whatever a government might (or might not) want to do regarding missile defence, it had little choice because of the Article 9 restrictions – and this, of course, was understood in Washington. SDI as a treaty-compliant research activity had slipped through the cracks with all the attendant domestic and international furor. Research was one thing, and the United States had made it clear that allied engagement would be determined according to Agreed Statement G limits prohibiting the transfer of blueprints to third parties. Development and deployment were something else, and alongside the treaty the government could look to the Thatcher criteria, alliance statements, and Canadian policy pronouncements in the House that placed the matter of development and deployment as a question for negotiations with the Soviet Union.

Unfortunately, a month after the September policy statement, the United States formally placed the issue of SDI development or testing on the international political agenda with a new – or, as some viewed it, reinterpretation – of the ABM Treaty. With it, the real extent of Canada's safety blanket emerged as a potential, and naturally unwelcome, issue, even though SDI opponents and critics in Canada failed to see the linkage clearly. Instead, like much of the SDI and earlier ABM debates, the opposition was framed in strict international security terms under the implicit assumption that Canada's national political and strategic security and defence interests and requirements were identical with, and followed from, international political and strategic security and defence interests, whatever they really were and whoever truly defined them – a point clearly made in the 1987 Defence White Paper, which explicitly defined Canadian national security interests as strategic stability, no matter its real meaning, to the drafters and the government.[4]

After more than two years of work, the Reagan administration had little to show for its significant investment in SDI research. To maintain momentum and public support, and ensure continued funding from Congress, evidence of progress was needed, and this need took two forms. First, some notional SDI architecture was required, and this would be subsequently dominated by two notional ideas/architectures from the multifaceted and multidimensional SDI research program – Brilliant Eyes and Brilliant Pebbles. The former, Brilliant Eyes, was to be an advanced space-based surveillance and warhead tracking and target discrimination system, which would evolve roughly a decade later into the Space-based Infrared System – Low (SBIRS-Low, space-based infrared sensors in low earth orbit), subsequently renamed the Space Tracking and Surveillance System (STSS) for the current Ground-based Mid-course Defense (GMD) system.[5] The latter, Brilliant Pebbles, was to be the primary architecture of a space-based defence entailing thousands of small kinetic-kill satellites on orbit, capable of independently identifying a target, and manoeuvring to collide with it in outer space during the mid-course phase. This idea would become the space-based component of the GPALS architecture and would re-emerge with the 2004 proposal by the Missile Defense Agency (the successor to the BMDO in 2002) to deploy a space-based kinetic-kill test bed in 2008.

Second, the notional architecture had to be more than simply on paper, and this meant demonstrating progress by the actual testing of SDI technologies. However, the ABM Treaty contained strict prohibitions on testing but not laboratory research.[6] One of the original US motives behind the prohibitions on "testing in an ABM mode" (Article 6 of the ABM Treaty) was concerns that extensive high-altitude Soviet SAM defences might be modified for use against ballistic missiles, thereby enabling the Soviet Union to break out of the treaty very quickly. To do so, the systems would have to be tested against ballistic missile warheads, and hence the United States sought and obtained strict limits on testing. With SDI, however, the situation was reversed. Now the United States sought a legal means to move forward with development (outside laboratory work) and testing. This required either a new or a reinterpretation of the treaty, depending upon one's view as to whether the line between all research and legal and illegal development and testing was clearly drawn and agreed upon by the two parties. Regardless, in October 1985, Abraham Sofaer, a State Department legal advisor, brought forward an interpretation of the treaty that would open the door for SDI research to move out of the laboratory and into development, testing, and evaluation.[7]

Sofaer argued that, during the original negotiations, the United States had sought a ban on any and all missile defence technologies beyond the formal, explicit reading of the definition of an ABM system and its components in Article 2: "(a) ABM interceptor missiles, which are interceptor missiles constructed and deployed for an ABM role, or of a type tested in an ABM mode; (b) ABM launchers, which are launchers constructed and deployed for launching ABM interceptor missiles; and (c) ABM radars, which are radars constructed and deployed for an ABM role, or of a type tested in an ABM mode."[8] The Soviet Union had rejected this ban. On this basis, drawn from the secret negotiating record, Sofaer thus argued that testing covered only ABM technologies that existed at the time of negotiations and ratification as spelled out in Article 2. Sofaer also referred to Agreed Statement D, which accordingly made little sense if the treaty covered any and all missile defence systems and components: "The Parties agree that in the event ABM systems based on other physical principles and including components capable of substituting for ABM interceptor missiles, ABM launchers, or ABM radars are created in the future, specific limitations on such systems and their components would be subject to discussion in accordance with Article 13 and agreement in accordance with Article 14 of the Treaty."[9] Sofaer's position, which was adopted by the Reagan administration, became known as the broad interpretation.

Quickly following the new interpretation of the treaty, missile defence opponents argued that Sofaer was in error and presented what would quickly become known as the narrow interpretation. Accordingly, any and all missile defence systems and components were under the treaty's prohibitions as a function of the Article 5 commitment "not to develop, test or deploy ABM systems or components which are sea-based, air-based, space-based or mobile land based" and the preamble of Article 2, which ended with the phrase "currently consisting of." This phrase broadly read did not limit the scope of the Article 2 definitions but meant to include any future systems as well (and indeed much was debated about the interpretation of the meaning of the comma that ended the phrase). As such, Agreed Statement D in this context was meant only to allow for discussions and possible agreement to allow other ABM systems, but in the absence of agreement they were prohibited by default. In other words, all was prohibited unless both parties agreed otherwise.

There was significant opposition to the broad interpretation in Congress and among US expert opinion. In the context of this debate, Democratic Senator Nunn, who had been present at classified briefings during the ABM Treaty ratification process, offered the most politically sophisticated argument of the day. At the time of the treaty's submission to the Senate for advice and consent, the Nixon administration, drawing from the secret negotiating record, had communicated the narrow interpretation. If Reagan wished to reinterpret the treaty, he would have to submit the treaty again to the Senate for advice and consent. Knowing full well that a resubmission could fail to obtain the necessary two-thirds majority, Reagan faced three options: to accept on constitutional grounds the narrow interpretation, potentially face a Supreme Court challenge (whose implications might go far beyond just the ABM Treaty interpretation issue), or withdraw from the treaty.

As the debate unfolded in the United States, both External Affairs and National Defence were concerned about the implications of the broad interpretation. Initial legal opinion within External Affairs cautiously sided with the broad interpretation in the absence of any direct knowledge of the private negotiating record.[10] Regardless, in not being a party to the treaty, the Canadian government had no role to play in deciding the question, even if a new interpretation might have direct implications for Canada. It was also obvious that if development and testing were permitted, the deployment issue would naturally follow. Even if deployment remained prohibited, the future looked bleak for arms controllers. The USSR might withdraw if the United States proceeded to act upon its own interpretation of the treaty. Having successfully tested the new technologies, the United States might invoke Article 15 and withdraw itself. Finally, the treaty might well become simply irrelevant, as the United States and possibly the USSR developed and tested new exotic missile defence technologies enabling either or both to break out from the treaty easily and quickly.

The broad interpretation – at least theoretically – meant that the scope of the Article 9 prohibitions on third-party involvement, even if Sofaer had not gone as far as arguing that deployment of non-ABM systems and technologies was permitted under the broad interpretation, would be radically changed. If Brilliant Eyes and Brilliant Pebbles were not covered by the Article 2 and 5 prohibitions and only required discussion (which implied the possibility of agreement to disagree), they would not be included under Article 9 and Agreed Statement G, as they pertained to the role of third parties. These, and potentially other relevant SDI technologies under the broad interpretation, would not be ABM technologies, and the development and testing of any systems based upon these technologies would not fall under ABM. Certainly, Article 9 would still bear on future Safeguard-like systems, such as the future development of the ground-based component GPALS. However, other ground-based technology systems, such as lasers, would not, and certainly any space-based systems or components, such as the second element of GPALS, would be outside Article 9.

Suddenly, the blanket might not provide much safety, and the Reagan administration appeared to be determined to push the issue. Even before the enunciation of the broad interpretation, administration officials had begun to back away from Thatcher's subject-to-negotiation criterion. With a new interpretation at play, the United States might seek allied endorsement, as it had done with SDI research, which had produced an alliance consensus that such research was prudent in light of Soviet efforts. In seeking another endorsement, the United States might return to Weinberger's original invitation, which had been framed around treaty-compliant research relative to Article 9. With the broad interpretation in place, the door was open much wider to the allies, and it would not be difficult for the administration to believe that the allies would naturally want to revisit their decisions under new circumstances. Regardless, the implicit combination of seeing missile defence as an ally loyalty test and an allied endorsement providing legitimacy for the US decision was now in place – if research was prudent, weren't development and testing as well? Besides, whether the allies liked it or not, they could not simply fall back on their status as non-signatories and thus avoid comment. Domestic forces, regardless of their views on SDI and missile defence, would drive a national response and were unlikely to accept no comment.

This was clearly a nightmare in the making for the government and senior officials in External Affairs and National Defence. There was no policy to cover this contingency. If this occurred, the conspiracy surrounding the *post facto* dropping of the NORAD ABM exclusion clause would reappear with a vengeance. The Opposition, which from the beginning linked SDI to weapons-in-space, would see this as proof positive of a government not only about to abandon Canada's commitment to the non-weaponization of space but also ready to sacrifice arms control *in toto* if the broad interpretation led the Soviet Union to withdraw under Article 15. It also meant the possibility of a major fight between the United States and Europe if the Thatcher criteria become inoperative. As one official noted early on in the 1985 SDI debate, whether SDI worked or not was irrelevant for the European allies.[11] If SDI led to a credible defence, it would be destabilizing, result in the Soviet Union developing its own system, and leave Europe as a vulnerable battleground between two secure giants. If it didn't, SDI would still be destabilizing and extremely expensive, draining resources away from other pressing defence needs. Either way, SDI moving forward and threatening the future of arms control, especially with improving East-West relations, would be nearly impossible for the European allies to swallow.

On the other side of the equation, if SDI was a loyalty test and the government failed, the prime minister's overall foreign policy strategy might collapse. Free trade might become hostage to SDI. The US administration, Congress, or both might turn against free trade, and other issues of vital concern to Canada (Arctic sovereignty and acid rain) might be sacrificed. In addition, the domestic climate would change as well. If Canada acquiesced to the new version of SDI, the image of subservience as Hockin had warned might undermine domestic support for free trade. Canada would also face a major threat to its strategic bilateral defence relationship if it stood aside from the most important new development in North American defence. Canada, after decades of balancing its international/alliance interests with its continental interests, might be forced to choose. In other words, the SDI nightmare could force a complete rethinking of Canadian strategy – a possibility that no government would look positively upon.

Fortunately, the worst nightmares of senior officials could find some solace in the waning of public and opposition attention to SDI. The issue of the ABM Treaty had drawn the attention of the Special Joint Committee of the Senate and the House of Commons on Canada's International Relations, whose initial mandate had included SDI, but the committee report went no further than a generic call for Canadian policy to take arms control into account in its policy considerations.[12] Following the Reagan-Gorbachev Reykjavik Summit in the fall of 1986, the interpretation issue was raised by Derek Blackburn (NDP).[13] Clark responded the next day with a reference to Agreed Statement D that the issue was not a question for Canada: "What is precisely intended in that treaty

[ABM] is for those two governments [US and USSR] which are parties to the agreement to work out."[14] Over the next several months, the interpretation issue would be raised a few more times by the Opposition, leading Clark to formally endorse the narrow interpretation on February 13, 1987.[15] On March 6, an NDP motion calling for the end to cruise missile testing in Canada was linked to the SDI question. In the debate, Clark stated that US policy was to abide by the narrow interpretation, while believing in the broad one. In a traditional refrain of Canadian policy makers, he argued that Canadian influence would be destroyed if Canada cancelled the testing agreement, and alliance unity was the most appropriate means to deal with these questions.[16]

In reality, a Canadian government was again saved by outside forces. Just before the February 13 House exchange, the High Level National Security Planning Group chaired by Reagan met to consider three options: (1) deploy a ground-based system in 1988, followed by a space-based system in 1993; (2) defer a deployment decision but adopt the broad interpretation of the treaty; or (3) defer entirely for a period of time in light of improving relations with the Soviet Union. Most of the senior members of the planning group were in favour of moving forward, except Schultz, who preferred option three.[17] In the end, the Schultz position won. The Reagan administration continued to endorse the broad interpretation, while acting within the parameters of the narrow one. Officials, including Nitze and Iklé, repeatedly stated that the United States would continue to act within the parameters of the narrow interpretation, as Clark stated on March 6.

The US decision recognized that the technology was still too premature to move to a full-scale development, test, and evaluation stage essential if a 1998 deployment schedule was adopted. There was no likelihood that the technology would be ready even if a very robust development, test, and evaluation program was adopted. SDI was still a research program. Brilliant Eyes and Brilliant Pebbles existed only on the drawing board. It made little sense to engage in a major political fight, at home and abroad, for a system that didn't exist in any meaningful way. Thus, there was no need to push the broad interpretation forward if the technology to cross this barrier did not exist. Besides, the barrier would be a problem for the next administration, and there was little political sense in generating international conflict in light of the dramatic changes occurring in the Cold War, for which Reagan could and would take credit (along with Mikhail Gorbachev, the new leader of the Soviet Union).

East-West relations continued to show signs of improvement, though no one expected that the end of the Cold War, or the Soviet Union, was near. At the 1986 Reykjavik Summit, NATO's INF double-zero option was agreed to, and the INF Treaty was signed the next year. At the summit, Reagan and Gorbachev also tentatively agreed to massive nuclear reductions and muted the idea of eliminating nuclear weapons entirely, much to the horror of Reagan's advisors (likely Gorbachev's as well). However, Gorbachev's condition was Reagan's agreement to sacrifice SDI, which he refused to do. Nonetheless, Reykjavik signalled the return of good East-West relations, if not the beginning of the end of the so-called Second Cold War, with arms control again as the political symbol.

In the end, Canada and all the allies were saved from confronting the implications of the broad interpretation made operative, and the interpretation question faded into obscurity. Although Clark had added a new element to Canadian policy in endorsing the narrow interpretation in the House, no formal government policy statement followed from the prime minister or anyone else, leaving speculation that Clark had stepped out on his own, as he had done in February 1985 in his blanket condemnation of SDI, and no reference to the interpretation issue was found either in the 1986 Foreign Policy Green Paper or the 1987 Defence Policy White Paper. Instead, the government spoke only of maintaining a close watch on missile defence developments in which "the nature of defences cannot be precisely determined ... Future decisions, on Canada's role, if any, in ballistic missile defence will depend upon these developments" and "will have to be considered in light of the impact ballistic missile defence could have on strategic stability and Canadian security."[18]

The government's luck in sidestepping the interpretation issue, and thus keeping SDI and missile defence off the public agenda, was more or less replicated in the definition and scope of the government official no to SDI research participation. Two days after Mulroney's SDI policy announcement, Pauline Jewitt (NDP) directly raised the issue in calling for no government research-funding support to Canadian industry engaged in SDI research. Deputy Prime Minister and Defence Minister Nielsen replied that government support to industry research projects was "subject to a decision-making process which is in place and will be utilized when any new project comes up."[19] In effect, Nielsen sidestepped the issue, and the government would continue to do so when it was raised several times again in the coming weeks, before it, too, faded from public view. However, it was not an issue that could be easily ignored by senior officials within the bureaucracy over the next couple of years.

Defining the meaning of government research had not been prominent during the SDI decision-making process. It had been largely assumed that no official government involvement meant no money and no government laboratory research on SDI. However, the assumption contained two, somewhat interrelated, problems. The first was the existence of objective criteria that delimited

SDI research from other military research. The second was the question of government research programs for industry – the Defence Industry Productivity Program (DIPP). Did no official government involvement mean that companies engaged in SDI research for the US Department of Defense were ineligible for DIPP funds? If the answer was yes (no meant no across the board), someone still had to decide what research was SDI research or, to avoid the criteria problem, perhaps simply assume that any contract issued by SDIO was by definition ineligible. However, the definition of SDI research, despite its inherent complexities, had wider implications. Defining SDI research touched directly upon the actual definition of missile defence and thus the meaning of participation if or when SDI moved out of the research phase and onto the path to deployment. If SDI research or technologies were defined very narrowly, the meaning of participation too would be very narrow. The ABM legacy had put early warning outside participation; direct ABM system components (as per Article 2 of the ABM Treaty) clearly as participation; and command, control, and communications as undefined. Implicitly, the SDI research question provided an opportunity for Canada, to the extent possible, to define the meaning of participation to its advantage. Unfortunately, the immediacy of the issue itself resulted in a missed opportunity – the triumph of short-term over long-term thinking – the long-standing Achilles' heel of government. Of course, there was no guarantee that doing so would actually be remembered in the future, or if remembered that it would prove a successful gambit.

Regardless, defining what was and was not SDI research could not be ignored. Differentiating between technologies that were specific or unique to SDI and/or missile defence and technologies that were multifunctional (and not unique to missile defence) was not an easy task. It was confounding the already complicated dual-use phenomenon in which technologies have both military and civilian functions. For example, radar provides for civilian air traffic control and, among other things, directs fighter interceptors, missiles, and anti-aircraft batteries to targets. For SDI and missile defence in general, a range of new radars, sensors, micro-electronic computing, and advanced communications was essential to the development of alternative non-nuclear intercept systems, such as kinetic-kill, laser, and particle-beam technologies. Advanced radars and other imaging sensors were vital to distinguish a nuclear warhead from a decoy or dummy. With the speeds and short response times involved in any missile defence system, computing speed, power, and capacity had to be dramatically improved and miniaturized. Alternative deployment schemes for sensors and interceptors, along with new interceptors, were also part of the multifaceted SDI research agenda. These same technologies had wider military and civil applications. Improved radar and sensing technologies could be applied for other non-missile defence functions in support of conventional military operations. The same technologies could, for example, enhance space-based remote sensing for a range of civilian uses and scientific purposes.

As the French feared, the real threat of SDI was not missile defence but the broader and deeper military and civilian spinoffs, with their direct impacts upon military and industrial/economic competitiveness. SDI played a significant role in exacerbating, and in some places creating, the gap between US military technology and capabilities and the rest of the world that continues to this day. What came to be known as the Revolution in Military Affairs was in large part because of SDI research – even technologies that predated SDI itself received a significant boost as a result of SDI investment, including areas such as imaging technologies and computer miniaturization.

With the immense range of technologies engaged in SDI, any government's ability to draw a line was problematic, and government policy was not intended to disengage from all areas of research just because they might have SDI applicability. Accordingly, National Defence officials felt that SDI technologies didn't exist per se. Instead, SDI emphasized certain technologies.[20] In addition, officials felt that nothing should prevent them from pursuing initiatives in areas essential to Canadian security, especially as they concerned NORAD's aerospace early warning mission (which included air-breathing bombers, cruise missiles, and ballistic missiles) and air defence mission – a position consistent with the air defence option presented to Kroeger in May 1985.[21] External Affairs fully agreed, arguing that SDI should be defined in architectural rather than technological terms. It also felt that the line regarding activities incompatible with ABM or other arms control agreements needed to be made clear.[22] In the end, both departments recognized that the parameters of the 1985 decision had to be better defined.

Existing and future cooperative research projects with the United States were a case in point. Even before the 1985 decision, National Defence had been concerned about Canadian involvement in US research projects that became subsumed under the SDI umbrella. If such projects became public, the government was liable to a charge of duplicity. Of course, officials knew that outside knowledge about Canada-US defence research cooperation, as well as the complicated defence industrial and technology relationship, was in short supply. One could count on the fingers of a single hand the number of outside experts on these questions, and, at best, critics might raise a general point of contention. This was to be the case for one project, Teal Ruby.

Teal Ruby was originally a US Defense Advanced Research Projects Agency (DARPA) experimental program on high-altitude sensing and target tracking that came to include Canada, Australia, and the United Kingdom. It initially had ballistic and cruise missile and aircraft tracking functions, although by 1985, a year before the first experimental satellite launch by the USAF Shuttle program, it had been reoriented to cruise missile and air defence.[23] Nonetheless, it was swept up by the SDI program, and as such Canadian involvement could be interpreted as an end run around the 1985 policy, even if it predated SDI. On June 16, 1986, Teal Ruby was indeed raised in the House. The associate defence minister was fully prepared to respond. He simply noted that the project had begun under the previous (Liberal) government in 1980 and that there was no reason to cancel it.[24] Except for a brief opposition statement the next day arguing that Teal Ruby violated the ABM Treaty, the question of bilateral research cooperation was never raised again.[25] The government found no need to make the more complicated argument about SDI architecture versus technologies or how Teal Ruby engaged fundamental Canadian defence interests.

However, while the question of government research support to companies engaged in SDI work through the DIPP quickly faded from public view, within the bureaucracy the issue acquired a crisis-like quality. In the wake of his June report, Kroeger had suggested that the government make it clear that DIPP and related government research funding should be applied in the case of SDI, and this was the implicit intent of Nielsen's response on September 9.[26] However, the government had not been forthcoming on the issue, and, as a result, any government support to industry involved in SDI research through the DIPP could be viewed as an end run around existing policy, leaving the government vulnerable to charges of misleading the House and the public. Of course, one would be hard-pressed to think that such charges on the basis of the DIPP, or other government-funded activities, would carry much significance, especially as the focus of the Canadian defence debate moved from SDI to the controversial nuclear submarine program of the 1987 White Paper. In addition, the traditional lack of interest and knowledge about defence within Parliament, the media, and the public at large, particularly on something as specialized as defence research activities, made it unlikely that these issues would get much play. Nevertheless, in the fall of 1986, the development of guidelines for private sector involvement in SDI engaged the bureaucracy and the ministers involved.[27]

The guidelines issue was concerned not only with the question of the government contradicting its own policy but also with the need to generate a favourable atmosphere to get Canadian companies and research groups to look for opportunities south of the border. With the leftward turn in Canadian policy during the late 1960s and the negative publicity surrounding contracts with the US military during the Vietnam War, Canadian companies had become increasingly sensitive to public exposure. One result was a growing reluctance on their part to engage in controversial research endeavours in the absence of active government support, and SDI certainly fit this bill. In addition, SDI compounded existing difficulties facing Canadian companies in exploiting the Defence Production Sharing Agreements, particularly in gaining access to top secret or black specifications necessary to bid on US defence contracts.[28] Original US concerns about the SDI invitation, and its retreat from the Weinberger invitation, had been largely driven by opposition to allowing allies access to top secret data and associated technology transfer questions. If the United States was concerned about allies in general on these counts, it would be equally if not more concerned about foreign companies whose governments had said no to SDI. In such an environment, it made little sense for companies to invest significant resources in pursuing SDI contracts – a situation implicitly recognized by officials who had argued that any significant economic and technology payoffs for Canada would demand government engagement and investment. With the Canadian no, Canadian companies were likely to sit on the sidelines unless directly approached by SDIO or other US research agencies for possessing technology of direct interest or potential. Thus, by reducing the risk of failure, DIPP funding was one possible way to get Canadian companies to engage.

Sending a signal to industry while managing the potential fallout of government research support programs was essential to the development of guidelines. However, developing the guidelines was one thing, distributing them was another. Certainly, a case could be made to disseminate the guidelines widely to the public at large. But bringing SDI back onto the political agenda through wide public dissemination might generate significant negative public reaction.[29] Thus, the government faced a conundrum. To limit distribution to only those concerned could result in significant negative fallout should they become public. To distribute publicly could be interpreted as the government violating its own policy of non-involvement by providing direct assistance to companies and research institutions. In the end, DND pushed for full disclosure, arguing that to do otherwise would lead to a charge of duplicity on the part of the government should the guidelines become known, while failing to do so could lead some companies to miss out on potential opportunities.[30]

The guidelines, which effectively meant a clarification of policy, became clear in response to reports that the Atomic Energy Commission had obtained a particle beam accelerator subcontract as part of a larger SDI contract from the Los Alamos Laboratories.[31] Existing policy had not meant to prohibit assistance to companies simply because the research might have SDI applications, nor was it intended to prevent scientific research. Regardless, the resulting publication of the policy clarification had no discernible public impact – it passed unnoticed. It also failed to alter the reality of Canadian company engagement in SDI-related research. But then official government support of SDI research did not prove very beneficial either. As suspected in the lead up to the September 1985 decision, expectations of significant economic benefits and technology transfer were not met, whether a state officially signed on or not. Nonetheless, there was a correlation between official participation and contracts. Canada ranked well below official participants and what would be expected given the long-standing defence industrial and technology relationship with the United States.[32] Moreover, industry certainly believed that the Canadian no had been a significant barrier. Gérard Lalonde of the Aerospace Industries Association of Canada stated, "we have been told by non-official sources that there is not a snowball's chance in hell of getting any SDI contracts."[33]

Of course, the actual extent and value of Canadian participation are extremely difficult to establish. With many firms owned by US parents and undertaking a variety of subcontract research in the context of the broad SDI research agenda, it is possible that many more companies were involved than meets the eye. In fact, Canadian companies could have easily been involved in SDI-related research without knowing themselves. Regardless, Canadian firms did not respond well to the 1985 policy or its clarified guidelines. They would await direct government engagement, and in the future they would not interpret the 1994 White Paper as a signal of direct engagement. If it was designed as such a signal, it failed to be understood or acted upon as such by industry. It was not until 2001 that the director general for the International and Industrial Program within National Defence made a concerted effort to inform industry of potential missile defence opportunities.[34] Even as late as 2002, Defence officials remained uncertain whether the 1994 White Paper prohibited Canadian industry from participation. Moreover, the signals from the United States about the meaning of the Canadian no were also mixed.

Reagan saw the Mulroney decision as unproblematic.[35] Senior Pentagon and SDIO officials had blessed the decision beforehand. At the research level, however, the relevant US decision makers and Canadian executives took the September decision as an absolute no to SDI, despite the general reluctance of these decision makers to share classified information (a reluctance partially driven by the belief held by some that any secrets provided to NATO allies meant they were in Moscow before all of NATO knew). This view of the Canadian decision extended further, to the primary actors in the day-to-day operation and management of the Canada-US defence relationship. For them, the decision was also a clear no to SDI, and as a no this meant no Canadian access to SDI-related thinking and activities. Most importantly, it was these middle-ranking actors that operationally defined what was and was not SDI related. Unless senior US defence decision makers intervened frequently and forcibly to impose their definition of SDI thinking and activities, Canadian access, especially through NORAD, would be determined by decisions at lower levels.

Almost immediately after the Mulroney decision, Canadian military officials faced the issue of how far they could go before hitting the SDI policy barrier and how far their US counterparts would let them go with SDI activities that butted up against the existing NORAD missions. As a result, Canada confronted an even deeper problem that struck at the heart of its long-standing defence strategy – whether one defines it as defence against help or in other terms.[36] Recognizing that the United States would defend Canada whether Canadians liked it or not – the original unsolicited 1938 President Franklin Roosevelt commitment that had never been retracted – Canada's strategy was to obtain knowledge of, access to, and influence over US thinking and plans for the defence of North America. In so doing, Canada would be able to manage inadvertent and unintentional US thinking and action that affected Canadian sovereignty and national security interests. In other words, the manner in which Washington went about defending the continent (and thus Canada) could, in effect, threaten Canada.[37] As part of this strategy, cooperation through NORAD provided the means to obtain this information, inform the United States of unique Canadian interests and concerns, and, ideally, influence US defence plans.[38] However, by virtue of the SDI decision, National Defence officials in general and Canadian NORAD officers in particular feared that one major aspect of US defence planning was potentially off the table for both sides: SDI.

For senior Canadian Defence officials, there was no clear guidance about how far they could go in engaging their US counterparts on SDI without being open to a charge of violating government policy. The brilliance of the SDI air defence option was its openness to engaging (participating) without full engagement (participation). But the prime minister had rejected the option, even though on the research front the caveat of Canada's defence interests and requirements applied. In other words, Canadian officials could take some latitude, particularly given the inherent difficulties in defining what was and was not SDI. The real question was on the American side. The United States had committed itself to SDI consultations with its allies in general and with Canada in particular. Ostensibly as part of the guideline or policy clarification process, US defence officials sought guidance on how far Canadian officials could engage the United States, especially relative to the North American air defence modernization program. Weinberger directly enquired whether Canadian policy foreclosed the Canadian military from any involvement in US outer space or missile defence planning.[39] However, this did not necessarily resolve the dilemma of how far American officials would be willing to go. As far as the French were concerned, consultation with the United States meant the United States telling the allies only what it wanted.

Close personal relations between Brian and Ronnie prevented high-level political fallout from the 1985 decision. But, at lower levels, where perhaps 90 percent of the relationship is actually carried out, US military officials in general (and SDIO and US NORAD in particular) took the Mulroney decision as a straightforward no and acted accordingly. In so doing, Canada had neither the right nor the need to know about SDI. As non-participants in SDI research, Canadian officials had no right to access any information about SDI – and if they had no right to know, then obviously they had no need to know. SDI was research and thus had no immediate impact on existing operational binational defence requirements via NORAD. However, research decisions, as a function of thinking about SDI in theoretical and operational terms, could not help but affect continental defence. Nitze, among others, had recognized that SDI would generate incentives for the Soviet Union to compensate by developing its offensive air capabilities. Moreover, SDI had subsumed a wide range of aerospace defence activities, and this extended into research that had significance for the NORAD mission. For example, the dividing line between early warning (NORAD) and missile tracking (SDI) was a fine one, and this component of SDI research could well fall into the category of need to know. Moreover, there was an ongoing bilateral air defence modernization program that would be affected by SDI research thinking and developments. In other words, the need to know might be much more complicated than simple access to research, and if there was a need might there not be a case for the right through a broad reading of the NORAD agreement?

In fact, this was merely a symptom of a more pervasive, fundamental problem. In saying no to official government participation in SDI research, the government had not actually said no to participation in missile defence. It had left the issue hanging, as it had done with the ABM question. If SDI or some other variant of missile defence for the defence of North America proceeded, the government would need access to a range of technical information upon which to base a decision – the dilemma officials had confronted during the ABM and SDI decision-making process. As technical development went hand in glove with strategic and operational planning, the government or its experts also needed this information. Both the technical and operational planning would evolve over time and influence each other with implications for the defence of Canada. Canadian officials had a need to know in order to prepare for the future and make an informed decision – no to research did not mean no to needing information on future missile defence planning and thinking. Defence officials clearly recognized this dilemma in making their case for saying yes to the Weinberger invitation.

Of course, it would have helped if the government privately or publicly made this clear. But to do so could look like duplicity. Done privately might suffice, but like the industry guidelines, if it became public, the government would have a hard time communicating the nuance of its approach to the public. The Opposition would certainly see it as an end run around the 1985 decision. Besides, the US government might well reject a private clarification and demand a public one to obtain allied endorsement, and a public one would resurrect the SDI debate, which the Canadian government had no interest in doing. In the end, no public clarification occurred. Privately, however, the new defence minister, Perrin Beatty, did seek clarification in a March 29, 1987, letter to his American counterpart.[40]

The context for the letter was not SDI but, rather, North American defence cooperation that predated SDI. The United States had divided its North America aerospace defence modernization into two streams, a strategic (missile defence/space) stream and an air defence stream. Each represented two parts of the SDA 2000 study. In Canada's experience, this division had been confirmed in the development of the Joint Parliamentary Committee report and the NORAD renewal process in 1986. Beatty sought continued cooperation in the areas of concepts, and research and development, of interest to Canada. He also wished to ensure Canadian access to the SDA 2000 study and continued Canadian engagement in the Aerospace Advanced Technology Working Group in areas compatible with the Canadian 1985 decision.

Compatibility, however, was left to the Americans to define. South of the border, particularly within SDIO and among US military officials engaged with Canada, any possible nuance or empathy relative to the Canadian position was lost. Although American officials on the Permanent Joint Board on Defence (PJBD) would subsequently release a version of the SDA 2000 Phase II Report to Canada, the PJBD referred the issue of strategic defence cooperation to the Military Cooperation Committee. At the committee's May 1989 meeting, USAF officials rejected Canada's request for full participation in the aerospace strategic defence planning process. The justification for this rejection was the Canadian position on SDI.[41] As far as the United States was concerned, when Mulroney said no to research, Canada said no to everything. The door was closed.

The case for excluding Canada from access to aerospace strategic defence planning followed logically from the SDI decision. Being on the outside of SDI research, Canadian officials had no right to SDI research information, and this information would be fed into any planning for future strategic defence.

Regardless of whether the actions of USAF officials were a spiteful attempt to punish Canada for the government's failure to endorse SDI – as some Canadian officials believed – numerous efforts to change the American position proved unsuccessful. However, even more damaging was a significant decline in co-operation and collaboration on the air defence front where Canada ostensibly had a right and need to know. Canada was denied membership on the US Inter-Agency Air Defense Steering Committee responsible for program oversight, even though air defence was a binational mission in North America.

Perhaps even more significant was the shift in USAF attitudes toward its Canadian counterparts in NORAD regarding the no-foreign-disclosure rule. Historically, NORAD had fostered very close relationships between American and Canadian air force officers. Working side by side, officers on both sides of the border established close personal relationships and came to share as much, if not more, in common with each other than their fellow officers outside NORAD. In some respects, this was the foundation of the relatively independent NORAD perspective, as distinct from the defence perspective in Ottawa or Washington. Opportunities also abounded for Canadian officers to distance themselves privately from the Ottawa policy line in the company of their American compatriots, as there were for US officers to do the same with respect to Washington. As a result, for Canadian officers in many circumstances, the no-foreign-disclosure rule had been readily ignored. Canadians were treated as Americans. This all ended after the SDI no. Canadians had become foreigners, even in NORAD. Any hopes that these close lower-level relationships might insulate Canada from the implications of the no for Canadian access were dashed.

In addition, as a result of the SDI decision, US Space Command began to draw lines between itself and NORAD. Even though the two organizations were tied at the hip as a function of sharing the same commander and the former providing the assets (data) in support of the latter's early warning mission, concerns that the no to SDI might produce a deep split between the two were pronounced in the Canadian Department of National Defence. Moreover, to some Space Command officials, like attitudes in SDIO and elsewhere, the Mulroney decision was not just a no but actually opposition to SDI. Canada had failed the loyalty test, despite attempts by Canadian officials, especially in NORAD, to clarify the Canadian position. The fallout was the basic closure of the space component to Canadians, who had long been posted into the Space Control Operations Center at Cheyenne Mountain. Overall, NORAD and with it Canada, even in the air defence sector, appeared to be on the quick road to marginalization. Canada was increasingly in the dark about current and future United States' strategic and North American defence thinking and plans.

The loss of Canadian access was not the product of a high-level decision. Gottlieb might have been right about SDI being a loyalty test, but he was wrong in believing it applied to senior levels of the Reagan administration. In the summer of 1985, Iklé told Canadian officials to ignore decisions coming out of the USAF and Colorado Springs on SDI activity.[42] He noted that such decisions communicated to Canada through NORAD as final had been withdrawn. Echoing comments of senior US Defense officials about decision making between Canada-NORAD and Ottawa, Iklé emphasized that decisions were made in Washington. A year later, Weinberger sought to open the door more to Canadians in looking for clarification on how far the United States could go in providing Canadians access to, and information about, SDI. The problems were at the lower implementation levels, which were the real drivers in the defence relationship more often than not.

Of course, the problem of SDI might have been resolved if the government had been bolder in clarifying the exact meaning of no. But the government quickly moved on and had little interest in reopening the SDI question, which appeared to have been resolved to its satisfaction. At least for the prime minister, the SDI no did not appear to have any negative implications, as close relations continued and the policy priority of free trade moved forward quickly on both sides of the border. Indeed, two weeks after the no, Reagan had gone to Congress to obtain fast-track negotiating authority for a free-trade deal with Canada. Key elements of Canada's defence strategy might have been unravelling, but then defence was not the government's top priority.

The problem of clarification reflected concerns among officials in External Affairs who worried about too much access, especially when GPALS emerged. Too much access could be viewed as an attempt to participate and thus a reversal of policy.[43] Moreover, the more Canada requested information about SDI or missile defence, the more it was feared that Canada became vulnerable to US pressure to reverse Canadian policy. Nonetheless, Defence officials looked for ways to break out of the SDI dead end.[44] In so doing, three options were identified in late 1989. The first was to seek a new policy on SDI by getting the defence and external affairs ministers to bring the requirement to the attention of the prime minister and cabinet. This was viewed as a risky option, with uncertain payoffs. The second was for senior defence officials to attempt once more to clarify the government's view on SDI and air defence cooperation to the new Bush administration, in the hope that a positive response might filter down into the US bureaucracy. However, no one expected likely success, as this option had failed in the past.

What emerged as the preferred option was to take a long-term perspective and try to link the question of Canadian access to the 1991 NORAD renewal process that would begin shortly. To do so first required Canadian officials to identify specific areas of Canadian interests and needs. This option was premised on the need for sufficient time for the administration and Congress to move the US bureaucracy in the appropriate direction. It contained the implicit idea of Canada buying its way back into American good graces – a forerunner of the mid-1990s asymmetric contribution to missile defence. Specifically, defence officials highlighted a Canadian contribution of $1 billion to a $10 billion US project on wide-area space-based surveillance that was earmarked to be operational by 2006.

This was about more than buying one's way back in simply by throwing money at the problem. Instead, Canada would become a real player, bringing valuable expertise and technology to the table. Despite US rhetoric of allied participation, there always remained the suspicion that, given the stakes and cutting-edge technologies involved, the United States really didn't want too much allied involvement. Engaging directly in a formal cooperative program that was important to the United States, while bringing expertise and funding to the table, made it harder for the United States to deny access. In other words, with its own resources, Canada could force the door open to pursue its interests.

What stood in the way was Canada's poor record of following through. In 1986, defence officials had suggested to the United States that Canada would develop a radar satellite constellation with global utility in support of the North American defence mission.[45] In the ensuing years, even before the defence budget cuts of 1989, little had been done on that front (though the Canadian Space Agency was moving forward with its radar satellite program – RADARSAT I – which the United States would launch at no cost in return for access to imagery). There was also the question of internal support from Canadian senior military decision makers and the absence of any formal high-level group within the Canadian Department of National Defence tasked with military space.

By 1989, the ambitious 1987 Defence White Paper was dead, gutted by defence cuts set out in the federal budget. This budget marked the onset of repeated deep reductions in funding that would last a full decade. In this environment, obtaining support from the three services for a large investment in any non-traditional military dimension such as space was highly unlikely, even if greater attention had begun to be paid to space within the department. Above all else, the world had suddenly and unexpectedly changed almost overnight. The Cold War was over, and the strategic context had changed. The spectre of global nuclear war was disappearing and with it the need for costly investment in strategic defence. Instead, a new rationale for missile defence emerged, reminiscent of the emerging Chinese ICBM threat that had driven Sentinel – the proliferation of weapons of mass destruction and ballistic missiles made real by the 1991 Gulf War.

Yuri Andropov succeeded to power in the Soviet Union upon Leonid Brezhnev's death in 1981. He recognized that dramatic steps were needed to deal with the stagnant Soviet economy, but his early death placed such steps on hold during the brief rule of Konstantin Chernenko. Upon taking power, Mikhail Gorbachev, a protege of Andropov, moved forward with two radical initiatives – perestroika as a means to rejuvenate the stagnant economy by decentralizing economic decision making, and glasnost to create an environment of openness essential to decentralization. At the same time, Gorbachev recognized that the success of internal reform hinged upon obtaining a respite from the Cold War rivalry and the requisite overinvestment in military forces that had played a major role in the stagnation of the economy.[46] To obtain this respite, dramatic foreign policy steps were needed, of which the INF Treaty was one. More important was the decision to apply glasnost to the Soviet Union's relationship with its Warsaw Pact allies (or satellites). This meant, in effect, the end of Soviet control over Eastern Europe with the renunciation of the 1968 Brezhnev Doctrine in thought and deed.[47] Beginning with Poland, which had been for some time under the threat of Soviet military intervention following the emergence of the Solidarity Movement, Gorbachev made it clear that the East European socialist states were on their own. In 1988, Communist regimes attempted to reform but quickly collapsed peacefully, with the exception of violence in Romania. In the fall of 1989, the primary symbol of a divided Europe and the Cold War, the Berlin Wall, figuratively collapsed, followed shortly by the fall of the East German regime. A year later, on October 3, 1990, Germany was reunified with the absorption of the East German state by West Germany and became a united state within the NATO alliance.

The Cold War was over, though many would continue to have doubts for some time, even after the creation of the Commonwealth of Independent States by Russia, Ukraine, and Belorussia on December 8 and Gorbachev's resignation on December 25, 1991. Following the decision to let the East Europeans go, the Red Army began the process of withdrawing back within its borders at the same time as the borders of Russia began to shrink.[48] The strategic debate about deterrence or warfighting in US nuclear doctrine moved into the dustbin of history. SDI no longer really mattered, and concerns about the relationship between deterrence and defence and strategic stability became politically irrelevant. Within a span of about four years, an exciting new world emerged full of growing optimism about the future of international cooperation. As arms control developments had ended the so-called First Cold War and ushered in détente, and INF had signalled the beginning of the end of the Second Cold

Table 3.1

US and Soviet Union/Russia strategic force levels, 1991

	United States		Soviet Union/Russia	
	Launchers	Warheads	Launchers	Warheads
ICBMs	550	2,000	934	6,034
SLBMs	480	3,456	832	2,792
Bombers	209	3,844	102	376
Totals	1,239	9,300	1,868	9,202

Source: Natural Resources Defense Council, Archive of Nuclear Data. www.nrdc.org.

War, so two major arms control agreements – START I and the Conventional Forces in Europe (CFE) Treaty – would signal the start of a new world.

After ten years of little progress, despite the lost opportunity at Reykjavik in 1986, Washington and Moscow signed START I on July 31, 1991. The treaty limited both parties to 1,600 launchers (ICBMs, SLBMs, and bombers) and 6,000 warheads, of which only 4,900 could be deployed on ICBMs and SLBMs. Reflecting earlier concerns about heavy ICBMs, the agreement also limited the Soviet Union/Russia to 1,540 warheads on their SS-18 missiles and 1,100 on their new mobile ICBMs.[49] The agreement represented a significant reduction in strategic force levels (see Table 3.1). With the collapse of the Soviet Union, START I was amended with the signing by Russia, Belarus, Kazakhstan, and Ukraine, the four Soviet Union successor states with nuclear weapons of the Lisbon Protocols in 1992, that committed the latter three to transfer nuclear weapons stationed on their territory to Russia and sign the Nuclear Non-Proliferation Treaty.[50]

Alongside START I, the CFE Treaty was signed on November 17, 1990. Two decades earlier as part of détente, the Mutual and Balanced Force Reduction Talks had been established. For a variety of reasons, including a fundamental disagreement over which forces to include, especially concerning the inclusion of US forces in the continental United States, no progress had been made in reaching an agreement on NATO and Warsaw Pact conventional force levels in Europe. These discussions were subsequently linked to the Conference on Security and Cooperation in Europe talks, which had been institutionalized after the 1975 Helsinki Conference.[51] With the events of 1988-89 in Eastern Europe, the barriers to a conventional forces agreement quickly evaporated, leading to the CFE Treaty. With the collapse of the Warsaw Pact, the treaty, being based upon the Warsaw Pact and NATO, was out-of-date almost immediately upon its signing. Nonetheless, it facilitated the peaceful withdrawal of the Red Army from Eastern Europe and was a valuable confidence-building measure in the process of ending the Cold War division of Europe and the Cold War itself.[52]

The dramatic progress on the arms control front was largely a function of the acceptance by Gorbachev and his ostensible successor as president of the new Russian Republic, Boris Yeltsin, of long-standing Western/US arms control positions – something that would come back to haunt the next iteration of arms control issues in the mid-1990s. For the time being, whether Westernizers or just seeking respite for domestic economic reform and access to Western economic support and aid, a remarkable convergence of political interests between Moscow and Washington occurred, and this convergence was no more evident than in the United Nations Security Council. After decades of respective Moscow or Washington (Paris or London) vetoes, the Security Council began to operate on the basis of Great Power consensus as hoped by the original framers of the institution in 1945. Indicative of Great Power cooperation was the council's quick response to the Iraqi invasion of Kuwait in August 1990, leading to the first and in some sense only use of the UN's Chapter Seven, Article 42, collective security provisions legitimizing the US-led military coalition operation to expel Iraq from Kuwait that began on January 16, 1991.[53]

Hussein's invasion of Kuwait represented more than just a threat to this new world order of international cooperation, as Bush called it. It also brought to political centre stage one of the two dominant security features of the post-Cold War era until 9/11 – the proliferation of weapons of mass destruction and ballistic missiles.[54] Concerns about Saddam Hussein's proliferation efforts existed long before Iraq's August 1990 invasion of Kuwait. Even more disconcerting, Hussein had demonstrated no reluctance to use these weapons in the 1980-88 war against Iran. He had employed his short-range Scud missiles, acquired from the Soviet Union, and indigenously modified medium-range Scud missiles (Al-Hussein) to attack Iran's major cities and employed gas on the front lines against Iranian troops and the Kurds. With these tactics combined, Hussein created the image of a decision maker willing and in the future able to use weapons of mass destruction married to ballistic missile delivery systems.

Hussein and Iraq became the occasion for the emergence of the concept of rogue state – a state whose decision makers act outside the rules and norms of international conduct. In terms of strategic weapons, this meant the tacit rules and norms governing nuclear deterrence. In other words, Hussein (and quickly joined by North Korea, Libya, and subsequently Iran) was not only undeterrable according to the standards developed in the Cold War relationship but was likely to use the weapons if the capability and opportunity appeared.[55] The United States and its coalition partners prepared for chemical and possibly biological warfare, and eliminating Hussein's weapons of mass destruction development programs, especially nuclear weapons, became one of the primary goals of the war. Correctly predicting that Hussein would employ his medium-range mobile ballistic missiles, possibly equipped with chemical or biological warheads, against Israel in order to bring it into the war and thus alter the political situation, with numerous Arab states participating in the coalition, the United States rushed into service its first-generation non-nuclear tactical missile defence system still under development – the Phased Array Tracking to Intercept of Target (Patriot) Advanced Capability – 2 (PAC-2).

The first-generation Patriot was a high-altitude air defence surface-to-air missile system originally designed for use on the central front in Europe.[56] In 1985, Raytheon, a major US defence contractor, began work on a tactical anti-missile variant designed to intercept the Soviet SS-21 and SS-23 short-range ballistic missiles (SRBMs). Because of Iraqi missile capabilities, this variant, the PAC-2, was deployed in Israel and Saudi Arabia, with civilian engineers working on the system to deal with the ballistic missile threat it had not been designed for.[57] In deploying PAC-2, the United States hoped that, by providing some degree of effective defence, the Israelis would be dissuaded from retaliating against Iraqi missile attacks. PAC-2 was also to defend Saudi cities and American and coalition bases in Saudi Arabia.[58] Alongside PAC-2, the United States devoted significant air assets in an extensive and largely unsuccessful Scud-hunting campaign designed to destroy the mobile Scud launchers either before they launched their missiles or afterward to prevent reloading.[59]

Hussein had not yet developed nuclear weapons, even though the subsequent United Nations Special Commission investigation as required by the Gulf War ceasefire agreement discovered that Iraq was much closer than thought. Iraq did not employ chemical or biological weapons during the war. Nor had it developed IRBMs capable of threatening Europe or long-range ICBMs capable of threatening North America. Finally, its modified Scuds proved of little military value, as most broke into pieces upon re-entry due to poor engineering. Regardless, Hussein's clandestine proliferation programs and the Gulf War served as the foundation for a new set of strategic missile defence arguments divorced from the arguments tied to the strategic nuclear Cold War relationship between the United States and the Soviet Union. These arguments would propel missile defence in a variety of directions and to the first tier of US defence priorities in the post-Cold War era.

Central to the more extreme variant of these new strategic arguments was the first principle that rogue state leadership – the new proliferators – was undeterrable. As ruthless dictators, the threat to destroy their large population centres in a retaliatory nuclear strike was meaningless. As undeterrable, this

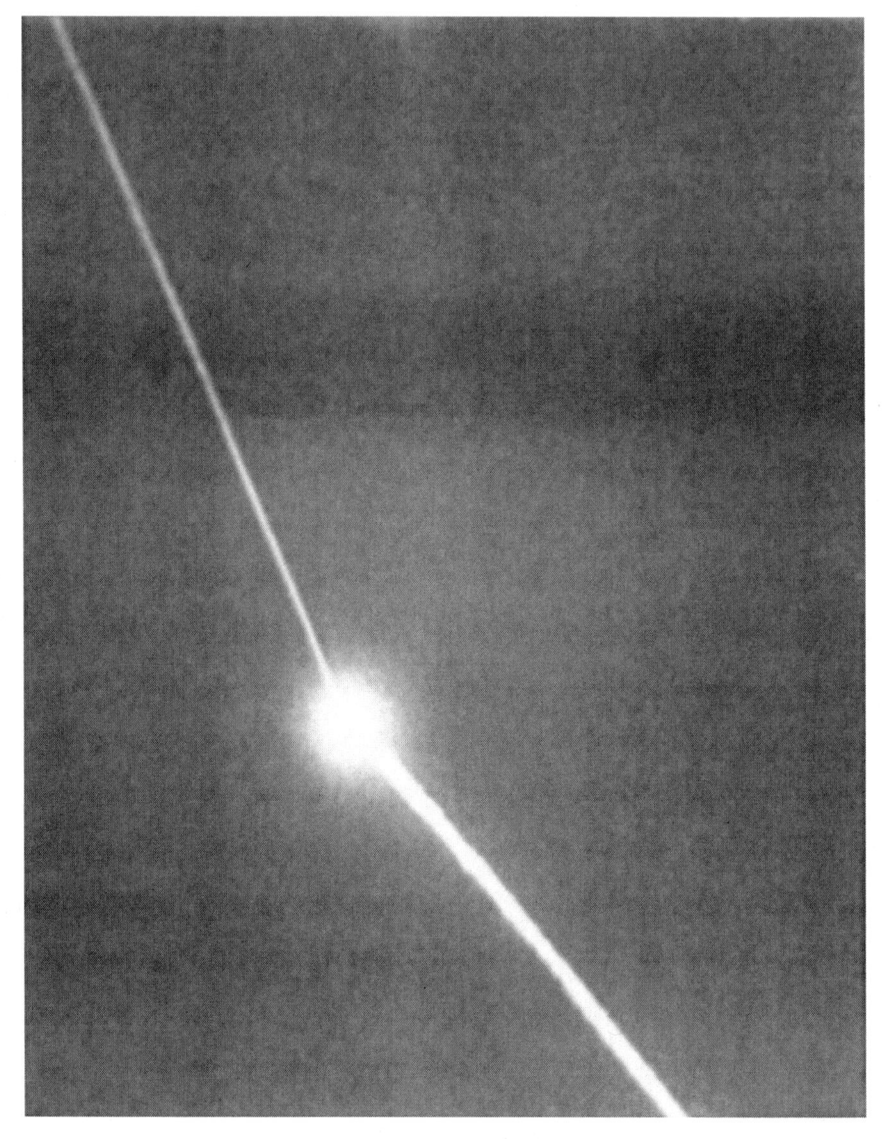

Intercept by a Patriot Advanced Capability 3. *Courtesy US Missile Defense Agency*

leadership had no compunction about using nuclear (or chemical and biological) weapons against US forces and their coalition partners. Nor, once it acquired ICBMs, would it hesitate to use these weapons against the United States directly. Furthermore, political, diplomatic, and legal attempts to prevent these states from acquiring weapons of mass destruction and ballistic missiles were doomed to failure.[60] Indeed, they had already failed, as clearly demonstrated by Iraq, which was a signatory to all of the international non-proliferation agreements.

The less extreme variant put aside the question of whether this leadership could be deterred and focused upon the capacity of these states to deter the West. In the context of the Gulf War, would the United States have intervened if Iraq had been able to threaten credibly US forces with a nuclear attack (the chemical and biological threat could be managed through passive defensive measures such as chemical suits deployed with US forces in the Gulf)? Would the Saudis have hosted US forces, and would other Middle Eastern states have joined the coalition if Hussein were able to threaten credibly their cities with a nuclear attack? Would the United States have acted if Hussein was able to threaten US cities? Was Kuwait worth it? These questions, in effect, were simply the post-Cold War equivalent of the Cold War problem of US extended deterrence in which missile defence during the SDI debate had been seen as strategic decoupling. But this wasn't the Cold War, and rogue states were not the status quo Soviet Union.

Both variants came to the same conclusion. New policies and capabilities were needed to deal with the proliferation challenge, and these emerged under the umbrella concept of counter-proliferation. Integral to these new policies was missile defence to defend against rogue state use of weapons of mass destruction married to ballistic missiles and eliminate the capacity of these states to deter the international community. In an inverse replication of the Cold War seamless vertical deterrence requirements that dictated tactical theatre and strategic nuclear forces linked within a coherent escalation strategy, seamless vertical missile defences were needed – tactical defence for military forces in the field against SRBMs, theatre defence for allied cities against medium-intermediate-range ballistic missiles, and strategic defences to defend the homeland against ICBMs. On the surface, priority investment should follow the same pattern – tactical to theatre to strategic – reflecting the proliferation pattern and process. Most rogue states already possessed SRBMs as a function of the Soviet Union selling these weapons during the Cold War. These provided the basis for indigenous re-engineering to develop longer-range missiles. Even though existing SRBMs put a priority on tactical defences such as Patriot, the case for simultaneous investment in theatre and strategic defences drawing upon SDI research was strong.

First, there was no significant technological case to favour one over another. Short-range missiles might appear easier to intercept because they were slower. However, the time available from launch to detonation on target was very compressed (roughly three to five minutes). The speed-time ratio changed as the range of the missile increased – the longer the range, the faster the missile but the more time for intercept. In addition, the basic technology required was the same such that development in one area could be transferred to others – the argument that led President George W. Bush to merge the tactical, theatre, and strategic defences into a single stream after Clinton had split the US effort into two streams, theatre (including tactical) and national (strategic).

Second, the here and now of SRBM threats and future of IR/ICBM threats actually provided time to develop strategic defences ahead of rogue state capabilities. It would be many years before the rogue states would acquire long-range ballistic missiles, and the United States could reasonably expect to field an effective strategic defence before the proliferators were successful, if the United States moved forward quickly. In so doing, the offensive-defensive unit cost-ratio argument, made by Nitze in 1985, potentially shifted in favour of the defence. If sunk research and development costs were included relative to the same costs facing proliferators, it was possible that the cost of a unit of defence would be significantly less than a unit of offence – reversing the situation most held as true during the Cold War in making the case against SDI.[61] In these circumstances, seamless vertical missile defences could also dissuade other potential proliferators from acquiring such weapons.[62] Why invest heavily in missiles for weapons of mass destruction when missile defences stripped them of their deterrent or military value?

Third, there was a simple political reality to consider. Defending forward-deployed military forces and coalition or allied cities was problematic if the United States could not defend its own cities. And there remained the possibility of an accidental launch, especially with the situation in the Soviet Union/Russia, where the leadership's control of its nuclear arsenal was in some doubt. In some ways, politically it was 1967 again, with rogue states replacing Mao's China and Russia providing fears of an accidental launch. Overnight, opposition to SDI was replaced by a growing congressional and public consensus on the importance of developing and deploying a limited strategic defence for the United States. Although the United States would begin to develop a range of tactical and theatre missile defence projects, the primary answer was GPALS, and like its ABM predecessors it too would consist of two types of interceptors. Unlike Sentinel and Safeguard, however, the ground-based terminal intercept component would not employ a nuclear warhead. Instead, it would deploy a kinetic-kill warhead. The other component would be located in space – SDI's Brilliant Pebbles brought to life to intercept warheads during the boost and mid-course phases. The kinetic-kill warheads or intercept process for both components had been initially tested in the pre-SDI Homing Overlay Experiment, which had also been applied to the US F-15 ASAT. In support of both layers, GPALS was to deploy a new generation of space-based satellites, Brilliant Eyes from SDI days, to provide tracking, cueing, and target discrimination.[63]

Following President Bush's announcement in January 1991, GPALS was divided into three phases. Phase One was to see the deployment in 1996 of the first ground-based component at the old Safeguard site in North Dakota (and thus be ABM Treaty compliant and subsequently consistent with the NMD Act). Phase Two was to follow with additional ground-based sites and the deployment of Brilliant Eyes. Phase Three was the space-based component. No specific dates were set for Phases Two and Three. Nonetheless, a Phase One deadline had been set, and to meet it numerous steps and decisions would have to be taken quickly, particularly in developing and implementing a robust testing envelope. A concept of operations would need to be developed, which because of new technologies could not simply be a modernized version of Safeguard procedures. Developing a new concept of operations naturally was predicated upon the specific capabilities of the system. A basic Phase One architecture was one thing. Details about the range of the interceptor; new tracking and cueing radars/sensors; the future of the existing old ABM radar at Cavalier, North Dakota; and command, control, and communications, among other system specifications, had to be established.

All these were vital for the Canadian government to make an informed decision on Canadian participation once an invitation was extended, and there was little if any doubt that at least an informal invitation would follow. If past practices were any indicator (ABM and SDI), there would likely be a year lag as the United States got the GPALS project up and running. So officials had some time to consider the basic issues facing Canada, and these emerged in an External Affairs–National Defence consensus in September 1991. The departments agreed that a formal US commitment not to deploy weapons in space and not to violate the narrow interpretation of the ABM unless renegotiated or as part of some other mutual agreement between the United States and the Soviet Union was the *sine qua non* of Canadian support.[64] Beyond this, Canadian participation would depend upon Canadian security interests and financial costs. It would also turn on exactly what aspects of GPALS would be open to Canada. In other words, Canadian participation would depend upon the United States deciding on what roles or functions were open to Canada. This was condensed into two possibilities: the minimalist possibility of the United States seeking Canadian endorsement with some participation in an operational command linked to NORAD, and the maximalist possibility of full participation along with the requirement to invest resources. Either, of course, depended in some part on the meaning and interpretation of participation relative to the

ABM Treaty's Article 9 prohibition, which had never apparently been a major issue of discussion between the two signatories in the Standing Consultative Commission.[65]

This basic agreement on Canada's GPALS position was essential with a US GPALS briefing delegation arriving in Ottawa on September 27, 1991. The meeting gave the United States the opportunity to provide further details on the GPALS architecture, operational questions, and related technological issues and provide some idea or hint of how the United States might see, or define, Canadian participation. At the same time, the US delegation wanted as clear an idea as possible of how Canada was likely to react to GPALS. In effect, the meeting was the beginning of the next round of manoeuvring on the part of both parties after SDI, and for Canada it might serve as the end of the information dilemma that had plagued ABM and SDI. However, to communicate a clarified policy position directly to the United States was problematic. If officials communicated Canada's two preconditions (though it was questionable at least whether the narrow interpretation was formal government as distinct from bureaucratic policy), Canada had for all intents and purposes set the conditions for participation. If these were met, Canada would say yes, and if prepared to say yes the United States might provide full access to GPALS information as the means to negotiate some form of agreement. However, these preconditions might be interpreted as a no to GPALS because of the weapons-in-space issue. Certainly, US officials were well aware of the weapons-in-space problem for Canada. The key for Canadian officials was to try to finesse this issue in developing a new policy. Of course, neither the United States nor Canada could or would commit at this initial meeting with many issues, including the ABM Treaty question, still up in the air.

A month and a half later, senior officials recommended a formal GPALS position to the minister of external affairs, Barbara McDougall. The recommendation moved past the initial joint departmental position of September.[66] First, the government should clearly and favourably recognize the security benefits of GPALS in light of the new strategic environment. Second, GPALS needed at least the acquiescence of the Soviet Union/Russia to ensure that it would not prove destabilizing. Third, the minister should express reservations about weapons in space but not condemn GPALS Phase Three outright. Finally, the minister should signal Canadian willingness to cooperate in principle and concentrate on early warning, tracking, and cueing, including space-based systems.

A public communication strategy was also recognized as essential. Even though significant support had existed for Canadian participation in SDI, the highly vocal nature of public opposition dominated bureaucratic and political memory. Officials invoked the ghost of SDI by reminding the minister of public opposition to SDI and public support for Mulroney's SDI decision. Officials emphasized the difficulty the government faced in communicating the difference between SDI and GPALS – even more difficult than that faced by the government in 1985 in trying to communicate the simple reality that SDI was only a research program. The new optimistic post-Cold War environment, symbolized by the idea of reaping the benefits of the peace dividend, was also not conducive to a new major defence initiative. The Canadian public was woefully unaware of the dangers of proliferation by anti-Western states, even with the Gulf War and the subsequent work of the United Nations Special Commission. Government leadership was essential, and this would require an extensive program of background briefings to the media by senior officials, emphasizing the security, arms control, and proliferation dimensions of GPALS, while repeating Canadian concerns about space-based systems.

Whether with the initial September consensus position or the softer November position, the government faced two major problems. Bush had made it clear that GPALS would proceed regardless of the Soviet Union. Although he preferred an agreement to amend the ABM Treaty to permit GPALS, Bush was prepared to proceed without Soviet cooperation. Bush had also clearly stated that GPALS would not be ABM Treaty compliant. There was to be no resurrection of the SDI interpretation debate. The president was on the path, if necessary, to announce US withdrawal from the treaty as allowed under Article 15 by citing the threat of rogue state proliferation as having "jeopardized its supreme interests."[67] For Canadian officials, this meant potential disaster. If their reading of Canadian public opinion was correct, explaining the collapse of the ABM Treaty in the new post-Cold War environment of American-Soviet cooperation, with or without the GPALS question, would be difficult, to say the least. The opposition parties would have a field day, either calling for the outright condemnation of the United States or levelling charges of subservience to Washington.

It was, of course, only a potential disaster, as the likelihood of such a dramatic step was low. From CFE to the Gulf War to START I, prospects of a US-Soviet agreement on ABM grew. The Bush administration had publicly offered to negotiate with the Soviet Union and make significant concessions. Gorbachev had responded favourably with certain conditions: GPALS had to be non-nuclear and limited in size and scope. It had to be designed to complement deterrence. The Soviet Union had to be reassured that GPALS would neither neutralize their strategic nuclear deterrent nor drag it into a new costly arms race. In the midst of these discussions, on September 22, Bush announced possible Soviet participation in a ballistic missile early warning and surveillance system.

A joint US-Soviet Union early warning and surveillance system that complemented mutual deterrence was a valuable confidence- and security-building measure and an avenue into ballistic missile defence cooperation. Such cooperation also had symbolic value as another indicator of the end of the Cold War, which had previously been unthinkable. Finally, it had real practical significance. The collapse of the Soviet economy had led to serious cuts in defence spending, resulting in a significant loss of capabilities. One of the victims was the Soviet ballistic missile early warning network. Large gaps in the network raised the possibility that benign rocket and missile launches might be misinterpreted as an attack on the Soviet Union, leading to a decision to launch a nuclear strike with a launch on warning posture.[68] This problem was compounded with the collapse of the Soviet Union and the presence of Russian early warning facilities now on foreign territory. In addition to the presence of significant elements of Soviet strategic forces; test and production facilities in Belarus, Kazakhstan, and Ukraine; and attendant command and control issues that would not be fully resolved until the Lisbon Protocols six months later, early warning gaps made the prospects of a nuclear accident relatively high – another argument in favour of the stabilizing function of, and pressing need for, a limited missile defence such as GPALS.

Movement toward greater cooperation, and an amended treaty, continued with the independence of the Russian Republic under President Yeltsin, which was recognized as the formal successor state to the Soviet Union. Yeltsin had first raised the idea of a global protection system employing US and Russian missile defence technology at the UN in January 1992. The June 17, 1992, US-Russia summit set the stage for further cooperation and an international effort centred on US-Russian cooperation on missile defence for a global protection system. The two presidents agreed to establish a high-level working group to look at three initiatives: sharing early warning with the establishment of a joint early warning centre, cooperating with states on missile defence and missile defence technologies, and establishing a legal basis for such cooperation either through new treaties or the amendment of existing ones.[69] They also agreed to revise the ABM Treaty to allow GPALS to proceed and to further reduce their strategic forces as part of START II, which would be signed in January 2003 just before Bush left office.[70]

For Canadian policy makers and policy, the June 17 summit turned the missile defence–deterrence–arms control relationship on its ear. GPALS had been blessed and, contrary to SDI arguments, coincided with further cuts to strategic nuclear forces.[71] The door was open to a bold Canadian move forward on missile defence with missile defence opponents essentially disarmed. GPALS did not hinder strategic arms reductions but was a force for international cooperation. Indeed, Canadian participation through a multilateral global system was ideal.

Indeed, this new environment was being replicated in NATO. Following the Gulf War, the allies at the Rome summit on November 8, 1991, identified "the proliferation of weapons of mass destruction and of their means of delivery [ballistic missiles]" as a clear threat to international peace and security.[72] Among other things, this consensus initiated a series of studies on tactical and theatre missile defence, along with the establishment of the Senior Politico-Military Group on Proliferation and the Senior Defence Group on Proliferation two years later. The United States also subsequently briefed the North Atlantic Council on a three-phase approach to allied cooperation: a joint early warning centre, technology transfer to facilitate allied missile defences, and the development of an operational program that would include the linkage of US space-based sensors to allied defences. The US representative felt this would be undertaken under the auspices of a revised ABM Treaty. Regardless, he pointed out that the Russians did not have a veto on this cooperation, despite Article 9.

For the NATO allies, their missile defence requirements, tactical and theatre, fell largely outside Article 9, as the ABM Treaty had only dealt with strategic defences. The Article 6 reference to "testing in an ABM mode," originally designed to prevent the potential use of high-altitude air defences in a missile defence function, applied only to tests against strategic ballistic missiles, which travelled at much higher speeds than tactical and theatre missiles – short-, medium-, and intermediate-range ballistic missiles. There were no formal agreed criteria demarcating strategic from non-strategic missiles, which would become the centrepiece of the demarcation debate and the 1997 Russian-US agreement on demarcation. It was on this basis that the United States believed the Russians had no veto and cooperation could proceed.

Of course, the same did not apply to Canada. Because of geography, any direct Canadian participation in GPALS was strategic and thus under Article 9. Direct Canadian participation required an amended treaty, depending upon the ill-defined scope of Article 9. Regardless, Canada could engage in tactical and theatre missile defence development, which it would at a low level as the NATO effort moved forward. More importantly, the allies provided the political opportunity to place Canadian participation within a wider framework as more than just a bilateral initiative. Another piece of the puzzle was in place for Canadian movement – allied missile defence engagement alongside an international initiative, burgeoning US-Russian cooperation, and significant reductions in strategic forces.

All that remained was the weapons-in-space dimension of GPALS. As part of the Bush-Yeltsin agreement on missile defence, it appeared that Russia might accept the weapons-in-space component of GPALS. Regardless, this component had no future deployment date assigned to it. The potential existed that the weapons-in-space problem might be pushed off into the future and for another government, while the Mulroney government concentrated upon the more immediate Phase One ground-based component. Unfortunately, Canadian officials clearly understood through their meetings with US officials that GPALS was a package deal. Canada couldn't accept one piece and reject another. Canada had to announce its support for GPALS as a whole, and such an announcement appeared necessary in order for Canadian officials to get access to the details on GPALS vital for making the next decision on participation itself.

The November 1991 External Affairs recommendation to express reservations represented an attempt to finesse the question. At the same time, announcing support for GPALS would not mean Canadian participation per se. It opened the door to negotiations on participation. Nor did it mean that Canada would have to contribute to and participate in each and every GPALS component. This would be central to the negotiations, and there was no guarantee that the United States would agree to Canadian participation in any and every GPALS component. Regardless, in order for Canada to engage or receive an explicit invitation to negotiate participation, the government would likely have to accept publicly GPALS as a whole, including the future space-based weapons side of the equation.

Since 1967 and the signing of the Outer Space Treaty at least, Canadian policy had opposed the weaponization of space implicitly defined as the stationing of a weaponized satellite on orbit around the earth.[73] This policy was supported by all the federal political parties. The weapons-in-space issue had also been the dominant line of opposition attacks on SDI in 1985. If anything resurrected the ghost of SDI, it would be weapons in space. It would take a major communication effort to overcome the GPALS–SDI–weapons-in-space link. Even with the new strategic environment, US-Russian cooperation, and the threat of proliferation, overcoming this link would be difficult.

The problem of weapons in space was most clearly articulated by Peggy Mason, the ambassador for disarmament. To support a program that led to weapons in space was in direct conflict with the primary arms control objective of non-proliferation – an argument raised during the ABM debate. It would also dictate a fundamental shift in Canadian policy.[74] In contrast to other advice to the minister of external affairs, Mason felt that the government must clearly point out that endorsement of Phase One (the ground-based component) did not mean acceptance of Phase Three weapons in space. The minister should unequivocally state Canadian opposition to weapons in space and reject participation in any such element of missile defence.

At the same time, Mason pointed out, there was significant opposition within the United States to Phase Three. The congressional bipartisan consensus that had led to the 1991 NMD Act did not extend to Phase Three. Like the opposition to SDI in Canada, the Democrats in particular, along with a large body of expert opinion in the United States, were still opposed to weapons in space, as they had been during SDI. This opposition had also been implicit in the 1991 NMD Act, which contained no references to weapons in space and called for an initial ABM-compliant first phase. Bush might wish weapons in space, but it remained to be seen whether Congress would fund that part of the program. A major fight loomed over weapons in space. As such, a formal reiteration of Canadian opposition to weapons in space might have some impact on the US debate, especially if the Democrats won the 1992 presidential election.

But expressing reservations was one thing; openly opposing the administration's position was another. It would be a bold and potentially dangerous gamble for an inherently conservative bureaucracy to propose. Bush had made it clear that the United States would proceed with GPALS as planned, and any public Canadian siding with the opposition to weapons in space would likely prove counterproductive in getting access to US thinking and planning. It was also unlikely that a bold Canadian position would carry any weight in the US debate. Regardless, if Bush won re-election and the Republicans gained control of both Houses of Congress, weapons in space would proceed, and Canada might have to publicly reverse course. If Clinton and the Democrats won, weapons in space might go away, but not much would have likely been gained by staking out a pro-Democrat position early on. If there was a division between the presidency and Congress, who could predict how the issue might play out? One other factor made a bold statement null and void: the prime minister would not countenance it. His personal relationship with George was as close, if not closer, than it had been with Ronnie. In the end, time was still on Canada's side, and the primary issue of strategic concern still had to be considered – GPALS and the future of NORAD and with it North American defence cooperation and the fundamentals of Canadian defence policy.

In the spring of 1991, GPALS was raised in the memorandum to cabinet on NORAD renewal largely for information. The memorandum emphasized that the missile defence issue was not pressing, as significant time would be needed for the program to move forward, and there was no invitation looming in the immediate future. Nonetheless, in placing GPALS within the context of NORAD renewal, there was an implicit recognition of the existential threat GPALS posed. One thing the legacy of SDI had made clear, especially to National Defence, was that, one way or another, GPALS would have a major impact on North American defence cooperation. If saying no to a research program had marginalized Canada in NORAD, standing aside from GPALS meant complete Canadian exclusion from an important aspect of US defence policy, with implications for broader operational North American defence cooperation – and, with it, the possible end of NORAD. If so, Canada would have to rethink its entire national defence strategy and policy. Canada might suddenly have to undertake its defence alone, with all the political and economic costs this entailed. In this regard, the United States did not need to pressure Canada – the very existence of GPALS was sufficient.

It was not just the legacy of SDI that affected NORAD. The end of the Cold War had raised the question of its relevance. The Soviet threat had evaporated. Indeed, the Soviet Union itself had vanished. With it, the Soviet ballistic missile, bomber, and air-launched cruise missile threat had effectively disappeared. The Russian bomber force began to withdraw from forward operational bases, and bomber flights, which had been tracked and intercepted by NORAD, came to a complete end. There was simply no money for the new Russian government to continue these Cold War exercises, and, of course, politically they were counterproductive. With the air defence mission on the verge of bankruptcy, what would NORAD do now, and who would be willing to pay for it? One answer was the search for a new raison d'être, found in NORAD engagement in the US war on drugs by moving into drug interdiction. But such a role didn't justify the costs of NORAD. The bulk of Canada's commitment had little value for the drug interdiction role, and the United States had just paid for 60 percent of the new radar line, the North Warning System (NWS), with zero value for the war on drugs. It would just be a matter of time before Congress would begin to wonder why the United States was paying hundreds of millions of dollars for an institution with no significant function to play outside of being a symbol of binational continental defence cooperation. Indeed, the history of NORAD's creation spoke to its logical end. NORAD was created for functional reasons from the bottom up. As these reasons disappeared, NORAD had no function unless it adapted, and this adaptation was functionally possible in only one mission: missile defence.

Even within the Pentagon, Canadian NORAD officials feared that questions would likely emerge about the utility of NORAD expenditures as US defence spending began to fall in search of the Cold War peace dividend. If push came to shove, NORAD and continental defence were not a high US defence priority

– with the exception, of course, of missile defence. Even here, US military officials were much more concerned with tactical and theatre missile defences for their forces deployed overseas. Indicative of the situation facing NORAD, the US Joint Chiefs of Staff, in deliberations on the US command structure in 1993 as part of the biennial Unified Command Plan review, proposed to downgrade NORAD from a four-star to a three-star command.[75] This idea was met with horror in Ottawa. It not only implied the downgrading of the defence relationship with Canada but also potentially foreshadowed much larger changes in US thinking. Regardless of the actual reasons behind the proposal, the chief of the defence staff, General DeChastelain, lobbied the chairman of the Joint Chiefs, General Powell, to maintain the status quo. DeChastelain argued successfully that the United States could not make such a decision without either the consent of Canada or a revision to the NORAD Agreement. As NORAD had been renewed in 1991 and Canada would not consent, the United States at a minimum would have to wait until the 1996 renewal.

In the end, GPALS was at once the largest threat to, and potential saviour of, NORAD. From the onset, Canadian NORAD and National Defence officials recognized the seminal importance of GPALS for the future of NORAD and aerospace defence cooperation. Recognizing that the old distinction between air and space had become increasingly blurred, rejecting missile defence left Canada a spectator in the defence of North America, including Canada. Seeing aerospace as a single environment, if Canada stayed outside GPALS, the United States would shift its efforts into US Space Command. Except for the NWS, NORAD's aerospace early warning mission was supported entirely by US assets, and these assets came under Space Command and USAF Space Command. With no strategic air defence requirement, Canada on the outside of GPALS meant being on the outside of the major aerospace threat and response to North America. Defence officials also recognized that GPALS' Phase Two space-based sensors potentially rendered NORAD's surveillance function via the NWS obsolete – something that had been predicted by officials during the ABM debate.

In effect, decades of few contributions, limited investments, and reliance upon the United States were now coming to threaten the future. If Canada stood outside, it would return to the 1950s, when the government lacked the resources to control Canadian airspace and had to rely upon the United States. Canada faced three potential choices.[76] First, it could focus all its efforts on the aging air defence system able to deal only with a non-existent air threat and eventually quit the North American defence business entirely. Second, it could develop its own indigenous aerospace defence surveillance system at overwhelming expense.

Finally, it could accept missile defence and prepare to be a viable partner in an integrated aerospace defence relationship for the next century.

Even though this reality was obvious and existential, the United States made these choices very clear to Canada. On February 12, 1992, Bush wrote to the prime minister extending an invitation for Canada to engage in discussions on GPALS. George also informed Brian of his intention to raise the issue at the next meeting of NATO's North Atlantic Council. In response, External Affairs prepared a series of talking points for the prime minister to use in a call to the president. These notes included clear support in principle of GPALS, along with the aforementioned international stability link and reservations on weapons in space, which Mulroney more or less communicated. Following this exchange, at the spring 1992 PJBD meeting in Charleston, South Carolina, US officials provided a detailed briefing on GPALS and proposed the next step in bilateral consultations through a US inter-agency team in April.[77] The US delegation noted that, as a function of past Canadian policy, the United States had gained a great deal of experience working around Canadians in NORAD on numerous missile defence issues. This time, however, the United States preferred not to, implying that this time NORAD would be fully removed, which potentially meant that the early warning mission for GPALS would go elsewhere. Instead, the United States needed to know Canada's interests in and intentions on the phases of GPALS before it began deliberations on command and control questions. Still, the US delegation concluded that the relationship with Canada required special consideration, regardless of whether a memorandum of understanding on GPALS was negotiated.

This exchange appeared on the surface as direct pressure for Canada to move forward. However, from a practical perspective, the United States had little choice. The Safeguard solution for early warning would no longer necessarily work. Technology had taken dramatic steps forward. The GPALS system, including the ground-based interceptor, could engage much earlier than the long-range Spartan interceptor of Safeguard. As the new system responded much earlier in the flight time of an incoming warhead, NORAD engaging only as an early warning agent was simply redundant. It didn't necessarily mean that NORAD's fundamental mission to provide warning to the respective national command authorities of a ballistic missile attack was irrelevant. Even with a missile defence capability in which release authority was delegated to the commander on-scene – either possibly NORAD if Canada was in or Space Command if not – someone would still have to give the president and the prime minister the bad news. Nonetheless, NORAD outside GPALS simply complicated the entire process, and there was little interest in engaging once again in debate on what would or would not be open to Canadians as non-participants. As Canadian defence officials knew quite well without being told by the United States, Canada as an outsider would have little say and involvement in North American aerospace defence, potentially raising the bigger question of whither NORAD.

Conversely, GPALS gave NORAD meaning and purpose, and NORAD had already demonstrated its potential value in this regard. During the Gulf War, NORAD had provided early warning of Iraqi Scud attacks to the US-coalition command, providing time for civil defence measures to be taken in Israel and Saudi Arabia as well as basic cueing for the Patriot batteries. In effect, NORAD had demonstrated the initial role it could play for GPALS, and it was not difficult from there to imagine a much broader role for NORAD in a North American missile defence system as envisioned by GPALS. Moreover, with Canada still basking in the glow from its overall military commitment in the Gulf War, and the close friendship between the prime minister and president, the prospect of NORAD becoming the missile defence agent for the defence of North America looked very promising.

While many issues of how GPALS would actually operate had yet to be even considered, it wasn't difficult to imagine its integration into the NORAD mission as a logical extension of binational air defence cooperation. NORAD already possessed the ballistic missile early warning mission supported by US early warning assets. As these assets were modernized and augmented by new sensors, they would be fed into NORAD through an integrated ballistic missile early warning and defence operations centre. Tracking, target discrimination, and cueing were an easy and logical extension of early warning. If, in turn, Canada contributed by engaging in the Brilliant Eyes research and development effort and then provided additional tracking and battle damage assessment radars, such as proposed by NORAD officials for Churchill (Manitoba), Goose Bay (Labrador), or Frobisher Bay in the Arctic, NORAD's case for obtaining the missile defence mission would be strengthened. Once these functions were integrated into NORAD, the logic of assigning overall command and control followed easily. NORAD assessors already decided whether North America was under ballistic missile attack and characterized the nature of the attack. Although its authority did not extend further, the possibility of the command assessor giving the order to release missile interceptors was not unfounded. After all, the decision was not about releasing nuclear weapons.

All of this was possible, at least theoretically, whether the ABM Treaty was amended or not. A narrow reading of Article 9, relative to Article 2, did not provide any direct prohibition to such a role for NORAD. Indeed, the "testing in an ABM mode" referred to radars directly communicating and guiding an interceptor to target. As long as this role remained US-only, Canadian participation would not run afoul of the treaty. As for interceptors, like air defence fighters, they would be US-only, even if committed to NORAD's integrated command. They would remain American, with direct command being American. The treaty was also silent on the question of command and control. As long as no interceptors or specific ABM Treaty components resided on Canadian soil or were transferred to Canada, then Canada would not be a participant even with NORAD engaged – or at least the issue was open for discussion between the United States and Russia in a very favourable international climate.

Here also lay the solution to another one of the problems of Canadian participation, weapons in space. Analogous to Canada's arm's-length distance from US strategic forces, weapons in space were just a different type of interceptor. NORAD would not participate in the actual intercepts, whether ground- or space-based per se. Canada, regardless of where the interceptors were located, would contribute via passive sensors and radar, and personnel in a command and control role, while ensuring Canadian access to and input into operational missile defence plans and the execution of these plans, thus ensuring a key role in the defence of North America and Canada. At best, a Canadian officer in command in the Cheyenne Mountain Operations Center would provide release authority to the US commander of the missile defence batteries – not very different from the role NORAD played in the process by which US strategic forces would be released, *sans* the role of the president. Thus, Canada would not actually participate in the intercept side unless the government agreed to deploy interceptors on Canadian soil or Canadian interceptors in space, which it could not do without placing the United States in violation of the treaty. Canada could in effect stand aside in a manner similar to its relationship to US strategic forces and its nuclear disarmament policy as a member of a nuclear alliance.

The government might have to accept GPALS as a whole, but it might be able to pick and choose the nature and meaning of participation. This complicated yet simple logic was not without precedence. The Pearson government had never explicitly said no to ABM. It had only agreed that ABM would be decoupled from NORAD. Even though Trudeau had expressed serious reservations about ABM, he had not explicitly said no either. Instead, the Trudeau government had faced the question of ABM command and control going to NORAD. In so doing, there had existed an implicit definition of the meaning of participation, and, on the surface at least, participation was limited to interceptors. Furthermore, Mulroney had never said no explicitly to missile defence either. He too had sidestepped the question by supporting SDI research as prudent while keeping the government officially out. Thus, the government had the opportunity, if it so desired, to define the issue of participation in such a way that allowed Canada to participate without participating.

While all this might prove possible, it would require bold leadership on the defence front from a government that had weathered both the bilateral free-trade and subsequent North American free-trade debates. It had also recently weathered the debate over Canada's involvement in the Gulf War and was confronted by new overseas military commitments in Yugoslavia, as it collapsed, and Somalia. Moreover, it continued to be deeply embroiled in the Quebec constitutional question, after the failure of Meech Lake, and by the end of 1992 the failure of the Charlottetown Accord as well. Meech Lake, in particular, had seriously damaged the prime minister's public standing and, alongside the Goods and Services Tax decision, made him arguably the most unpopular prime minister of the century. Playing cute on the meaning of participation was too dangerous to contemplate in such a difficult domestic political environment. Besides, SDI lurked too closely in the background, and raising GPALS promised to resurrect all of the acrimony of that debate. It would no doubt be seen as SDI through the back door. The potential charge of government duplicity could not be ignored. In this sense, the Mulroney government had become handcuffed by its own policy – one perceived by many to have been a no, period, and not just a no to official government participation in SDI research.

By 1992, the Mulroney government looked exhausted, and the prime minister was increasingly unpopular. There appeared little political payoff to be had by taking the lead and placing missile defence on the public agenda. Having decided to retire, it was unreasonable to leave his successor with such a contentious political issue, with the general, if mistaken, view that SDI had been widely unpopular in Canada. As for Mulroney's successor, Kim Campbell, missile defence also appeared as a vote-loser file to be left alone if at all possible. None-theless, the government did not have the luxury to stay completely quiet on GPALS. For the third time in the missile defence saga, when pushed the government would express an understanding of US security concerns. On March 9, 1992, in response to a query in the House on GPALS, the associate minister of defence, Mary Collins, expressed concerns about the threat of proliferation.[78] She added that the US Congress had placed explicit limitations on missile defence, GPALS was not Star Wars, and no discussions were under way. This would be repeated in the government's response to a second query two months later: "Beyond these general points, Canada has not taken a decision regarding possible participation in the proposed US ballistic missile defence programs. We have indicated our willingness to engage with the United States in consultations concerning the details of the American proposal and its implications for Canada."[79]

Of course, all of this came to naught. Bush was defeated in the November presidential election. With the Democrats in control of the White House, GPALS would disappear. However, this did not mean that missile defence would. All indications pointed to some form of restructuring of the US missile defence effort, as the Democrats sought distance from SDI. Besides, the new defence secretary, Les Aspin, had been one of the architects of the NMD Act. For now, Canada could breathe a sigh of relief, having obtained some respite from the inevitable, as history would prove. Above all else, hidden from the glare of public discussion, the legacy of SDI and GPALS had established the next government's new policy on missile defence, one that opened the door to Canadian participation.

Act 4

National Missile Defense: Let Sleeping Dogs Lie (1993-2000)

On February 16, 1996, President Clinton's secretary of defense, William Perry, announced the National Missile Defense (NMD) Deployment Readiness Program – also known as 3 plus 3. It was designed as a robust, compressed three-year research, development, test, and evaluation program (the first "3"), to be followed by a decision on whether to proceed to deployment. Once, or if, the decision to deploy was made, an operational system employing only ground-based interceptors would be in place three years later (the second "3"). Six months later, the Joint Requirements Oversight Council (JROC) within the Office of the US Joint Chiefs of Staff agreed to give operational command either to NORAD, if Canada agreed, or to US Space Command, if it didn't.[1] In 1997, the prime system integrator contract for NMD was awarded to Boeing and the kinetic-kill warhead contract to Raytheon. The next year, the first flight and intercept tests began. In the spring of 1999, Congress passed and Clinton signed into law the second National Missile Defense Act, which mandated deployment once the technology was available. After the failure of the third live intercept test on July 8, 2000, Clinton announced on September 1 that the technology did not warrant deployment, thereby passing the decision onto the next administration.

Until 1998 or so, the developments that presaged and followed Perry's announcement met with deafening silence in Canada. Little different from the silence between ABM and SDI, and then SDI through GPALS, by all accounts missile defence did not exist as an issue for Canada. Although the Chrétien government took a major step in the direction of participation in the 1994 Defence White Paper, which formally announced a new missile defence policy actually put in form by the bureaucracy in 1992, no one paid any attention, including the government.[2] Two years later, few paid attention to the 1996 NORAD renewal when one of the most virulent opponents of missile defence, Foreign Minister Lloyd Axworthy, accepted a new clause in the treaty that provided a mechanism for Canadian participation.[3] Two years later, with NMD momentum growing, Foreign Affairs and National Defence engaged in an acrimonious debate on the meaning and nature of consultation with the United States on NMD. By the next year, with a critical deployment decision point drawing ever closer, concerns rose that Clinton might well have no choice but to proceed forward in 2000. Such a decision would force the government to decide finally on Canada's position.

Despite these developments and concerns, the government and cabinet avoided the missile defence file with the connivance of Foreign Affairs and National Defence. With the government, cabinet, and the Liberal caucus divided, and with Foreign Affairs and National Defence on opposite sides of the issue, neither department could predict, if push came to shove, which way the government might go on NMD. For opposite reasons, both departments hoped and prayed that the Democrats in the White House, the ABM Treaty, Russia, the NATO allies, technology, costs, or any combination therein would prevent a positive NMD deployment decision. Their prayers were answered, leaving intact the informal four no's policy developed by National Defence in the late 1990s – no US architecture, no US deployment decision, no US invitation, no Canadian decision. But it was only a temporary respite. The missile defence writing was already on the wall and in many ways had been for four decades.

With the defeat of Bush in the 1992 presidential election, GPALS was to disappear, if only because it was a Republican program. But missile defence did not. During the election campaign, Bill Clinton had gone on record as a supporter of missile defence. However, the future would show that his support was lukewarm at best, just as Johnson's and McNamara's had been. Domestically, no presidential candidate could afford to appear weak on defence and security, even if the economy dominated the 1992 election. Defence and security had long been the Achilles' heel of the Democrats. Besides, there was widespread bipartisan support for missile defence in Congress. Senior Democrats had supported the 1991 NMD Act, and Clinton's defence advisor during the campaign and his first secretary of defense, Les Aspin, was one of its architects.

The Democrats, including Aspin, had been fundamentally opposed to SDI, and the 1991 NMD Act contained provisions that had implicitly ruled out the SDI-inherited space-based component of GPALS. Nonetheless, it was not immediately clear from the campaign exactly how the new administration would proceed. Phase One of GPALS, the ground-based system earmarked for the old ABM site in North Dakota, fit the NMD Act and was nearly ABM Treaty compliant, except for the absence of an ICBM field to defend.[4] Canadian officials suspected that Phase One might proceed under a different program. However, none of the three US military services was a formal sponsor of what would become known as NMD, which was the norm for new development and deployment projects. Yet all three were involved in missile defence development programs, focused primarily on the tactical and theatre levels and designed to defend military forces deployed in the field and rear support bases.

The US Army had begun research on the next generation of the Patriot tactical missile defence – the kinetic-kill Advanced Capability 3 (PAC-3) – and a more advanced theatre-level defence, which would become the Theater High Altitude Area Defense (THAAD) program sponsored by the US Army Missile Defense Command in Alabama.[5] The navy had become involved with a lower-tier Patriot equivalent and an area theatre-wide defence THAAD equivalent. The air force was moving toward initiating an airborne laser development program, which as a boost-phase intercept capability had tactical, theatre, and strategic value.[6]

Overall, then, indications led to an obvious conclusion that Canada would face the missile defence question in some form or another, even with a Democrat in the White House. Although tactical and theatre defences dominated the services development programs, one would be hard-pressed to believe that any administration would ignore the missile defence of the United States entirely and simply proceed with defences for forces in the field (and, thus, allies). Of course, a national defence inevitably raised the ABM Treaty question. During the GPALS interlude, officials were concerned that the SDI-ABM Treaty interpretation debate might resurface as a function of the robust development and testing envelope required by the GPALS timeline. If this occurred, External Affairs officials advised the government to finesse the issue without committing itself by only expressing reservations about weapons in space, if it had to. With the Conservatives in Sussex Drive and the Republicans in the White House, anything more forceful was simply unacceptable. The possibility of a public announcement endorsing the narrow interpretation of the ABM Treaty as a precondition for Canadian participation was simply not on. Nonetheless, both External Affairs and National Defence concluded long before Clinton's and Chrétien's elections that the narrow interpretation was essential to disarm domestic opposition and smooth the process toward Canadian participation.

For the time being, as long as Bush let that sleeping dog lie, there was no need for the government to make any statement about GPALS, especially if such a statement might awaken opposition voices within the Canadian public that had fallen silent with the passing of the Cold War. Besides, even an expression of a reservation might be interpreted in US circles as a no. At least within National Defence, the memory of the fallout from the SDI decision was fresh. For the time being, a wait-and-see posture appeared to be the only option – to wait and see whether the United States would proceed with or without Russian agreement and to wait and see if Bush and GPALS would survive.

Nor was it wise to pre-empt the new Clinton administration with any announcement, private or public, even if the Democrats were long-time supporters of the narrow interpretation. No one knew how the administration might react, and in the absence of such knowledge, caution had always been the Canadian way. Nonetheless, External Affairs and National Defence agreed that Canada needed to engage the new administration as early as possible on missile defence and flesh out some real options for the current government – and, of course, prepare the ground for whatever government emerged after the forthcoming federal election.[7] National Defence recommended that the defence minister at the time, Kim Campbell, raise the issue directly with Aspin at a visit scheduled for the end of February and look to engage in more substantive talks beginning in March. By May, serious concerns arose that an invitation from the United States was likely and could require a cabinet decision by the fall.[8]

These concerns were unfounded. As one would expect from previous experiences, a new administration would take some time, in part simply because of the complicated process of staffing key positions, before it would begin to move forward on the policy front or, in this case, make missile defence its own. In effect, it took until the lead up to the next presidential election for this to be fully achieved with the announcement of 3 plus 3 in early 1996. Immediately after taking office, the Clinton administration undertook a review of US missile defence policy, followed by a series of incremental steps. Its first policy decision, announced on May 13, 1993, restructured SDIO by emphasizing theatre missile defences and renaming it the Ballistic Missile Defense Organization (BMDO).[9] It also split the US missile defence development effort into two independent streams: a theatre stream that included tactical defences and a national stream that would be labelled a Technology Readiness Program. A month later, the Clinton administration, following the pattern of ABM and SDI, made NMD safe for Canada. The administration fully endorsed the narrow interpretation of the treaty.[10] The way was now open for Canada to revisit publicly missile defence, especially with the election of the Liberals on October 25, 1993.

The Liberals had long viewed the ABM Treaty as a sacrosanct document. Even before the interpretation question was born, they saw the treaty in very strict terms. Their statements inside and outside the House during the SDI debate suggested that even laboratory research was a violation or at least undermined the spirit of the ABM Treaty. Regardless, any possibility of Canadian participation or simple acceptance of any US missile defence program had to be within the clear and unambiguous narrow interpretation of the ABM Treaty. Canadian officials also wondered whether an amended or renegotiated treaty might even be acceptable to the new government. Missile defence for the Liberals, and especially Chrétien and Axworthy, who had led the opposition to SDI, had to include full adherence to the prohibition on space-based weapon systems. Otherwise, their past threatened to come back to haunt them. No signal to the United States of Canadian willingness to consider participation in any form of missile defence was even thinkable without a restrictive view of the ABM Treaty that included no weapons in space.

Welcoming the US endorsement of the narrow interpretation, as was ultimately done in the 1994 Defence White Paper, was not, by itself, sufficient to undo the damage of the Mulroney SDI decision. US officials below the administration and senior levels of the Pentagon interpreted the 1985 Mulroney policy as an unequivocal no to SDI rather than a qualified yes as intended and blessed by the Reagan administration and senior US defence officials. Undoing this interpretation was essential to gain access to US missile defence thinking and planning. With a GPALS Phase One deployment date set for 1996, this was pressing even before Clinton. Numerous architectural and operational decisions had to be made concurrently through the development process, and these might affect Canadian interests. Paramount among them was the decision on command and control of the ground-based Phase One component of GPALS, which might well survive regardless of who won the White House in 1992, and the operational employment strategy for the system, especially in terms of coverage and intercept priorities.

Of course, with most of the technology still in the research phase, it was unlikely that the original 1996 date would be met. The probability that, from a standing start, an operational ground-based missile defence capability could be built in North Dakota by 1996 was remote. The new kinetic-kill warheads had just begun testing.[11] But this was a plus rather than a minus for Canada. It provided more time not only to obtain access to US thinking and planning but also to potentially influence US thinking and planning in the early stages and make the United States at a minimum aware of, and sensitive to, Canadian interests. This became pressing after the 1991 GPALS announcement and led External Affairs and National Defence to recognize early on that the SDI policy was truly out-of-date. In fact, it stood as a default no. Moreover, access was essential for any future decision on Canadian participation – access to the technology, architecture, operational plans, and US thinking about possible Canadian roles.

Along with Canada's need to clearly signal to the United States an openness to undo the SDI decision, industrial and technological considerations were at play. Although the SDI policy guidelines for government support to private industry had been made public, there was little in the way of a payoff. Canadian companies had not done well in accessing SDI contracts and had demonstrated little interest in looking for them. Neither the relevant US officials nor Canadian industry interpreted the guidelines as a signal of government support. With GPALS and the subsequent possibility that the new administration would proceed with a near-treaty-compliant component, and rapidly expanding tactical and theatre missile defence research and development programs, the range of industrial and technological opportunities to industry had expanded greatly. Moreover, tactical and theatre systems were clearly outside the ABM Treaty. A new signal was needed, and this required government to at least clarify the 1985 policy or reverse it.

The need to go beyond simply welcoming the new administration's commitment to the narrow interpretation and make a clear signal of Canadian interest in possible participation resonated on the military/NORAD front. The fallout of SDI had mostly affected Canadian access through its NORAD–Space Command relationship. By 1992, the relationship in some ways had reached its low-water mark. As part of the congressionally mandated Unified Command Plan (UCP) review, senior US military officials concluded that NORAD should be downgraded from a four-star to a three-star command. The initial decision was taken to bring some consistency to the US command structure as it continued to move down the path laid out from the Goldwater-Nichols reforms rather than as punishment of Canada for its missile defence policy or anything else.[12] Nonetheless, senior Canadian military officials worried that missile defence was somehow in the background. With its old Cold War defence missions now bankrupt, and Canada on the outside of the only potentially new defence mission, NORAD was marginal. If marginal, it didn't need four-star status. The centrepiece of the Canada-US defence relationship, the institutional symbol of Canada's special and unique relationship as the United States' only binational command, and the most important conduit of Canadian access could be on its way out. In the end, Canada was successful in blocking the unilateral US decision. But unless something further was done, it might be a short-lived victory: the 1996 NORAD renewal loomed on the horizon.

All was not entirely foreboding on the NORAD front, however. The Canadian access issue had, surprisingly, taken a move in the right direction, thanks to the appointment of Lieutenant General Horner as the commander of NORAD and US Space Command. Horner had commanded the air component in Operation Desert Storm, where the participation of Canadian F-18s left him with a very favourable view of Canada. He understood the value of NORAD in providing early warning to the coalition and Israel during the Iraqi missile attacks. Moreover, upon arriving at Colorado Springs, Horner had seen little value in keeping Canadians out. Under his command the doors swung open again, and Canadians returned to the full range of activities, including the Space Control Operations Center. Even so, the viability of Canadian policy could not rely upon an individual personality. Horner would leave, and his replacement might not see Canada in a similar light. His replacement might also be a missile defence zealot.

Thus, well before the election of the Chrétien Liberals, the stage was set on several fronts for a fundamental change to Canadian missile defence policy. The environment after the defeat of Bush was conducive to forward movement, especially after Clinton's public endorsement of the narrow interpretation. The need to move forward was also clear to prepare for what appeared to some, but not all, as inevitable – a deployed missile defence system for North America. Finally, all was quiet on the domestic missile defence front, and the anti-missile defence forces in Canada were further lulled into complacency by the election of the party that had led the opposition to SDI. To no one's surprise, missile defence played no role whatsoever in the fall 1993 election campaign. Indeed, the term "missile defence" was not heard. Nor, in the months to follow, did the term see the light of day. Instead, public discussions on defence were initially dominated by the new government's decision to cancel the EH-101 helicopter purchase, and then cuts to the defence budget, as well as the very active operational tempo of the Canadian Forces overseas.

With the new strategic environment that accompanied the end of the Cold War, the new government announced the first major defence review since 1987.[13] This included the creation of the Special Committee of the Senate and House of Commons on Canada's Defence Policy, which would conduct a study, including public consultations. With the deficit dominating government policy objectives, further significant defence cuts loomed, and this meant that decisions about future fundamental roles, missions, and capabilities logically dominated the process – decisions also made possible and necessary by the post-Cold War strategic era. Missile defence was obviously marginal to fundamentals. Regardless, whatever little chance existed for missile defence to be injected into the process was eliminated entirely with the release of the *Canada 21* report in early 1994.[14] The Canada 21 committee consisted of Trudeau-era Liberals, giving the report a degree of legitimacy and relevance rarely seen in Canadian politics. Regardless of its actual contents, the committee was seen to endorse a fundamental shift in Canadian defence policy away from combat missions and toward constabulary ones. For National Defence, the fundamental goal was to drive a stake through the heart of the constabulary force idea, which proved successful with the endorsement of multipurpose combat capable forces in the 1994 White Paper.

When the public consultations were complete, and the Special Joint Parliamentary Committee released its report, missile defence was absent. The only reference appeared in the minority Bloc Québécois committee members' report: "That Canada renew its intention to oppose ... the placing of any anti-ballistic missile system on Canadian territory or in aerospace above Canada."[15] As far as Canadian policy was concerned, however, the White Paper released less than a month after the parliamentary report went much further and in a different direction. The government essentially opened the door to Canadian participation. Neither cabinet nor Foreign Affairs, much to its later regret, paid attention to the two missile defence paragraphs inserted in the White Paper.[16] When cabinet met to discuss and approve the White Paper, no one raised a word about the new missile defence policy.

The new policy did not commit the government to participation, but it did signal the government's willingness to engage. Citing Clinton's adoption of the narrow interpretation – Canada's safety blanket – the government committed itself to consultation with the United States and European allies on participation and the possibility of direct participation with two caveats. Participation had to be cost effective relative to Canadian defence needs. Canada's direct role was to be in the areas of surveillance and communication – areas identified by every government whenever missile defence was under consideration, beginning with ABM.[17] From the onset and throughout, even the thought of Canadian engagement on the sharp end intercept side was rejected.

The signal was sent. The defence minister, David Collenette, was formally told by senior Defence officials that the new policy was intended as a signal to the United States. At home, it passed virtually unnoticed. In Washington, the primary target of the new policy, the signal was apparently understood. NORAD renewal discussions, begun a year earlier in 1995, concluded with a new clause that enabled NORAD and Canada to undertake the NMD mission without opening the agreement. Following the announcement of 3 plus 3, the JROC expressed its desire to assign the NMD mission to NORAD.

The 3 plus 3 program naturally followed from Clinton's initial decision to limit NMD to a technology readiness program; the 1995 revision to the NMD Act, which removed 1996 as the deployment date in favour of technological availability; and the politics of the approaching presidential election. Although announced in early 1996, the first three years of research, development, testing, and evaluation didn't fully get under way for about a year, largely for organizational, administrative, and budgetary reasons. This set the decision point for the second three in 2000. However, technology alone would not determine the decision to deploy in 2000. The Clinton administration made it clear that technological capacity was only one of four decision points. The state of the ballistic missile proliferation threat to North America, the cost of a deployed NMD system, and the impact of a deployed NMD on international peace and security were also to be considered. Clinton had raised the bar substantially. To the Republicans, who now controlled both the Senate and the House, this was tantamount to further evidence of Clinton's essential duplicity. Labelling the program "3 plus infinity," Republicans concluded that Clinton had no intention of deploying NMD, regardless of the technology. Congress would have to force Clinton's hand, and missile defence became a major part of the vicious fight between Clinton and Congress for the next several years. The four criteria also served to frame the elite debate in the United States as well as the subsequent elite or attentive public debate in Canada. In neither case was there a commensurate public debate. In the United States, the public supported missile defence. In Canada, missile defence was merely irrelevant at this time.

NMD was simply Phase One of GPALS. Like its predecessor, NMD would employ all the new technologies stemming from the SDI research effort within the limits of the ABM Treaty.[18] There would be no space interceptors as envisioned in GPALS, but a new generation of space sensors was to be developed and deployed. NMD was to employ a kinetic-kill warhead on a new high-speed interceptor. Whereas the GPALS architecture placed the interceptor field at the old Grand Forks ABM site outside Cavalier, North Dakota, by 1997, a competitor emerged – Fort Greely, Alaska. Throughout the life of NMD, no formal decision on location was ever taken, and as long as the ABM Treaty was in effect the United States was limited to a single site. However, neither Grand Forks nor Fort Greely was treaty compliant. Neither was colocated with an ICBM field as required by Article 3(b), and no one in the United States was going to consider rebuilding the Grand Forks field, closed through START I reductions, or constructing a new field in Alaska. Of course, the Safeguard precedent might provide a way to sidestep the colocation provisions for Grand Forks. Regardless, by 1999, Fort Greely had become the favoured site, and Grand Forks was relegated to a possible second site.

The reason for Fort Greely's triumph was North Korea. In 1993, North Korea created an international crisis by announcing its withdrawal from the Nuclear Non-Proliferation Treaty. Even though the announcement was clearly legal within the limits of international law, it was met by widespread horror and condemnation. North Korea was one of the undeterrable rogue states identified by Bush – renamed "states of concern" by the Clinton administration. Some speculated that North Korea's decision was simply an attempt to blackmail the West for economic and energy aid. Others speculated that North Korea already possessed several nuclear warheads. Regardless, North Korea had an aggressive ballistic missile program, including re-engineered two-stage Scud ballistic missiles, and was rumoured to have exported its technology to Iraq and Pakistan. Fort Greely, by virtue of its location and the curvature of the earth, was the optimal site to intercept future North Korean ICBMs. It provided time for several intercept opportunities. It also could intercept a missile on an east-west

US Secretary of Defense Robert Gates examines GMD interceptor, Fort Greely, Alaska. *Courtesy US Missile Defense Agency*

First GMD interceptor being loaded into a silo, Fort Greely, Alaska, July 22, 2004.
Courtesy US Missile Defense Agency

trajectory launched from the Middle East and defend Hawaii and thus meet the requirements of the 1991 NMD Act for a system capable of defending all fifty states.

The Clinton administration was well aware of the implications of North Korean activities as well as those among the other states of concern for the missile defence file. In November 1995, the hitherto classified 1995 National Intelligence Estimate (NIE) was leaked to the public to limit pressures on the administration to move more quickly on missile defence and to manage the potential political implications of missile defence for the upcoming 1996 presidential election.[19] The NIE predicted the acquisition of long-range ballistic missiles married to weapons of mass destruction by states of concern – North Korea, Iran, Iraq, and Libya – no sooner than 2010. The 3 plus 3 program provided more than ample time to meet the threat.

Following the election, the Republican-dominated Congress, in an initiative that echoed the Committee on the Present Danger's approach in the 1970s, decided to challenge the administration's estimate. In 1997, it appointed an independent commission to examine the threat time line, chaired by the future George W. Bush secretary of defense, Donald Rumsfeld. The Rumsfeld Commission report, issued on July 15, 1998, not surprisingly raised doubts about the NIE's time line and argued that proliferators might acquire long-range ballistic missiles much sooner – especially North Korea, which had already tested a two-stage missile.[20] The report cautioned about the amount of advance warning necessary to ensure adequate time to deploy defences if the initial intelligence estimates were wrong and whether decision makers would be willing and able to respond quickly enough.

The 1995 NIE and the Rumsfeld report debate were destined for future history to resolve, except for the actions of the North Koreans and then India and Pakistan. Right on the heels of the Rumsfeld report, North Korea tested a three-stage missile, estimated to have the range to strike Hawaii and possibly the western extremities of the continental US. Shortly thereafter, India and Pakistan both tested nuclear devices, to the surprise, at least publicly, of Western intelligence agencies. Suddenly, everything predicted by the Rumsfeld Commission had come true. The threat box had been checked, and the jig was up for the Clinton administration. It faced little choice but to recognize the inevitable. On January 20, 1999, Secretary of Defense Cohen announced: "We are affirming that there is a threat, that the threat is growing, and that we expect it will soon pose a danger not only to our troops overseas but also to Americans here at home."[21] In September, the 1999 NIE confirmed the Rumsfeld Commission's conclusions.[22] Above all else, in the same speech announcing the threat, Cohen noted that "technological readiness will be the primary remaining criterion."[23]

Immature technology had given Congress no choice but to alter the conditions of the 1991 NMD Act on deployment in the 1995 Department of Defense budget. Until 1998, little testing had actually occurred, not least of all because NMD came out of SDIO and its successor the BMDO, whereas the services had concentrated upon tactical and theatre systems. Regardless, 3 plus 3 demanded a very robust and, by new weapon system development standards, compressed test and development schedule. Decisions would have to be made quickly and steps taken in a concurrent manner rather than through a linear step-by-step process. The integrated system would have to be developed at the same time as the system's components and be flexible enough to adapt to engineering changes and different components coming on line at different times. For example, intercept tests had to get under way relatively quickly and employ an existing rocket (launcher) not designed for missile defence. The new dedicated missile defence interceptor was not expected until sometime in the first part of the next decade. The very nature of the process made the project ripe for trouble – technological and political.

One of the first vital steps was to select the prime contractors for NMD. Boeing and Raytheon/Hughes had each developed a kinetic-kill warhead, known as an exo-atmospheric kill vehicle (EKV). In June 1997, the Boeing EKV successfully demonstrated its tracking and target discrimination capabilities in a fly-by test. It was able to discriminate successfully between the re-entry vehicle (warhead) and decoys launched by a Minuteman III ICBM out of Vandenberg Air Force Base. In January, the Raytheon/Hughes EKV demonstrated the same capability.[24] Shortly thereafter, Raytheon/Hughes received the EKV contract. Boeing was awarded the system's integration contract. With the two primary contracts awarded, the testing schedule moved toward actual intercepts. A year later, on October 2, the first actual intercept test took place. The Raytheon/Hughes EKV launched from Kwajalien Atoll in the South Pacific successfully intercepted a dummy warhead launched by a Minuteman III from Vandenberg during its mid-course phase in outer space.[25] The test included no decoys, and shortly thereafter reports emerged that the test had been rigged. The dummy warhead emitted a signal that the EKV used to guide it to target. Thus began the great technology feasibility debate.

Unfortunately, the debate was as much informed by the past and a different system as it was by the present. During ABM and SDI, the majority of the US scientific community argued that whatever system was envisioned would not work. In the case of SDI, scientists believed that the millions of lines of computer

GMD exo-atmospheric kill vehicle. *Courtesy US Missile Defense Agency*

software code necessary for a strategic intercept capability were impossible. Of course, dramatic advances in computing speed and processing capability by the mid-1990s had eliminated most of that line of reasoning. Now NMD opponents pointed to the Gulf War Patriot experience as a sign of technological unfeasibility and Pentagon duplicity. Immediately following the Patriot-Scud duel, the Pentagon and the US Army made bold claims about the success rate of the PAC-2 system. George Lewis and Theodore Postol from the Massachusetts Institute of Technology, after closely examining camera footage of the intercepts, declared that Patriot had failed to intercept a single Iraqi Scud – a conclusion that the US Army finally publicly agreed to.[26] It didn't matter that PAC-2 had been rushed into service while under development with civilian contractors still working on it in the field and successfully provided an image of defence important in dissuading the Israelis from retaliating. All that mattered was that PAC-2 was a failure and that the Pentagon and the BMDO were willing to hide the truth. Every failed intercept test, as in the case of the second and third tests in January and July 2000, were viewed by critics as examples of a technology that would never work.[27] Like the first intercept test, subsequent successful tests after 2000 were viewed as dubious and rigged.[28]

Whether intentional or not, the technology critics of NMD in the United States and Canada demanded that intercepts from the onset replicate reality – including dealing with sophisticated decoys and counter-measures designed to confuse the interceptor's on-board guidance system. However, the ability of proliferators dealing with first-generation ICBM technology and nuclear warheads to develop even simple decoys and penetration aids was difficult and costly. Although technology had diffused, engineering decoys and penetration aids required basic knowledge of multiple warhead technology. The British during the 1970s had spent over a billion dollars on the Chevaline program to defeat the Soviet Galosh missile defence surrounding Moscow.[29] North Korea was still dealing with the fundamentals of a three-stage rocket, never mind the engineering demands of developing and deploying a nuclear warhead that would work. Even China, which had developed nuclear weapons and long-range ballistic missiles in the late 1960s, was only just beginning the transition to solid-fuel propellants and multiple warheads.

Even if one accepted the possibility that some crude form of decoys might be possible, no development process attempted to deal with the complexity and reality from the onset. Instead, small incremental steps were taken in a step-by-step process, looking as much for failure as success and moving from simple intercepts employing, for example, a target-emitting signal for the interceptor to track and intercept to test initial components to more difficult tests with additional elements of the system being integrated together. Moreover, if earlier new technologies had been judged on the basis of their initial tests and subsequently cancelled, few, if any, new weapon systems would have been fielded. For example, the US ballistic missile program had witnessed failure after failure before demonstrating the technology. To suggest otherwise was to demand, for instance, that the first test of Apollo be a moon landing. However, the time-compressed NMD development, test, and evaluation process did lead to numerous significant development problems, poor planning, and insufficient testing, as exposed by the Welch Panel Report in February 1998 – what one observer called the rush to failure.[30]

Following the release of the report, the BMDO restructured the NMD development program, which, nineteen months later, the panel found to be much improved, though not ideal.[31] For the Republicans, the first report confirmed their suspicions of the Clinton administration's lack of commitment by failing to fund NMD appropriately. As early as 1995, they argued that the NMD program was not being funded at a level adequate to ensure a viable testing program vital to fielding an operationally capable system by 2003. In response, the 1997 US Quadrennial Defense Review called for the doubling of testing funds to $4 billion over the next five years and an increase in scheduled flight tests from eight to thirteen by 2003. In 1997, Congress increased NMD funding by $474 million.[32]

Although the Republicans saw the BMDO's NMD funding requests as evidence of Clinton's lack of commitment, NMD opponents on both sides of the border saw the funding issue in an entirely different light. Even though Cohen had stated that the funds were available, opponents suggested that any effective defence would break the bank. As with the technology criterion, the ghost of SDI was at play. Then, critics looking at the idea of a space-based umbrella over the United States argued that such a defence was unaffordable, especially with the exorbitant costs to launch payloads into outer space at over $10,000 per pound. However, NMD was not SDI, and the costs of a limited ground-based system were not exorbitant. Throughout the life of NMD, expenditures never exceeded 2 percent of the defence budget, and these included development costs for a new generation of space-based sensors. According to one set of estimates, an operational NMD system would cost between $4 and $13 billion (see Table 4.1), spread over many years. The Congressional Budget Office estimated the total costs of NMD, including deployment and operating to 2015, for the initial Capability 1 at $30 billion, with Capability 2 and Capability 3 costing an additional $6 and $13 billion respectively.[33]

Naturally, final costs depended upon the overall size of the system. NMD was originally designed to deal with a very small number of single warhead missiles from proliferating states. It only initially required a comparatively small number

Table 4.1

National Missile Defense options and costs

Proposed system	Name	Interceptor				Anti-ballistic missile (ABM) radar			Cost
		Kill vehicle	Type of booster	Number	Location	Name	Type	Base radars	
Candidates for the administration's 3 plus 3 system									
Army system	Ground-based interceptor (GBI)	Exo-atmospheric kill vehicle (EKV)	New or off the shelf	100	Grand Forks, ND	Ground-based radar (GBR)	100-GHz phased array	10-GHz phased array advanced early warning radars (much like GBR)	$9 billion
Air force system	Minuteman interceptor	Kinetic kill vehicle	Minuteman III (modified)	20-100	Grand Forks, ND	Have Stare	10-GHz dish (based on existing radar)	10-GHz dish (same as ABM radar)	$4-6 billion
Other possible ground-based defences[a]									
4 plus 2 weeks system	Payload launch vehicle	EKV	Payload launch vehicle (existing)	20	Grand Forks, ND, but in a crisis at sites in Alaska and Hawaii	Have Stare in Alaska	2- to 4-GHz and 100-GHz dish	Unknown sea-based radar (Cobra Judy?) or Have Stare in Japan	Unknown – a few billion?
Multiple site defence	GBI	EKV	New or off the shelf	200-300	Northeast, northwest, and perhaps Grand Forks	Ground-based radar	10-GHz phased array	10-GHz phased array advanced early warning radars (much like GBR)	$13 billion+

a The 4 plus 2 weeks system was envisioned as a system that would essentially be in storage and capable of being deployed in two weeks for the defence of Hawaii and Alaska. David Mosher, "The Grand Plans," *IEEE Spectrum*, September 1997.

of interceptors located in a single site. As, or if, the threat expanded, the number of interceptors would increase, and with the increase came the possibility of expanding NMD to a second site. Thus, NMD was conceptualized in three stages: an initial operating capability (Capability 1) at a single site of about twenty interceptors to be followed if or when necessary by an increase in interceptors up to roughly the ABM Treaty limit of one hundred at one site (Capability 2), followed by an expanded system with two sites (Capability 3).[34] The final size, and thus cost, of NMD depended upon the size of the threat, the location of the threat, the operational employment strategy, and assessments of NMD effectiveness, and it was the effectiveness consideration that opponents suggested drove costs far beyond existing estimates, while at the same time arguing that the technology wouldn't work and thus was a waste of resources.

The emergence of Fort Greely as the preferred initial site by 1998 was a function of identifying North Korea as the first state of concern likely to acquire an ICBM capability and the legal requirement to provide a defence for all of the United States, not just the lower continental forty-eight states. Optimized for both when compared with Grand Forks, North Dakota (actually Cavalier), there was also the question of future threats from the Middle East, whose trajectories tracked over the eastern Arctic and down the eastern seaboard to targets such as New York and Washington, DC. Fort Greely had the potential to intercept warheads launched from the Middle East but was not optimally located to do so. Here the system had to take a more difficult trailing shot, where the interceptor had to catch up to the warhead in flight. In contrast, an interceptor launched from Grand Forks, or perhaps an alternative site in the northeast, would take a more optimal straight-ahead shot – as the Alaska site afforded for a North Korean launch.

The necessity of a future second site was also a function of the possible requirement to take more than one shot at an incoming warhead, which was highly problematic for an interceptor launched from Alaska against a Middle East warhead destined for the northeast. A second or even third shot was central to the planned operational NMD employment strategy of shoot-look-shoot. An initial interceptor was to be launched at the incoming warhead early in its mid-course phase, followed by a rapid assessment of success or failure. If the intercept missed, a second interceptor was to be launched and, if necessary and time permitting, a third. Shoot-look-shoot relative to size and location of the threat thus determined site and interceptor requirements. Basically, three interceptors might be needed for each incoming warhead. If the threat was estimated at ten warheads (ICBMs), the system needed at least thirty interceptors.[35]

Although the actual ratio of interceptors to warheads remained classified, this ratio became the heart of the opponents' cost critique. Asserting that any national missile defence system had to be perfect, the number of interceptors required for perfection given the immature and dubious technology required so many interceptors as to raise costs substantially. Strangely, only the NMD system among all missile defences was held to this standard – a standard no defensive system had met or would ever meet. Moreover, even with a probability of 60 percent intercept, with three intercept chances increasing the probability to over 90 percent, the possibility that only one warhead might leak through was deemed unacceptable to opponents in the United States and Canada – echoing views during ABM that George Lindsey effectively addressed during his testimony to the Standing Committee on External Affairs and National Defence in 1972.[36] Although these interrelated cost-technology arguments carried little political traction in the United States, the requirements for an effective NMD system as the threat grew put the United States on a direct collision course with the ABM Treaty, Russia, and, on the surface, NATO-Europe.

With the Clinton administration committed to the narrow interpretation, both Grand Forks and Fort Greely violated the treaty, although the absence of an ICBM field within a radius of 150 kilometres appeared to be a marginal violation. Nothing in the numbers of interceptors until one got to Capability 3 fully violated the treaty. However, in reality, NMD was potentially a fundamental violation of a treaty principle prohibiting a national defence, even though within the limits of Article 3 a national defence was legally possible: "Each Party undertakes not to deploy ABM systems for a defense of the territory of its country and not to provide a base for such a defense, and not to deploy ABM systems for defense of an individual region except as provided for in Article III of this Treaty."[37] When negotiated, neither Safeguard nor Galosh – the Soviet system defending Moscow – had the capability to defend the entire nation. Both systems were limited to a relatively small defensive footprint due to the limited range of the interceptors. Safeguard, for example, could engage targets only at a range of five hundred kilometres, employing the Spartan missile. Galosh had similar characteristics. These limitations, along with other technological constraints of the day, made the provisions governing the capability of an ABM site, and the distance between interceptor sites, national capitals, and ICBM fields, meaningful.[38] NMD, however, promised national coverage from a single site.

It wasn't just these treaty issues that confronted the Clinton administration. There was also the question of the new generation of advanced sensors essential for NMD to be able to undertake early shots at a warhead during its mid-course phase. Alongside the new interceptors and kinetic-kill warheads, NMD entailed the modernization of both the early warning infrared Defense Support Program (DSP) satellites and the ground-based early warning radars at Fylingdales (United Kingdom), Thule (Greenland), and Clear (Alaska). The DSP was to be

US sea-based X-band radar. *Courtesy US Missile Defense Agency*

replaced by the Space-based Infrared System–High (SBIRS-H) of satellites in geostationary and polar orbits. The ground-based radars were to be replaced by new X-band radars. Neither violated the treaty.[39] NMD initially included an X-band radar in near proximity for tracking, target discrimination, and the cueing of the on-board in-flight guidance system, which also met the treaty's provisions – the role the phased array radar at Cavalier, North Dakota, had played for Safeguard.[40]

However, NMD contained another set of planned sensors that potentially violated the treaty: a constellation of infrared satellites in low earth orbit to provide an effective means to track targets and discriminate between warheads and decoys in the mid-course phase and cue the colocated ground-based radar. The treaty specifically prohibited space-based components under Article 5, despite the long-forgotten SDI interpretation debate. However, it was unclear whether this constellation fell under those provisions. In one of the agreed

Artist's representation of a Defense Support Program satellite.
Courtesy US Missile Defense Agency

statements governing treaty interpretation from the Standing Consultative Commission, the United States and the Soviet Union had agreed that "tracking alone is insufficient for a radar to be tested in an ABM mode; the presence of an ABM interceptor being guided by an ABM radar is also required."[41] US officials subsequently argued that these satellites did not communicate directly with the interceptor and that they had other non-missile defence military functions to perform. As such, it was not a system component but only an adjunct and thus outside the treaty's radar provisions.[42]

The radar provisions did not affect just the United States. With the collapse of the Soviet Union in 1991, key elements of its Galosh ABM system defending Moscow were no longer on Russian territory.[43] Beyond this, some voices in the United States argued that the ABM Treaty was null and void as a result of the collapse of the Soviet Union.[44] Even if one accepted the administration's argument that Russia in international law was the legal successor state to the Soviet

Union, both parties had to sort out the presence of Russian ABM radar on foreign territory. There was one other major issue facing the treaty: the status of tactical and theatre missile defence development programs. This was the issue of demarcation, which became central to the continuing fight between Clinton and the Republican majorities in Congress.

The Clinton administration had purposely separated tactical (at that time known as anti-tactical ballistic missile defence) and theatre missile defence development programs (TMD) from the national defence program that became NMD. As missile defence opponents in the United States and Canada focused upon NMD, they were virtually silent on TMD programs. It appeared that defending US military forces deployed overseas, as well as host nations and/or allies in the Middle or Far East, was acceptable, but defending US cities was not. At the same time, voices on the right, finding support from Republicans, outlined an alternative architecture for NMD, one that would employ the future US Navy's theatre-wide capability in a national role.[45]

The navy's theatre-wide development program was designed to use the area air defence Aegis-class guided missile cruiser and the new Aegis-class *Arleigh Burke* destroyer to provide an umbrella missile defence for US naval task forces at sea or directly supporting littoral operations at a host nation embarkation point.[46] Facing the potential threat of medium- to intermediate-range missiles (for example, an Iraqi-type capability threatening US forces and allies in the Gulf), the next generation of surface-to-air missiles with a kinetic-kill warhead – the Standard Missile-3 – not only might provide such theatre defences but, if upgraded for speed, would also be able to intercept an ICBM warhead.[47] Deploying these cruisers and destroyers around the United States, including Hawaii, might result in an effective defence against ICBMs and SLBMs.

Clearly, this would violate the ABM Treaty Article 5 prohibitions and reflected a much larger problem – the possibility that US TMD programs might be employed in a crisis to break out of ABM Treaty constraints. But the United States was not alone. The Russians also possessed a fledgling tactical and theatre system – the S-300 family.[48] During the original negotiations, US officials had been deeply concerned about the possibility that the Soviet Union's extensive air defence network might serve as the basis for it to break out. They successfully pushed for the provisions on "testing in an ABM mode" to prevent such a possibility. If the air defence network could not be tested against strategic ballistic missiles, it could not provide a breakout capability that might threaten strategic stability. But no one had considered the possibility of TMD, and the treaty was conspicuously silent. Technology was on the verge of outpacing a treaty designed to last forever.

Standard Missile-3 launch from Aegis destroyer. *Courtesy US Missile Defense Agency*

This basket of ABM Treaty issues, led by the demarcation question, confronted an administration publicly committed without reservations to the treaty and its now dated and politically irrelevant narrow interpretation. It was also linked to the START arms reduction process, and together they were a key indicator for the final Clinton deployment criteria – international peace and stability. From their beginnings with McNamara, limits on strategic missile defences had been linked to reductions in strategic nuclear arsenals, even though opponents of the ABM Treaty had regularly pointed to the failure of this linkage to occur. Soviet strategic nuclear forces had increased in quantity and quality after the signing of ABM. Regardless, the linkage between limited or no missile defences and strategic arms reductions had become gospel in the Western arms control community. It had also become gospel in Soviet rhetoric and regularly part of Gorbachev's dramatic proposals for significant strategic nuclear cuts in return for Reagan abandoning or significantly limiting SDI, whether this was Soviet rationale or not. It also became gospel for the Yeltsin and Putin governments in Moscow.

The 1992 Bush-Yeltsin announcement of a cooperative approach to missile defence with the goal of creating a global protection system had effectively gone nowhere. Certainly, US-Russia cooperation existed in the areas of lasers, an observational satellite system, and space experiments, and Russian military officials were invited as observers of US tests.[49] Moreover, early warning had dominated cooperation efforts largely because of significant gaps in a deteriorating Russian early warning network. Russia lacked the resources to maintain its early warning capabilities. Both the United States and Russia had an interest in ensuring that the Russians didn't misinterpret a peaceful rocket launch as a hostile act, as occurred in 1995 when a Norwegian sounding rocket had placed the Russian Strategic Rocket Forces on high alert (though some officials doubted Russian claims about this).

As the grand design for cooperative missile defence stalled, so did START. Just before Bush left office, START II was signed in Moscow. Building on START I reductions and inspection procedures, START II committed both parties to reduce their strategic weapons to between 3,000 and 3,500 warheads by 2007 and the elimination of multiple warhead (MIRVed) ICBMs. In effect, START II achieved the long-standing US goal that dated back to the 1970s Committee on the Present Danger of eliminating the Russians' first-strike capability. Moreover, in allowing for MIRVed SLBMs, as well as other provisions, Russia had ostensibly agreed to structure its remaining strategic forces on the US model, even though this and the elimination of MIRVed ICBMs had long been rejected by previous Russian and Soviet negotiators and Russia lacked the funds to reduce its forces, never mind restructure them and maintain a SLBM capability at sea.

The United States with the Nunn-Lugar Amendment created the Cooperative Threat Reduction program to assist in eliminating Russian nuclear forces. It did not provide funds for restructuring.

START II was problematic from the beginning. Certainly, Russia had little if any choice but to reduce its strategic nuclear arsenal. The Cold War was over, and the Soviet Union was dead. There was simply no need to maintain existing levels of strategic nuclear forces. Besides, there was also no money to maintain these forces. But reductions were one thing, restructuring another. Moreover, START II had all the appearances of a US diktat, exploiting Russian weakness while holding out the hope of economic support that never really materialized. But START did have one great attribute as long as the Russian legislature, the Duma, did not ratify the agreement. It could be used as a means to threaten the West by evoking the arms control gospel. Unless the issues threatening the viability of the ABM Treaty were dealt with, START II would not be ratified. If START II was not ratified, START III would never happen.

By 1996, Yeltsin had made it clear that START II ratification in the Duma hinged upon the negotiation of a demarcation agreement or protocol to the ABM Treaty. It didn't matter that START II ratification had been invoked to get the West to revise the 1991 Conventional Forces in Europe (CFE) agreement and legitimize the Russian military presence on the territory of its former republics. Nor did it matter that START II had been tied to Russian opposition to NATO enlargement and NATO activities in the former Yugoslavia. START II hinged upon demarcation, and the Clinton administration, committed to ABM, had little choice but to negotiate demarcation.

In March 1997, Clinton and Yeltsin met in Helsinki, where the basic parameters of demarcation were agreed on.[50] Eight months later, US and Russian negotiators reached an agreement.[51] It consisted of a memorandum of understanding on Soviet successor states, two agreed statements and common understandings covering lower velocity tactical and higher velocity theatre systems respectively, a confidence-building measures agreement, and a joint and US unilateral statement. The First Agreed Statement limited tactical systems as identified in the agreement to interceptors, which did not exceed 3 kilometres per second. The Second Agreed Statement entailed theatre systems also identified in the agreement and including the Russian SA-12 system (the foundation of the Russian S-300 TMD system), with interceptors that exceeded 3 kilometres per second.[52] Although no upper limits were identified, the United States separately stated that it had no plans to test an interceptor whose velocity exceeded 5.5 kilometres per second for land-based and 4.5 kilometres per second for sea-based systems and had no plans to test non-strategic systems against MIRVed ballistic missile targets.[53] Finally, the parties agreed that "each party shall provide assurances that it will not deploy systems ... in numbers and locations so that these systems could pose a realistic threat to the strategic nuclear force of another party."[54]

Whether the demarcation agreement resolved the breakout problem remained open for debate. Some critics argued that any theatre system capable of intercepting intermediate-range ballistic missiles with a range of thirty-five-hundred kilometres could intercept longer-range ICBMs.[55] Regardless, demarcation satisfied no one. Whether Yeltsin, the military, or the Duma, demarcation did not satisfy Russian concerns about breakout.[56] Alternatively, perhaps, it was simply insufficient to give up Russian leverage and ratify START II in the short term.[57] For the remainder of the Clinton presidency, no agreement was reached to amend the treaty to permit NMD to proceed, reaching a point where even US defense secretary Cohen in 1999 threatened to withdraw from the treaty.[58] A year later as part of a shift in its diplomatic strategy, and with the NMD deployment decision point rapidly approaching, the Duma ratified START II, the demarcation agreements, and the Comprehensive Nuclear-Test-Ban Treaty (CTBT) as an alternative means to pressure the United States.

As for Congress and the Republicans, the demarcation agreements amounted to dumbing down US TMD capabilities. Such a view might have been irrelevant had the agreements executive status.[59] However, with the issue of the CTBT facing ratification, the Senate had squeezed a written commitment out of Clinton that he would submit any demarcation agreements for ratification in return for early consideration of CTBT. Both CTBT and demarcation were never ratified. Nor was the revised START II agreement, which extended the time for Russia to meet its reduction commitments. Whatever the merits of either, they fell to the unrelenting feud between Clinton and the Republicans.

The Republican desire to force Clinton's hand one way or another led to repeated attempts beginning in 1997 to pass new legislation mandating deployment. None had been successful until early 1999. On January 6, the National Missile Defense Act of 1999 was introduced to both Houses of Congress. In mid-March, the brief three-section bill calling for "the United States to deploy as soon as is technologically possible an effective National Missile Defense system capable of defending the territory of the United States against limited ballistic missile attack (whether accidental, unauthorized, or deliberate)" passed with overwhelming bipartisan support in both Houses and was then signed into law by Clinton, with funding subject to the annual authorization of appropriations and the annual appropriation of funds for NMD.[60] Immediately after the passing of the act, Secretary of State Albright cabled key US embassies overseas, including Moscow and Beijing, that the second amendment to the bill "confirms that U.S. policy with regard to the possible deployment of a limited [missile-defense system] must take into account our objectives with regard to arms control."[61] However, this amendment, the third section, had simply stated that it was "the policy of the United States to seek continued negotiated reductions in Russian nuclear forces."[62] No linkage with deployment was made. Nonetheless, a year and a half later, speculation abounded that the Clinton negative deployment decision had been taken long before the outcome of the missile tests, with most members of the administration opposed (the exception being Cohen), the emergence of the Russian stumbling block, and the traditional arms control gospel.[63]

For NMD opponents, Russian opposition was proof positive of the threat NMD posed to international peace and stability. Even though the Cold War was long over and the strategic, political, economic, and social conditions necessary for its return were entirely absent, critics raised the spectre of the Cold War and SDI. If the United States proceeded, the ABM Treaty would die, followed by the collapse of START, a new arms race, and the shattering of the entire international arms control and non-proliferation edifice. Any hope for an indefinite extension of the Nuclear Non-Proliferation Treaty would evaporate. If critics were to be believed, the ABM Treaty was the cornerstone of not just strategic stability as understood during the Cold War but of everything related to arms and peace. Of course, for proponents, the logic was difficult to follow. Why a limited strategic defence of twenty interceptors would threaten Russian strategic forces even at START II levels (3,000 to 3,500) warheads was hard to understand. If a three-to-one ratio of interceptors to warheads was necessary for a plus 90 percent intercept rate, the United States needed over 9,000 interceptors. Even if all the TMD systems were added together, they didn't come anywhere near that number. Moreover, NMD – even with the largest number of interceptors expected (Capability 3) – still provided ample head room for further large-scale reductions in offensive forces. Whether such reductions occurred was likely to be determined by a range of political factors outside of NMD numbers. NMD might be the whipping boy, but certainly one would be hard-pressed to see it as the cause.

Similarly, the Nuclear Non-Proliferation Treaty indefinite extension, and indeed the whole regime, might collapse, but why NMD should be held responsible was hard to comprehend. Indeed, why states in various conflict regions, such as the Middle East, seeking to reassure each other in order to remove an incentive to acquire nuclear weapons would eliminate a treaty in their interests was also hard to fathom. However, there was one nuclear state with the interest and capacity to respond, and it appeared as the likely, real target of NMD: China.

With strategic forces (ICBMs) numbering in the high twenties according to public sources, NMD appeared to threaten China's strategic nuclear deterrent.[64] With no sea-based forces, and its ICBMs still liquid fuelled, the initial NMD

Capability I likely provided a reasonable defence from retaliation before, or after, a US first strike. Besides, China had risen in some circles within the United States as the future threat to the West. Confronting NMD, China appeared to have no choice but to expand its strategic arsenals.[65] Of course, China had been proceeding with a strategic modernization program for some time. It was developing an initial SLBM capability, and reports began to emerge of a new-generation mobile, solid-fuel MIRVed ICBM – the DF-31. Here was the beginning of the arms race that missile defence opponents had always predicted.

Yet China seemed to be in no rush. Nor had China done much to counter its vulnerability to a pre-emptive attack when its relations with the Soviet Union were at a low ebb during the 1970s and 1980s. Indeed, like McNamara's rationale for Sentinel, the Soviet Galosh system around Moscow was designed to deal with its own rogue state – China. No one seemed to care about the dangers of strategic stability then, and in many ways China was even peripheral to the NMD debate. Certainly, the Chinese government condemned NMD in the language of the arms control and disarmament gospel. China supported calls for strict compliance with the ABM Treaty, voting, for example, in favour of the resolution calling for such in the First Committee and United Nations General Assembly in the fall of 1999.[66] But the Chinese appeared more interested in the weapons-in-outer-space issue. As for modernization, the Chinese also were more concerned about the implications of new high-technology military capabilities as demonstrated by the United States in the Gulf War – the so-called Revolution in Military Affairs. The Chinese were not indifferent, but they were never front and centre in the NMD debate. Instead, more attention was directed to the Europeans in their role as NMD opponents, at least diplomatically.

As with most issues confronting the transatlantic alliance, NATO opposition was not monolithic. European concerns about US missile defence developments were long-standing. Despite general public opposition to SDI, the European governments divided over participation in the research program. They were also divided over NMD, with most of the allies formally undecided.[67] Opposition was led by France and Germany. The Chirac and Schröder governments agreed that the WMD/ballistic missile threat was a distant one, and Middle East proliferation incentives were being driven by regional concerns rather than directed toward Europe. In contrast, the British and Italians were the closest to the US threat assessment, and the Turkish government already faced a regional missile threat from Iran and Iraq. The two European nuclear powers, France and Britain, also disagreed on the adequacy of nuclear deterrence. The French, at least publicly, believed deterrence would work against Third World proliferators and had clearly stated that it would retaliate with nuclear weapons against any nation employing biological, chemical, and nuclear weapons. The British were somewhat ambiguous but had rejected using nuclear weapons in response to biological and chemical attacks, at least against British military forces.

France and Germany were united in their concerns about the implications of NMD for cooperative relations with Russia. For all the Europeans, except NATO's new Eastern European members, the old Cold War fear of Europe being caught between Russia and the United States at odds remained. This fear extended into their views about the amendments to the ABM Treaty – a desire for any agreement, with the only significant non-starter being weapons in space. However, the French were also concerned about the implications NMD and an equivalent Russian missile defence response could have for their own strategic nuclear forces, which had been unilaterally reduced after the end of the Cold War. Finally, the old strategic decoupling arguments were resurrected. Like the strategic stability argument, they spoke to a different time and place. Nonetheless, this time decoupling took the form of differential security – a US defended and a Europe undefended – with all the political implications that might follow for the alliance. Of course, one solution was for NATO to become engaged in missile defence, which in fact it had already begun to do.

Even if proliferation threat assessments differed from those of the United States and among the Europeans, geography alone dictated that the threat to Europe from Middle East proliferators would precede the threat to the United States. In the ballistic missile proliferation process, short-range single-stage rockets preceded the development of medium- and intermediate-range rockets that could strike Europe, and these preceded intercontinental rockets capable of striking North America. Most of the Europeans agreed with the 1995 NIE, but the public version of the NIE had not dealt with Middle Eastern threats to Europe. Regardless, if 2015 was the date for proliferators to be able to threaten the United States, the threat to Europe would come much sooner. If defence was to be one of the responses, time was at something of a premium.

Immediately following the Gulf War, the NATO Rome Declaration (1991) identified "the proliferation of weapons of mass destruction and of their means of delivery" as a clear threat to international security.[68] Shortly thereafter, the proliferation of ballistic missiles and their threat to NATO territory, along with weapons of mass destruction, were identified as priorities in NATO's new Strategic Concept. In the June 1994 framework document, the allies agreed to "improve defence capabilities of NATO and its members to protect NATO territory, populations and forces against WMD use."[69] At the same time, the Senior Politico-Military Group on Proliferation and the Senior Defence Group on Proliferation were established.[70] The former was to examine a political approach to proliferation, whereas the latter was to investigate military requirements either to dissuade proliferation or to protect NATO territory and forces from attack.[71] In addition to these two groups, Supreme Headquarters Allied Powers Europe, the Defence Planning Committee, and the Conference of National Armaments Directors (CNAD) turned to examine missile defence requirements. The CNAD responded to the American offer to share ballistic missile early warning data with its NATO allies and established the Early Warning Inter-Agency Staff Group. In June 1999, phase one was declared operational, providing US early warning data from US Space Command to the Joint Analysis Centre in the United Kingdom and from there to NATO Headquarters and member nations as necessary. The CNAD also established the Missile Defence Ad Hoc Working Group on extended air defence and theatre missile defence, which produced the 1998 NATO Layered Tactical Ballistic Missile Defence (TBMD) Sensor Architecture Study. During this period, the NATO Air Defence Committee was asked to develop and implement a new air command and control system with modernized ground-based radars that could provide a detection and tracking capability against tactical ballistic missiles.

The numerous studies undertaken by various NATO groups culminated in 1999 with the release of the alliance's Strategic Concept at the April Washington summit: "The alliance's defence posture must have the capability to address appropriately and effectively the risks associated with the proliferation of NBC weapons and their means of delivery, which also pose a potential threat to the Allies populations, territory and forces. A balanced mix of forces, response capabilities and strengthened defences is needed."[72] At the same time, missile defence was identified as one of the elements of NATO's Defence Capabilities Initiative, and the North Atlantic Council – the alliance's highest decision-making body – approved the Stand Alone TMD Feasibility Study Project for tender, with a reporting deadline of 2003-04.[73]

Alongside NATO's efforts, several European tactical missile defence development programs began.[74] The NATO Medium Extended Air Defence System (MEADS) emerged in 1995 with the signing of a statement of intent by the four initial participants, France, Germany, Italy, and the United States. This cooperative program was designed to field an operational, manoeuvrable, limited area point defence against the full range of air-breathing threats, including cruise missiles, and short-range ballistic missiles using a kinetic-kill interceptor.[75] MEADS was earmarked for operational deployment in 2005, even after the French withdrawal from the program. The British, French, and Italians began work on the Horizon frigate program, which also included a tactical missile defence capability based upon the Franco-Italian Sol-air Moyenne-Portée/Terre (SAMP/T, Future-to-Air Family of Missiles) – the Aster interceptor – originally designed as a replacement for the American Hawk and Patriot systems.[76] In April 1999, the British withdrew from the Horizon program and proceeded with their own Type-45 area air defence destroyer replacement, which was still to employ the Aster. The Franco-Italian partnership continued.

Germany and the Netherlands were also engaged in missile defence. Both possessed the Patriot PAC-2, and discussions had begun to acquire the PAC-3.[77] The Germans and the Dutch were engaged in developing a new generation of naval radars that would have a missile-tracking capability. Canada was participating quietly in a Dutch-German advanced phased array radar (APAR) program, with an eye on its possible employment on its City-class frigates.[78] Of course, Canada had also participated quietly in the various NATO extended air defence/theatre missile defence studies, though no Canadian company joined the consortia bidding on the feasibility study. Not surprisingly, Canada's primary interest was the surveillance and communication side of the missile defence equation – an interest wholly consistent with policy outlined in the 1994 White Paper.

If the Chrétien government was hoping to use NATO to legitimize Canadian participation in missile defence, it was sorely disappointed. As the 2000 NMD decision point approached, European opponents and Russia escalated their opposition to NMD. Even with missile defence cooperation part of the Russo-NATO charter, and the strange Russian offer of an all-European missile defence capability employing Russian technology, diplomatically the Europeans appeared opposed to NMD. Even the British issued only cautious, non-committed statements. NATO and European interest in missile defence was conveniently ignored in Canada as the debate began to heighten in 1999. The government made no effort to reverse the image of Europe and Russia as fundamentally opposed to missile defence in general and NMD in particular. Strategically, however, there was at least one element of the European dimension that could not be ignored entirely – the early warning radar at Thule, Greenland.

The modernization of the US Ballistic Missile Early Warning System was essential to NMD effectiveness. Both the United Kingdom and Denmark had to agree to modernization. The British site at Fylingdales had no direct implications for Canada, except that it would be a political surprise, if not a shock, to have the Blair government say no. The Danish site was something else, however. Adjacent to Ellesmere Island, the Thule radar was essential to missile tracking and the cueing of the X-band radar for the interceptor.

Denmark was generally seen on the left of the alliance, and all indications in the late 1990s pointed to significant domestic opposition to NMD and thus Thule modernization. Indeed, the United States had not yet formally approached the Danish government with a request. Alongside domestic opposition was the thorny internal politics of relations between Denmark and Greenland. There were indications of Greenland opposition to NMD and the modernization of the Thule radar, but the real issue was the bigger question of Greenland autonomy from Copenhagen, in which the Thule radar might become a victim. Danish agreement might not help the Canadian case for participation, but it certainly wouldn't hurt it. But if Denmark said no, the United States might turn to Canada as an alternative, and the repercussions of a Canadian no to such a request were likely to be devastating. Thule was early warning, and early warning was NORAD's mission.

Whether the safety blanket of the ABM Treaty would help Canada, if this eventuality occurred, was unclear. The Russians complained about US early warning radar modernization as a violation of the treaty, but the BMDO and the administration had simply ignored the complaints. The radars were grandfathered according to their interpretation of the ABM Treaty, and this extended to their modernization. However, an alternative site in northern Canada violated the provisions as a new radar because both sides were limited to radars at the outer edges of their territories, facing outward. At the same time, though, Russia faced the same problem of radars being outside its territory. Regardless, Thule was a key sensor feeding information into the NORAD early warning mission. If push came to shove, the administration and especially the Republican-dominated Congress would not take a Canadian no well. It was one thing to stand aside when the United States was publicly stating that NMD was US-only, designed as such, and did not require a Canadian contribution (territory) to be effective. It was another thing if Canadian territory became essential to NMD effectiveness.

NORAD officials on both sides of the border had long believed that the command's future hinged upon Canada's decision on missile defence. It was the primary, if only, defence mission with the Soviet/Russian bomber and cruise missile threat gone and with it the strategic significance of Canadian territory for North American defence. The Canadian provision of territory for additional NMD radars was one solution in this new strategic environment. In its 1991 modernization study, NORAD officials suggested that any missile defence system for North America would be much more effective with the deployment of radars on Canadian soil, especially for missile tracking from launch points in the Middle East over eastern Canada. Three sites were identified: Churchill, Frobisher Bay, and Goose Bay, Labrador. These radar sites would provide additional tracking, cueing, and battle damage assessment and were of special significance if a second ground-based interceptor site was established in the lower forty-eight states. While costly, favourable sharing arrangements à la the traditional NORAD model would limit the burden on Canada's defence

Thule ballistic missile early warning radar. *Courtesy US Northern Command*

budget, and perhaps even better arrangements could be negotiated to exploit the US missile defence obsession. Such radar sites could be sold as a major new contribution to Canadian sovereignty, especially over the North. Of course, there remained the ABM Treaty question, but that was an American problem. Regardless, the offer was the most important thing, and such an offer was consistent with the basic parameters of the 1994 White Paper. Besides, it was in the area of early warning, where Canada in NORAD already played a role.

The early warning mission for Safeguard had been assigned to NORAD. It had sparked no major concerns about violating the ABM Article 9 third-party prohibitions. Early warning radars on Canadian soil connected to NMD might violate the treaty, but early warning data collection and assessment did not. Adding NMD as another end-user was no different from the US Safeguard command or Strategic Air Command as an end-user. Providing early warning to the American strategic nuclear deterrent was never seen as Canadian participation in nuclear weapons, nor would adding NMD be necessarily seen as participating in ABM. Even if Russia argued as such, precedent worked in NORAD's and Canada's favour. Thus, for NORAD and National Defence, the silent minimum goal, as much as there was one, was to lay the groundwork for

NORAD to acquire the early warning mission for NMD. Indeed, it could be sold to government as a means to defend NORAD and Canada's strategic interests without violating ABM. Moreover, just as Trudeau had not objected to early warning in the 1970s over Safeguard, and he had been opposed to missile defence as well, so might early warning for NMD be acceptable to the current prime minister, and foreign minister, who had spearheaded Liberal opposition to SDI.

The early warning mission also raised the possibility of participating without participation. Along with being a receptor and accessor for data from Space Command's infrared DSP satellites, and in the future from their replacement, the SBIRS (space-based infrared sensors in geosynchronous and polar orbits)-High satellites and the ground-based ballistic missile early warning radars, NORAD's early warning might require access to data from the planned low earth orbit satellite constellation – space-based infrared low. Although the primary missile defence mission of these satellites, having been cued by early warning assets, was target discrimination and cueing of the specific missile defence radar, they also had an early warning function. They facilitated early warning by potentially providing a more accurate and quicker identification of the incoming warhead's target. This, of course, was of value in programming decisions about intercept priorities for the automated intercept launch process. Acquiring access to these satellites would bring NORAD, and thus Canada, ever closer to actual participation and knowledge about the NMD interceptor process.

The potential problem again, however, was the ABM Treaty. Beginning in 1998, National Defence officials closely watched the debate surrounding the status of the planned SBIRS-Low constellation.[79] As outsiders, they had no official role to play in whether the constellation was to be interpreted as an ABM adjunct (non-treaty) or component (treaty). Besides Canada being non-signatory, Canadian officials were also not privy to the secret negotiating record, which had been important, for example, in the 1980s narrow-versus-broad SDI interpretation debate. Nonetheless, like that debate, officials within Foreign Affairs and National Defence had their own views, and, not surprisingly, they tended to disagree. But then there was also internal disagreement or at least uncertainty within National Defence ranks. The ultimate decision, of course, was an American and Russian one. But life would be easier in dealing with NMD if it didn't turn out as a fight, and after the formal US explanation of its adjunct view this appeared to be the case, or at least demarcation placed this disagreement at the bottom of the public disagreement list. In the end, the SBIRS-Low program became plagued with technological development problems and cost overruns. Initial predictions of a first launch around the turn of the twentieth century moved further and further into the future, such that the system was still far off when the current missile defence capability was declared operational in 2004.[80] Even the replacement for the DSP satellites slipped many years forward, such that the United States had to continue to replace the existing DSP satellites with the same technology. The adjunct-versus-component debate, like its predecessor, the narrow-versus-broad, disappeared into history.

If NORAD obtained NMD early warning, and this logically and functionally extended into the future space-based infrared side of the equation, command and control were not far beyond. As the line between early warning and target discrimination and cueing blurred with new technologies, so did the line between making an assessment decision and a release decision. If the actual release process was automated, where exactly did command and control begin and end? If missile defence in general and NMD in particular demanded a quick release decision right on the heels of an early warning assessment of North America under attack, with specific detailed knowledge of the nature and targets of the attack, the logic to integrate the entire process within a single system and process from early warning to intercept release made impeccable sense. In other words, if NORAD undertook early warning, it made sense for NORAD to take over command and control of the entire mission up to and including the release decision but not to include any direct involvement with the actual NMD interceptor site and interceptors.

After the agreement to the NORAD ABM exclusion clause in 1968, both senior American and Canadian officers raised the issue of the command and control of Safeguard, and the opportunity to decide was largely left in Canada's hands. The issue was dropped for three major reasons. First, a Safeguard intercept by a Spartan or Sprint interceptor entailed a nuclear blast, and the release of nuclear weapons was a presidential US-only decision. Second, the limited range of Spartan and Sprint put it some distance from NORAD command, and the US-only intermediary Continental Air Defense Command existed for this part of the mission. Finally, the government would look duplicitous in accepting the exclusion clause and then turning around to agree to a command and control mission that included nuclear weapons.

But NMD was different. No nuclear weapons were involved. NMD was a kinetic-kill system. With its longer range, NMD needed the delegation of release authority to an immediate command from the national command authority (the president) to allow a quicker decision and multiple intercept chances. No intermediary command existed. The United States had one of three choices: NORAD, Space Command, or another command yet to be established. Another command meant additional resources in a period of constraint. NORAD and Space Command were married at the hip anyway, and separating them would be costly. Besides, NORAD had the early warning mission that obviously would extend into NMD when needed. Moreover, with the new radar technologies coming on line, the line between early warning and tracking was further blurred. The ability to launch early on against a warhead demanded a centralized, integrated approach. Multiple command layers in the intercept process made little sense, and new command and control technologies, unavailable in the 1960s, supported a single centralized approach. Finally, there remained the ambiguity associated with the meaning of participation. The actual intercept system would remain under the US Army (National Guard). The ABM Treaty said nothing specifically about command and control. It was not specified by the Article 9 prohibitions on third parties and at worst might marginally violate Agreed Statement G on the provision of blueprints to third parties. If the United States and Canada agreed, command and control fit into ABM.

The mechanism to undertake Canadian participation, especially following the 1994 White Paper signal to the United States, was NORAD renewal. Since the dropping of the ABM exclusion clause in 1981, NORAD renewal had been largely pro-forma, and nothing on the surface in the lead up to 1996 portended any significant changes, despite some Canadian fears. Nonetheless, significant change occurred, with both parties agreeing to a new consultation clause. Canadian agreement implied the possibility of forward movement on missile defence and could be well interpreted as another quiet Canadian signal. Since the adoption of the revised terms of reference in 1981, NORAD's mission had been defined as aerospace warning and aerospace control. The former included air and space, which covered ballistic missiles. The latter, ostensibly the defence mission, however, was strictly limited to air only. Space and by default missile defence were excluded, and their addition required a formal change to the terms of reference. In 1995, with negotiations under way and NMD only a technology readiness program, it was premature to even contemplate negotiations on the possible role of NORAD or the terms of Canadian participation. But if NMD was to proceed to a deployment program, an architecture was needed, and decisions would have to be made on system elements such as early warning and command and control. Some mechanism was needed, and the agreed solution was to insert a consultation clause into the agreement: "The expansion of binational cooperation in other aspects of the aforementioned missions should be examined and could evolve if both nations agree."[81]

Nothing in the revised agreement explicitly spoke of a future missile defence mission to NORAD. Ballistic missile defence was not even mentioned in the document, not least of all because of Canadian sensitivities. Nonetheless, clues were present, such as references to "large residual nuclear arsenals capable of striking North America" and "other nations are covertly attempting to acquire nuclear-capable ballistic missiles and weapons of mass destruction."[82] The qualification to the consultation clause emphasized surveillance, communications, and data fusion, consistent with the policy parameters laid out in the 1994 White Paper. Besides, nothing else existed on the horizon that would necessitate such a clause for the five-year duration of the agreement. In fact, a five-year period ensured that NMD could be integrated even in a rudimentary form into NORAD after the release of 3 plus 3, especially with the next formal renewal occurring a year after the completion of the first phase of 3 plus 3. If the development, test, and evaluation phase proved successful, such that Clinton's hand was forced (preferably for some members of the government within the confines of an amended ABM Treaty), a Canadian decision became inevitable and a means for Canadian participation necessary. Moreover, as this phase progressed, architectural and operational decisions would have to be made, and these directly affected Canada. The clause as a signal opened the door to Canada obtaining access to the vital NMD information necessary for a decision. In effect, the 1994 White Paper, the new NORAD clause, and shortly thereafter the JROC announcement all amounted to Canada and the United States agreeing that Canada had a need to know about NMD.

In 1995, however, the potential fly in the ointment was the appointment of a new foreign minister, Lloyd Axworthy, who replaced Chrétien's Quebec lieutenant André Ouellet, inheriting the NORAD renewal negotiations. Axworthy came from the left wing of the Liberal Party, and many observers quietly suggested that he was actually more at home with the foreign policy positions of the NDP. He was also viewed as somewhat anti-American and, with Chrétien in 1985, had been one of the most outspoken Liberal critics of all things SDI. It was also clear that, in finally reaching his ambition of becoming foreign minister, Axworthy was to be an activist. With a government dominated by domestic concerns, the budgetary deficit, and Quebec, unless Axworthy went too far, he could expect little interference from the prime minister. For that part of the Foreign Affairs bureaucracy that had bit its tongue during the Mulroney years, Axworthy also promised a return to the Trudeau age of the recent past. The previous Foreign Affairs–National Defence convergence on relations with the United States and the missile defence question was placed on uneasy ground with Axworthy's arrival.

Naturally, there was no shortage of issues facing the new foreign minister, and arguably NORAD renewal was not necessarily at the top of the list. Regardless, renewal had to be dealt with, and this meant the missile defence issue could not be entirely avoided. National Defence officials immediately flagged NORAD renewal as a potential problem with the new minister because of missile defence. To the surprise of many, little opposition emerged from the new minister. Instead, NORAD fit well into Axworthy's worldview of relations between states being governed by rules-based institutions, as he stated in the House debate: "We [Canada-US] have differences and similarities. The way to deal with these is to have rules in place."[83] Axworthy understood the significance of the binational defence relationship and had not been a major critic of NORAD. Agreeing to a new clause was far removed from agreeing to Canadian participation or placing the government in a position of being pushed or trapped into participation. Besides, consultation on missile defence was official government policy.

At the same time, he made it clear to the House that there "is no anti-ballistic missile system in any way connected to this NORAD agreement."[84] Besides, Axworthy was one of the strongest defenders of the ABM Treaty and believers in the arms control gospel. As he stated on repeated occasions throughout his tenure, the ABM Treaty "is the cornerstone of strategic stability. It should not be undermined with changes that are incompatible with its intent. In the effort to accommodate the possibility of an eventual National Missile Defense, great care should be taken not to damage a system that, for almost 30 years, has underpinned nuclear restraint and allowed for nuclear reductions."[85]

With NORAD renewal completed, NMD on the Permanent Joint Board on Defence agenda, and the 3 plus 3 program unfolding, the direct possibility of NMD command and control being assigned to NORAD came under discussion. In 1998, NORAD and US Space Command released *Concept for Operations for Ballistic Missile Defense of North America*.[86] The study noted that the current US Unified Command Plan had assigned the ballistic missile defence mission to US Space Command. Under the assumption that the Canadian government would join in, NMD command and control would be assigned to NORAD as the supported command, with Space Command being the supporting command for execution of the mission. Defense Engagement Authority would be directly assigned to the commander-in-chief NORAD (or his deputy), with preapproved employment procedures or rules of engagement agreed upon by the US National Command Authority and the Canadian chief of defence staff. Such procedures were vital because of the short time lines involved in making an engagement decision. These, in turn, required the establishment of defence priorities, which were to be agreed upon jointly by the US National Command Authority and the Canadian chief of defence staff, so as to be embedded in the battle management software of what would be an automated release process.

With the release of the study, and reflecting the recovery of Canadian access to US thinking and planning, a range of new opportunities opened for Canadians in US Space Command and at the Joint National Test Facility at Schriever Air Force Base, just outside Colorado Springs and the home of US 14th Air Force.[87] The facility was the location of operational NMD modelling and simulation.

With direct Canadian involvement in developing the simulation models, Canadians gained direct access to current US NMD thinking. In addition, senior Canadian officials were invited to observe and if willing participate in missile defence gaming exercises. All signs pointed to an open door to Canadian participation. Nonetheless, residual concerns existed within National Defence about the costs if Canada signed on.

On the surface, it was hard to believe that a major Canadian role, especially in command and control, could be achieved at no cost. Historically, Canada paid nothing for NORAD infrastructure on US soil, and no part of the currently planned NMD architecture was destined for Canadian soil. As for Canadian infrastructure, costs had been shared, with the United States paying the largest portion. Canada paid only 40 percent of the costs of the NWS modernization, and perhaps a better arrangement could be struck if the idea of NMD supporting radars emerged. However, even these costs were problematic for National Defence, never mind concerns that the United States might start to ask for a greater contribution once Canada agreed to be in. Since 1989, defence spending had declined in Canada as a combination of the deficit and a recession. These cuts paled beside those instituted by the Chrétien government, with its primary goal of eliminating the deficit. The government announced that the defence budget was to be reduced by nearly a quarter, and this amounted to nearly a one-third reduction in actual purchasing power. As the department absorbed these cuts, the operational commitments of the Canadian Forces, driven primarily by the conflagration in the former Yugoslavia, rose to their highest levels since the Korean War. Simply meeting these commitments strained the defence budget to its limits.

Senior defence officials recognized the implications of being outside NMD, but there were simply too many other pressing demands for the forces in the field to invest any capital. As spending levelled off and increased slightly, National Defence faced no shortage of demands from all the services to modernize or replace aging, obsolete equipment. Missile defence stood as a potentially new demand for the services. Although senior military officials might sympathize with missile defence, this did not mean they would be willing to sacrifice other more core traditional pieces of equipment, such as armoured vehicles, ships, and planes, in order to participate in NMD. Canadian NORAD officers drawn overwhelmingly from the air force might place NMD at or near the top of their priorities, but this did not mean that the Canadian Air Force as a whole would, never mind the army or navy. Certainly all agreed that close cooperative relations with the US military were the *sine qua non* of its strategic plan, as evident in the soon-to-be-released Strategy for 2020 from the chief of defence staff's office, whether the government truly understood this or not.[88] But this did not necessarily mean money for NMD.

Over the decades, the armed forces of Canada and the United States moved closer and closer together. Each Canadian service had a close working relationship with its American counterpart – arguably closer than the relationship among the three services in Canada. Each sent the best and brightest, including senior staff, on a variety of bilateral exchange programs. The navy boasted of its unique ability to integrate ships into US carrier task groups. Army officers on exchange occupied significant commands within the US Army.[89] The air force, of course, had been attached at the hip to the USAF as a function of North American bilateral air defence cooperation long before NORAD. However, these close relationships did not necessarily work in favour of NMD, particularly since the US services also did not place a high priority on NMD. They faced their own spending problems during the Clinton years.

For the US Army, which had responsibility for missile defence as a function of the 1957 Eisenhower decision, NMD barely registered on its horizon, which was a National Guard mission. Instead, tactical and theatre defences were of much greater importance, but even these were well below traditional preferences for tanks, armoured fighting vehicles, and attack helicopters, among other equipment. Similarly, the navy had interest in these defences, but it too looked to missile defences for the blue water fleets, despite the Heritage Foundation proposal to employ the navy theatre missile defence system in a strategic, national role. Even the air force had little interest in placing NMD as a priority and had little role to play in tactical and theatre defences, despite the interests of Space Command being dominated by the USAF. Above all else, the creation of the SDIO, and its successor the BMDO, had placed national missile defence outside the ambit of the three services, and its advocates largely came from outside, even if commanded by a USAF officer.

Thus, to the extent that US military preferences could influence Canadian ones, little could be expected on missile defence and NMD. If one asked senior US military officers their preferences on what Canadian military capabilities should be brought to the table, missile defence would not be one of them. In the end, only NORAD had a truly vital organizational stake in the NMD outcome, in spite of the array of policy officials concerned about the implications of NMD for Canada-US defence relations. If NMD could not be sold as a priority to senior military officials keenly aware of scarce resources, never mind the government, the alternative was to find a means to make a Canadian contribution, if need be, within existing investment priorities. Thus was born the idea of an asymmetrical contribution.

Territory was always one option for a Canadian contribution, and, as noted, a NORAD case for radars on Canadian soil was made long before NMD emerged. But it was unlikely that territory alone was sufficient, even if it made NMD more effective. The United States was likely to seek Canadian funds – at least, senior military officials and the government were not convinced that territory was sufficient.[90] Besides, a territorial offer was believed for the time being as potentially untenable because of the ABM Treaty and certainly would awaken Canadian nationalist sentiments and suspicions. It wasn't difficult to predict that such an offer would be interpreted as the first step to interceptors on Canadian soil, and, if the public's confusion surrounding the 1986 nuclear submarine proposal was any indication, these interceptors might suddenly be seen as nuclear tipped.[91] The alternative to territory was to be found elsewhere, in another contentious domain: outer space.

Despite bragging about being the third nation in space and the Canadarm on the US Space Shuttle, Canada's actual effort bordered on dismal, especially in the defence and security area. Originally, space had been the responsibility of National Defence; its research arm, the Defence Research Telecommunications Establishment, was responsible for the first Canadian satellite, *Alouette*, launched in 1962. The 1967 Chapman Report, the only major governmental-level study on Canada and outer space, emphasized a civil science concentration and eschewed an independent launch capability for Canada.[92] In 1972, space was turned over to the Department of Communications. With space removed from its profile, the high costs of military space, and few resources, National Defence largely withdrew. There was some continuation of joint scientific ventures with the US military among the various Canadian defence research agencies or labs, for example, Project Teal Ruby. Operationally, space languished within the air force, and Canada's engagement almost exclusively occurred through NORAD and its link to US Space Command.

In the late 1980s, roughly coinciding with the creation of the Canadian Space Agency, and after nearly two decades of being essentially on the outside of military space, National Defence began to examine space again. Like most nations, this interest was reinforced by the significance of US military space assets in the Gulf War. In 1992, National Defence issued its first formal Space Policy, followed several years later by the establishment of the Directorate of Space Development and the creation of the bilateral CANUS Joint Space Project. This project identified six generic areas of space cooperation and investment, of which only one – missile defence – was off limits. The project in general and the surveillance component in particular became the venue for Canada's possible asymmetric contribution to NMD.

Canada's only contribution to military space had been the deployment of Baker-Nunn cameras at St. Margaret's Bay, Nova Scotia, and at Cold Lake, Alberta, for the surveillance of space as a contribution to the US Space Surveillance Network. This network tracked objects in space, which was essential to NORAD's early warning mission. The resulting database ensured that objects in space or on orbit would not be confused with missile warheads tracking through space, causing a false alarm, with potentially horrific consequences. The cameras had become obsolete, and the last of the two facilities at St. Margaret's Bay closed in 1992.[93] Replacing this contribution became a priority for the directorate. It proposed an initial interim low-cost ground sensor to be subsequently replaced by an optical sensor in low earth orbit for the surveillance of space.

Although not publicly trumpeted nor internally confirmed as the solution to the Canadian missile defence contribution problem, the sensor had all the required attributes and more. The surveillance of space had an important missile defence role in differentiating background objects in space from warheads in transit. Whether this sensor had the capacity to perform a specific part of the missile defence mission depended on several factors, including its orbital orientation. But whatever role the sensor played, it offered a significant Canadian contribution to the vital space surveillance mission. Moreover, it was far removed practically and perceptually from operational missile defence – much further than even the US space-based infrared system in low earth orbit. It was far removed, too, from the weapons-in-space problem, not least of all because it could serve a civil science function in surveilling deep space. It would require a significant investment of funds from a constrained budget, but it was a project on the National Defence agenda. Finally, it could also serve as a hedge against a US NMD deployment decision if Canada stood aside and possibly protect NORAD. It was saleable as a meaningful contribution to North American defence, something the high-profile RADARSAT-1 did not easily do, and hopefully put a wedge in the space door if US officials decided that non-participation in missile defence meant no participation in military space and the elimination of NORAD from the space function entirely.

In the end, space-based surveillance, renamed Project Sapphire, languished for nearly a decade despite being a priority. With other pressing demands, none of the military decision makers saw merit in moving beyond the general definition phase, even as the defence budget began to expand slowly after 1998. Nonetheless, the asymmetric contribution idea was at least another possible signal to the United States of Canadian interest in NMD participation, and such signals were necessary at a minimum to obtain access to the information vital for an informed decision. In some senses, the initial responses to GPALS, the 1994 White Paper, the 1996 NORAD renewal, and the growing Canadian engagement in NORAD and at the Joint National Test Facility were all signals to the United States that Canada had a need and right to know. It had a need to know because, at the end of the day, Foreign Affairs and National Defence officials had to be able to provide detailed answers to the range of questions their political masters might raise about NMD and the implications for Canada one way or another. They also had a right to know because, all things being equal, the government appeared to be leaning toward some form of participation, given the aforementioned signals. Whether Canada or the United States was truly leading the dance didn't matter. National Defence remained coy publicly, leaving it to proponents within the academic community to make the case for participation, while beginning in 1998 to articulate the four no's policy – no US architecture, no US deployment decision, no US invitation, no Canadian decision. For the United States' part of the dance, NMD was a US-only system, designed to operate without the participation or contribution (territory) of Canada, even though the United States made it clear that it was open to discuss the issue with Canada. Neither was truly pressuring the other, even privately. Rather, there was simple existential pressure on both parties created by the time line of the 3 plus 3 construct.

The United States' open door and Canada's positive signals centred upon NORAD and direct interaction between the relevant actors in the Pentagon and at National Defence Headquarters. There was one actor that Canadian defence officials had only informal access to, and this actor had the NMD research and development mandate: the US BMDO. With its mandate, this organization held many of the technical and specific architectural answers on the NMD system, components, and capabilities. It held many of the answers about the technical implications of NMD for Canada. In many ways, getting inside the door of the BMDO was the final step in the NMD puzzle. By seeking to obtain a liaison position in the organization, another major signal would be sent of Canadian interest and intent. By obtaining the position, the United States would lay the basis for negotiations that would inevitably follow, once the president forced Canada's hand by formally announcing a decision to deploy. Ideally, Canada would have inside access to the detailed architecture, the deployment decision would be made, and the invitation would coincide with a Canadian yes.

Of course, this logic existed in the minds of only some, not all, National Defence officials. Certainly, senior National Defence officials, civilian and military, strongly believed that Canadian national interests likely lay in Canadian participation. But they also knew that this decision was not theirs to make. It was their responsibility to provide as much information as possible to the prime minister, cabinet, and Parliament (even though it did not require a decision by

Parliament, only an order-in-council), and the BMDO possessed a great deal of this valuable information. It would have been a dereliction of duty not to obtain as much information as possible. Moreover, by the fall of 1998, many senior National Defence officials had come to the conclusion that some form of US missile defence was inevitable. In the case of NMD, if a deployment decision was made in 2000, the government would have to decide in one way or another – silence meant no. The decision needed to be based upon the best information available, and that meant access to the BMDO.

One might fear that National Defence would use information acquired through its liaison officer to push the government in the direction of participation. Naturally, this was their responsibility, if the officials concluded that it was in Canada's best interests. But there were other voices in government ready and willing to offer alternative information and perspectives, and chief among these on the missile defence file was Foreign Affairs. On nearly every element of the missile defence argument, which was constructed around the four Clinton criteria, Foreign Affairs and National Defence were at odds in some way or another. The proposal from the assistant deputy minister for policy to place a Canadian liaison officer at the BMDO headquarters in Washington, DC, was based upon the consultation clause in the 1994 White Paper. Foreign Affairs opposed such an act. Outside of the small group tasked with Canada-US defence relations, there was the strong view within Foreign Affairs that missile defence in general or NMD in particular would never work and never would (nor should) be deployed. If neither was likely, there was no pressing need for information and thus no need for a liaison officer. Moreover, a liaison officer went too far in sending a pro-NMD signal. As with ABM and SDI, if Canada sent too strong a signal, the costs of an ultimate Canadian no would be much greater than if Canada avoided any signal one way or the other. And as far as the minister and Foreign Affairs were concerned, the final answer would be a no. Finally, a liaison officer might in some way place the United States in violation of the ABM Treaty regarding third parties. Unfortunately, Foreign Affairs officials had not objected to the consultation clause in the White Paper and were poorly placed to object to a policy that was arguably not theirs. After nearly a year of debate, National Defence prevailed. The United States agreed, and a Canadian Air Force lieutenant colonel was appointed.[94] It was all for naught. The lieutenant colonel never got beyond the unclassified library in the lobby. For all the fighting between National Defence and Foreign Affairs, he was left on the outside, looking in, with access to nothing.

The failure of the liaison officer to obtain direct access can be interpreted in various ways. Formally, Canada was neither a participant nor had openly asked to participate and thus had no need or right to any privileged access. Why should

Canada be treated any differently from the other allies, for whom there was no specific desk within the organization? Furthermore, Canada was seeking access to the most highly classified information on the US ballistic missile defence programs. This was definitely no-foreign-disclosure information. US NORAD might turn a blind eye because of its close working relationship with Canadians, but this was the BMDO – the guardians of US missile defence. The organization also held the institutional memory of SDI, and Canada had been on the outside, not a loyal ally. To force the doors open required direct intervention from senior administration officials at the Pentagon, the State Department or National Security Council, or the president himself, and this was not forthcoming. Whether an effort was made is unclear. What is clear is that the failure of the liaison officer to obtain any meaningful access coincided with ominous developments elsewhere. Canadian NORAD access began to close down, with an increase in the formal strict application of no foreign disclosure. In particular, the Canadian replacement officer assigned to the Joint National Test Facility at Schriever Air Force Base in Colorado Springs in 1999 met the same fate as the liaison officer to the BMDO. He was moved out of modelling and simulation and into public relations, tasked with preparing unclassified PowerPoint slides for visitors.

US officials explained the relatively sudden, strict application of no foreign disclosure as legally driven and non-prejudicial to Canada. It certainly had no reflection on anything the Canadian government might have said or done. However, the strict application coincided with other developments. In 1998, the US State Department began to hint strongly that the Canadian exemption to the International Traffic in Arms Regulations, better known as ITAR, was at issue. For years, National Defence and Canadian companies had been largely exempted from ITAR as a function of the close working relationship between DND and the Pentagon and general US satisfaction with Canadian export controls. The Canadian exemption reflected the long-standing integrated nature of the defence industrial relationship, hinged upon the Defence Production Sharing Agreements, among other agreements, and the usual Canadian success in acquiring exemptions to US legislation, especially in the area of access to the US market, with the direct support of the US Department of Defense. For these reasons, Foreign Affairs and Defence officials ignored the warnings of the State Department. Whether a function of the Chinese spy scandal at the time, being caught in a wider net of a US decision to tighten up ITAR enforcement to prevent the diffusion of advanced military technologies, rumours of Canadian export control failures, or displeasure at the direction of Canadian foreign policy under Axworthy, in early 1999 the State Department formally eliminated the Canadian exemption.

The immediate damage to National Defence, Canadian Forces, and Canadian industry was tremendous. ITAR became a focus of government attention right up to the prime minister, who spoke directly to President Clinton (even during a golf game) to restore the exemption in some form. In early 2000, the United States began to soften its position, as Axworthy announced on the occasion of the public signing of the NORAD Agreement renewal.[95] By the end of the year, a government-to-government agreement was in place.[96] Regardless, from its beginning to its end, the US ITAR decision was seen by many in Ottawa as the State Department punishing Canada for certain activities contrary to US interests. Perhaps this punishment was also spilling over into Canadian access in NORAD and, of course, to NMD.

Like Mulroney and Reagan during SDI, Chrétien and Clinton enjoyed close, good relations, and their relationship played some role in the resolution of the ITAR question. They were golfing buddies. As a friend and in some ways ideological soulmate, Chrétien could hope to rely upon Clinton, as the Liberals could rely upon the Democrats, to manage the NMD problem – no different from Pearson and Hellyer being able to rely upon Johnson and McNamara during the brief life of Sentinel and Mulroney upon Reagan and Bush during SDI and GPALS. This created an unspoken, if not unconscious, sense that, even if Clinton's hand was forced, the fallout was manageable. For officials opposed to NMD, public doomsayers ignored this political reality. Unfortunately, Mulroney and the Conservatives had thought the same, to the regret of National Defence and Canadian companies. Close, good political relations at the highest level did not guarantee the same at the lower levels, especially among officials who had to deal with a growing foreign policy chasm largely attributed to the activist Axworthy.

Nevertheless, four particular issues came to signify Canadian foreign policy activism in the late 1990s – the land mines treaty, NATO nuclear strategy, the establishment of the International Criminal Court, and the non-weaponization of space. In all four issues on the international stage, the United States was on the opposing side internationally. Not only was this painting the United States as a direct target for isolation and humiliation, it was being done by its closest ally, Canada (or so concluded many American officials involved). More importantly, Canadian foreign policy under Axworthy seemed to be departing from its traditions of seeking to build bridges as a faithful Western ally. This was particularly the case with the celebrated land mines treaty and the trumpeting of the so-called Ottawa Process.

Although the process leading to the land mines treaty ban was started by non-governmental organizations, such as the Red Cross and the American-based International Campaign to Ban Land Mines, and moved forward in part by

Norway, Foreign Affairs successfully came to dominate the negotiating process and quickly pushed forward the conclusion of the treaty without support of the major powers – the United States, Russia, and China. The result was the creation of the Ottawa Process in which states with the active support and involvement of non-governmental organizations (known in Canada at the time as civil society[97]) stepped outside traditional negotiating fora, such as the Conference on Disarmament, to bring to a quick conclusion a new treaty with or without all states on board, including, implicitly, the major powers that were cast as obstacles to such agreements. In effect, the Ottawa Process at its core was the idea of medium and weak states standing up to the strong states in the interests of the greater global good.

But it was not just the process leading to the land mines treaty that was so troubling to the United States. The US administration had joined the negotiations and signalled a desire to sign on to the treaty. However, it needed concessions allowing for some time to implement the treaty, particularly as it faced a hostile Republican Congress. There was an expectation among US diplomats that Canada, following its traditional diplomatic practices, would work to accommodate these needs and make a US signature possible. But they were sorely mistaken, and many US officials could not help but think that there was a strong streak of anti-Americanism among the Canadian officials involved (which, of course, they denied). No exceptions and no delays were allowed. Of course, it was an open question whether a hostile Congress and senior active and retired US officers speaking of the threat to US troops, especially on the Korean Peninsula, would have recommended signature and ratification even with more time. Regardless, Canadian diplomats and the foreign minister were completely unsympathetic – it was take it or leave it.

Part of the Ottawa Process spoke to a radical shift in Canadian coalition behaviour. The process emphasized the role of medium and small states in negotiations. At roughly the same time as the Ottawa Process, talk about Canada's new diplomacy started to percolate around Ottawa. The new diplomacy spoke of a move away from traditional Canadian allies and international venues to a new set of like-minded nations. Among the nations targeted as partners in the new diplomacy were South Africa, Brazil, and Mexico, with the related idea of human security as the central goal of Canadian policy.[98] For the United States, the Ottawa Process and the new diplomacy led by Canada's foreign minister appeared as moving America's closest ally in a radically different policy direction, one more akin to that of the non-aligned states than an ally.

If the process and the new diplomacy boded ill for the relationship (something several senior Foreign Affairs officials quickly recognized in coming to the conclusion that both should be avoided in the future[99]), Axworthy's run at the nuclear weapons question in NATO added fuel to the fire. In the lead up to the fiftieth anniversary of NATO in April 1999, nuclear weapons and the NATO first-use doctrine were not on the agenda for the strategic review and update. Whatever the merits of changing NATO nuclear doctrine, there was no appetite, especially among the three nuclear powers of the alliance – the United States, France, and the United Kingdom – the new NATO members, or the basing countries to open the nuclear question that had lain dormant since the end of the Cold War. Axworthy thought otherwise, bringing the question to the Standing Committee on Foreign Affairs and International Trade for review in 1998 and obtaining an ally in the German foreign minister, Fischer, to push for an examination of the doctrine in the strategic review leading up to the Washington summit. Despite deep fears among senior US officials at the US Embassy in Ottawa that the committee might take a bold step in formally rejecting first-use, if not NATO's nuclear status, the committee basically endorsed the status quo and called for regular consultations with civil society to monitor the issue, along with other steps toward disarmament.[100] The NATO review that followed also resulted in no change. Nonetheless, the entire episode only reinforced US concerns about the direction and reliability of Canada as an ally.

The foreign minister also picked up the call for the establishment of a permanent International Criminal Court to replace the ad hoc war crimes tribunals established for Rwanda and the former Yugoslavia. For the United States, this idea was deeply disconcerting, especially with respect to provisions that in the minds of administration and State Department officials would expose the United States to the equivalent of a frivolous lawsuit – such as the long-standing attempt to indict Henry Kissinger as a war criminal for the bombing of Cambodia during the Vietnam War. US officials watched closely the Spanish attempt to extradite former Chilean president General Pinochet from the United Kingdom for war crimes. In response to the process leading to the establishment of the court, US policy makers made it clear that the United States would take retaliatory action against any nation that arrested US personnel for war crimes trials before the International Criminal Court (and promptly and successfully negotiated a series of side agreements with other countries).[101]

These three actions led in one way or another by Canada and its activist foreign minister involved US diplomatic and military interests. But the fourth – the push for a global treaty banning the weaponization of space – came closest to the NMD file. Whereas National Defence was looking to space in terms of a Canadian role (early warning) and making an asymmetric contribution, Foreign Affairs was moving in the opposite direction. Coinciding roughly with declining Canadian access in NORAD and elsewhere, Foreign Affairs undertook a mini-diplomatic offensive around the Conference on Disarmament to ban weapons in space.

For decades, the Conference on Disarmament's annual work plan had included an agenda item establishing a Committee for the Prevention of an Arms Race in Outer Space. Little progress had been achieved, particularly because neither superpower had been greatly interested in moving toward a formal treaty ban, which External Affairs had expected to be the logical next step after the 1967 Outer Space Treaty and used as part of its argument against ABM. Beginning in 1994, US policy shifted under Clinton. It now opposed the establishment of the committee and the inclusion of discussions on space weapons in its work plan on the grounds that the Cold War was over, there was no arms race in outer space, and the existing legal regime was sufficient. As the work plan was established by consensus, unless all agreed to the US position, the United States could and did simply stop the adoption of the entire work agenda for the year, effectively bringing the conference to a dead stop. Attempts to get around this block, including US willingness to adopt parts of the work plan, were in turn blocked by other states.

Meanwhile, signals from parts of the Pentagon and US Space Command suggested that the United States might be moving in the direction of weapons in space. To a certain extent, the systems involved were a legacy of SDI and its successors – the kinetic-kill warheads of NMD – and the by-product of the idea of space-based kinetic-kill interceptors concept of Brilliant Pebbles. In an ironic sense, some of the references to the Son of Star Wars made sense.[102] However, what was really behind the most pronounced advocacy on space found in Space Command's *Vision for 2020* and *Long Range Plan* was not missile defence but the recognition of the growing significance of space-based assets for US military activities and the US economy.[103] This recognition emerged initially out of the 1991 Gulf War, which many observers informally labelled the first space war.[104] Constellations of sensors, such as the Global Positioning System, and associated communication systems played a key role in the tremendous success of US military forces in destroying the Iraqi military in Kuwait at negligible cost. This was evidence of the next revolution in military affairs – an idea current in US military thinking, inside and outside the Pentagon. The result was twofold. First, space was identified as a key military enabler for terrestrial warfare – the ultimate high ground. Second, the US military recognized its growing dependence upon space, and dependence translated into vulnerability. And a third conclusion would subsequently follow. The US and Western economies were also increasingly dependent upon space-based assets. Space had become a critical military and economic centre of gravity.

During the Cold War, both superpowers had pursued anti-satellite programs. But their activities were limited by strategic calculus, costs, and technology. All this began to change by the late 1980s as costs began to decline, technology improved, and more states acquired the capability to launch satellites into space. As ballistic missile proliferation occurred, more states acquired the potential capacity to launch payloads into space. Whether as a missile or launch vehicle, a rocket was a rocket. If a nation could throw a warhead a long distance, it could launch a payload into space, and if it could launch a payload into space it could target payloads (satellites) already in space. Moreover, payloads in space were relatively easy targets, following predictable orbital paths, as China formally demonstrated in January 2007. In addition, as long as a state was not dependent upon satellites in space, precision strikes against an opponent's satellites were not necessary. In other words, unless one's military and economy were dependent upon space, a crude intercept method was acceptable and indeed possible – nuclear warheads.[105]

For US military officials, especially in Space Command, US satellite dependence and vulnerability required a response, and among the various options was to defend satellites by preventing this crude form of attack. This meant, along with a range of passive defensive options, intercepting anti-satellite warheads before they reached target. However, destroying these warheads via a kinetic kill in space threatened to leave a debris field that could interfere with other satellites. A solution was to destroy the launchers on the ground or the rockets as they ascended from their launch pads by placing defender satellites in space. The former was equivalent to the largely unsuccessful Scud hunting campaign of the Gulf War. The latter, of course, was a boost-phase intercept long seen as the most effective means for missile defence. Defending satellites and defending the national homeland could be achieved by the same means – killing two birds with one stone. As such, any reference to defending US military, civil, or commercial satellites naturally led to missile defence, and for anyone with an eye on conspiracy NMD was simply the forerunner of weapons in space.

For Foreign Affairs officials, this was the foundation of a future arms race in the pristine environment of outer space. US defensive measures, as seen by others, were also seen as an offensive measure. Actively defending satellites in space by orbiting a weapon system able to intercept rockets or missiles during the boost phase also meant the system was able to strike at a range of terrestrial targets, not just missiles. Who needed bombers, air-breathing cruise missiles, or artillery when targets could be destroyed from space? One didn't even necessarily need high explosives with the kinetic energy generated by the speed of objects launched from space. Beyond this, deploying these constellations in space meant the United States effectively controlled space itself and determined unilaterally who would be allowed access. Space ensured the US Empire and fit nicely into explicit US plans to maintain and increase its military superiority over any possible competitor.

But despite the military logic involved and the vigorous advocacy of Space Command and others inside and outside government, the Clinton administration was not convinced. It had eliminated funding for space-based weapons and had shown no appetite for moving in the direction of weapons in space. Even Space Command recognized that the technology was decades off and the decision beyond their purview. Regardless, Foreign Affairs felt that the window was closing for a negotiated prohibition on weaponization. Reflecting the obsession of Foreign Affairs, the Liberal government, and in many ways the Canadian addiction with all things multilateral, the Canadian ambassador in the Conference on Disarmament tried to push the space weapons issue forward, and Foreign Affairs issued a non-paper outlining the basic elements of a negotiated international treaty prohibiting the deployment of on-orbit weapons in space. In so doing, the prohibition was not to include terrestrial-based anti-satellite systems as a means to defend US assets. Such systems, of course, implied a tacit acceptance of missile defence, and, indeed, a kinetic-energy capability such as NMD could readily be turned against satellites. This did not mean, however, that Foreign Affairs was motivated to legitimize NMD and thus create favourable conditions for the government to participate such that the proposal amounted to a signal to the United States. A treaty might eliminate the ghost of SDI entirely, but it did not mean that the foreign minister or the government planned to move forward on NMD.

American officials were not impressed. It demonstrated a complete ignorance of the realities of US politics. As a matter of strategy, it ensured American satellite vulnerability by legitimizing terrestrial-based anti-satellite capabilities – the very capabilities that threatened US satellites. Above all else, the proposal rested upon a feel-good case – saving space as the last realm of human endeavour not polluted by weapons. This had little traction in a country that placed a major emphasis on national security and had serious global responsibilities for maintaining international peace and security. Like much of Canadian policy in the strategic field, it smacked of a sanctimonious idealistic moralism that Dean Acheson, secretary of state for Harry Truman, would have understood in referring to Canada as the "stern daughter of the voice of God."[106] A feel-good policy was a luxury that Canada might afford, but it made no sense to the United States. Indeed, much of Canada's activist foreign policy smacked of such a luxury. Besides, when it came to weapons in space, Canada was aligned with the position of China and Russia – perhaps another element of Canada's new diplomacy.

Of course, Canadian foreign policy was not completely out of step with that of the United States. Although the government had disagreed with the United States over the bombing of Serbian forces in Bosnia prior to 1995, broadly, Canada was in step with the United States and its NATO allies on most issues concerning the dissolution of Yugoslavia. Canada also acted with NATO in the Kosovo campaign, sending eighteen CF-18 fighters as its contribution to the war from the air. Canada joined the United States in support of UN actions in Haiti and East Timor, among other places. Nonetheless, it was impossible not to see some connection between the actions undertaken during Axworthy's years, anti-Americanism, and a growing sense of concern in Washington about Canada. Indeed, throughout the period, many government officials and academics speaking with US government officials were exposed to an unofficial harangue about what was going on in Canada, and these were the officials who managed the day-to-day relationship.

Regardless of the actual impact of Axworthy's foreign policy initiatives on Canadian access, NORAD officials in particular had by 1999 become more concerned that the NMD train was leaving the station and that Canada was not on board. Time was disappearing and with it the opportunity to ensure Canada had a meaningful role to play and a reasonable possibility to influence architectural and operational planning. Even more, it appeared that, in the near term, the government might have to make a decision based upon very general information. Even though on occasion US officials raised the question of Canada and NMD, usually having been prompted by questions from the Canadian press, official US policy of NMD as a US-only system remained. For example, at a press conference in Ottawa on May 15, 2000, John Hamre, the US deputy secretary of defense, called upon the Canadian government to shed its Cold War thinking and join in an effort to convince Russia to agree to amend the ABM Treaty: "We'd like to be able to modify the treaty to save the Treaty."[107] If this and other press reports are to be believed, Axworthy was the stumbling bloc, as the prime minister remained uncommitted.

By 1999, US officials, including Secretary of Defense Cohen, had begun to target the ABM Treaty and Russian opposition to any amendments to permit NMD to proceed. The administration was bound by the 1999 NMD Act that firmly mandated deployment as soon as technologically possible, and the possible was to be partially informed by the political – the next presidential elections. If the United States proceeded, it might just have to walk away from the ABM Treaty, and even within the White House legal advisors began to contemplate what stage in NMD construction violated the treaty.[108] For Canada, its safety blanket was at stake, and here, on the surface at least, Canada shared one clear interest with the Russians – an amended treaty with some constraints on the United States was better than no treaty at all. But the new president of Russia, Vladimir Putin, never came to this conclusion. Nor did the Chrétien government and its foreign minister.

Hamre's suggestion, or public signal, smacked in some ways of an attempt to manipulate Canadian hubris about its role on the international stage. As the NMD debate in Canada started slowly to gain steam in 1999, the media regularly spoke of US pressure on Canada to join. To the question of why the United States wanted Canada in, especially if no Canadian money or territory was involved, the answer was legitimacy. Canada's status as the good, moral middle power gave it significant influence. If Canada stood in favour of NMD, the world would follow, or at least this was a dominant view among many Canadian proponents, with the inverse for opponents to NMD. Even Moscow, on occasion, played on this hubris in seeking Canada to condemn NMD. Of course, why exactly Moscow, Beijing, Washington, or any other nation would heed the Canadian call is unclear. Nonetheless, a foreign policy asymmetrical contribution, if Canadian commentators were to be believed, was to endorse an amended ABM Treaty to permit NMD to proceed. But this contribution might also be interpreted as the government leaning in favour of participation. It was tantamount to a decision by setting the basic condition for participation. But this government, unless forced by a US deployment decision, was not going to decide. Nor was relative silence about the ABM Treaty supposed to be viewed as a decision against NMD, though it certainly could be interpreted as such. The government and Foreign Affairs remained uncommitted and said nothing except to reiterate the importance of the treaty as "the cornerstone of stability."

Perhaps Chrétien simply believed that either Clinton would not proceed or Russia would eventually give in. In his memoirs, he invoked the standard views of critics about NMD. He doubted that it was possible, affordable, or necessary, and he feared a new arms race and the weaponization of space.[109] Regardless, the government was not going to show its hand until it had no choice. Proponents and opponents in the attentive Canadian public looked to their respective champions – National Defence and Foreign Affairs – for clear direction. To everyone's disbelief, both departments sought to convince everyone that they were neither champions nor divided, despite regular press reports about significant differences between the minister of national defence, Art Eggleton, and Axworthy. Even when the deputy commander of NORAD, Lt. Gen. George Macdonald, linked Canadian participation as vital to the future of NORAD and cooperative continental defence relations, National Defence acted quickly to get *Jane's Defence Weekly* to retract and tone down its report of the speech.[110]

Nonetheless, on every contentious issue over NMD in a Canadian debate that employed the US criteria of threat, cost, technology, and international peace and stability, there were subtle yet clear differences. Perhaps indicative of the subtle differences was the testimony of the two ministers to a joint meeting of the Foreign Affairs and National Defence committees on March 23, 2000. The occasion was early NORAD renewal, but the focus of attention was on NMD and NORAD. Both Axworthy and Eggleton stressed that NORAD and NMD were separate and would remain separate unless an agreement was reached to include NMD in some form in NORAD. Axworthy, however, took great pains to emphasize that NORAD would continue regardless of the outcome on NMD – something Eggleton sidestepped in his opening statement.

Both agreed in their opening statements that there was insufficient information for a decision, one way or another, at the time – faithfully adhering to the unofficial four no's policy. However, the defence minister focused in a matter-of-fact manner on the architecture issues at hand within the context of official government policy as laid out in the 1994 White Paper. The foreign minister, in contrast, left a clear message to all present of opposition to missile defence, NMD, and Canadian participation, without directly saying so. Employing the Clinton criteria in the form of rhetorical questions in his opening statement, the answers he provided were consistently negative. As far as the threat was concerned, he stated many didn't think NMD addressed it. As far as the technology, there were major challenges in a "high risk development" strategy. There were strong reservations among allies. If ABM was broken, an arms race would result, and "if the Americans are able to come to an agreement on amending the ABM Treaty, deployment of an NMD system could still have implications for China's position, or set up an arms race with rogue states as they race to build offensive weapons that could break through the NMD shield."[111]

Of course, there was nothing necessarily incorrect or misleading about Axworthy's (or Eggleton's) testimony, and one might believe that the differences were only superficial or merely reflected the different mandates and responsibilities of their departments. But in the absence of any clear positive reference to NMD in the foreign minister's statements (not just at the meeting but anywhere), it was clear where he and the Department of Foreign Affairs stood. Indeed, even an amended ABM Treaty was problematic, and the opposition was perhaps clearest in Axworthy's concluding statement: "We will address these NMD issues on their merits at the appropriate time, when the information is clear and many of these questions can be answered."[112] In reality, much of the information and many of the answers would not be available for years and decades as a function of the nature of the questions. The conclusion was tantamount to a no.

If it had been necessary, the likelihood that the two ministers and departments would have been able to agree on a memorandum to cabinet on NMD, especially in terms of policy recommendations, was pretty close to zero – similar in some ways to the problem in 1985 over SDI. They agreed on the 2001 NORAD renewal, begun in 1999 and concluded in 2000. It was in both their interests to keep the renewal and missile defence questions separate, and on this the United States concurred as well. What might happen if Clinton announced deployment in the fall of 2000 was anyone's guess. This extended not just to the differences between Foreign Affairs and National Defence but also to deep divisions on NMD in cabinet and in the Liberal caucus.

On NMD and Canada-US relations, the Liberal big tent contained both members with views closer to the NDP and those with views closer to Conservatives (Alliance/Reform).[113] The split was evident in the House and on the two parliamentary committees tasked with responsibility for these issues – the Standing Committee on Foreign Affairs and International Trade and the Standing Committee on National Defence and Veterans Affairs. In the House, which largely ignored NMD until 2000, one Liberal member asked the foreign minister to provide assurances that "Canada will decline the invitation to join the North American missile defence system, one of the most insane ideas emanating from Washington since Ronald Reagan's star wars proposal."[114] As for the Standing Committee on Foreign Affairs, NMD did appear in testimony for the nuclear weapons/first-use study. But, save a passing reference in the final report to the possible necessity of missile defence because rogue state leadership might be undeterrable, and a rejection of missile defence if it meant weapons in space, the committee largely ignored the topic. The National Defence committee, however, took up the missile defence question directly as part of its examination of the revolution in military affairs. In the process of examining missile defence and NMD, it was clear that there was significant Liberal support. Moreover, the process revealed the position of one of the government's most influential interlocutors on Canada-US relations, the Canadian ambassador in Washington, who also happened to be the prime minister's nephew.

In the spring of 2000, the National Defence committee travelled to Washington to obtain a better understanding of NMD and hold discussions with its US counterparts. The group consisted entirely of Progressive Conservative, Alliance, and Liberal members, including the chair. The first stop was the Canadian Embassy and a meeting with the Canadian ambassador, Raymond Chrétien. At the meeting, the ambassador was blunt in warning of the potential consequences of Canada not participating in NMD – not dissimilar to Ambassador Gottlieb's warnings in 1985. In recognizing that NMD was inevitable in some form and at some time, if Canada stayed out or, even worse, condemned NMD publicly, the impact could spill over onto economic and trade issues. Whatever doubts committee members had (and these were very few to start with), discussions with the ambassador and others affirmed the view that NMD was going to happen and that Canada needed to get on board, and sooner rather than later. When the ambassador's comments were raised in the House, however, Axworthy initially rejected their validity and subsequently sidestepped the issue.[115]

Even though the Canadian debate was dominated by the American debate, the question of the defence of Canada did emerge, if only briefly. National Defence and Foreign Affairs avoided as best as possible any direct reference to the potential value of an operational NMD system for the defence of Canadian cities. The deeply held belief that, regardless of what Canada did or did not do for its defence, the United States would defend Canada was simply extended into NMD.[116] Implicitly, the United States would use its interceptors to protect Canadian cities as American cities. Attempts to raise doubts appeared to fall on deaf ears, at least publicly. The most explicit reference came in May 2000 when the deputy commander of US Space Command, Vice-Admiral Herbert Browne, warned that, if Canada was not a participant, the United States "would have absolutely no obligation" to defend Canada. The reaction in Canada was incredulity that any US official would dare make such a statement.[117] Remaining committed to the four no's, the government's only response was to state that it would not be blackmailed.[118] In the press, there was a short-lived expression of disbelief that the United States would not defend Montreal – the city cited in the controversy.[119]

Even attempts to restructure the missile defence and NMD debate by proponents had little, if any, currency.[120] Arguing that missile defence in some form was inevitable, international security issues were largely irrelevant to a Canadian decision. Whether an arms race resulted or not, Canada would have to deal with the bilateral implications of a US decision to proceed. These implications touched directly upon the future of NORAD and the overall nature of the continental defence. Furthermore, if the United States was going to defend North America with NMD interceptors, it was the Canadian government's responsibility to ensure that the decision to defend Canadian cities was not left solely to the United States – the likely outcome if Canada said no to missile defence or said nothing at all.

Not surprisingly, the strategic national interest argument had little resonance. This was the legacy of decades of Canadian rhetoric about Canada the good, compassionate, neutral peacekeeper, the idea of Canada having no self-interest save the interest of the global community, and the prevalent notion (especially in Foreign Affairs and the foreign policy academic community) that Canada acted on the basis of values, not interests – the opposite to the United States, of course. Starting with Trudeau and ABM/Safeguard, and through SDI and GPALS, the idea of distinct and unique Canadian strategic and defence interests that might lead to a consideration to defend Canadian cities simply did not play significantly. Perhaps the whole idea was simply too American. Whether the Canadian public thought this way is unclear, despite National Defence polls that indicated widespread support for missile defence. But then polls had indicated public support for missile defence during the SDI episode.[121] For some, the polls in both cases were to be ignored. The public simply did not understand the complexities of the issue (unless they supported one's preferences, as evident in the fall of 2004 when the polls shifted against missile defence). Public views did not matter – missile defence was an elite debate.

With division everywhere (or so it appeared) and a government divided (even though it was unlikely that a decision on NMD one way or another would have torn the Liberals apart), all kept to the script devised by National Defence policy officials – no architecture, no deployment decision, no invitation, no answer. Even on the occasion when NDPer Svend Robinson referred to an Axworthy speech condemning missile defence, there was no deviation from the party line.[122] No one was going to force the prime minister's hand. Canadian policy was effectively paralyzed. The same was not true, however, south of the border. Chrétien, his government, and the Liberal caucus may have wished for an Al Gore presidency in part to manage the NMD file. Instead, the government turned to face the return of the Republicans and the Bush family. Both were ardent missile defence advocates, and the end of the four no's was just a matter of time. Clinton's decision not to deploy was likely to be fleeting. In between, as if in preparation for the inevitable, Axworthy left the foreign ministry and retired. His replacement, John Manley, immediately sought to reinvigorate Canada's relationship with the United States, and missile defence looked as one likely means to do so.

Act 5

Ground-Based Midcourse Defense: Is This the End? (2001-05)

On December 12, 2002, President George W. Bush announced that the United States would deploy the Ground-based Midcourse Defense (GMD) system at two sites – Fort Greely, Alaska, and Vandenberg Air Force Base, California. Operational command of the system was assigned to the commander of US Northern Command, who had also been dual-hatted as the commander of NORAD after the integration of US Space Command into Strategic Command. In the fall of 2004, GMD became operational, with an initial deployment of four interceptors at Fort Greely, two interceptors at Vandenberg, and a sea-based X-band tracking and cueing radar stationed in Alaska. Over the next year, additional interceptors were deployed at both sites, and the United States began negotiations with Poland for a third interceptor site and with the Czech Republic for a forward X-band tracking radar site. During this process, Canada and the United States in early August 2004 agreed to assign the early warning mission for GMD to NORAD – replicating its role during the brief operational life of Safeguard in 1975. The past had indeed been repeated, except for one vital difference. The door to Canadian participation left open in 1969, 1985, 1992, and 2000 was formally closed in 2005. On February 24 in the House of Commons, Canada's foreign minister, Pierre Pettigrew, announced that Canada would do no more than early warning. For the first time, the government issued a blunt no to missile defence.

This apparent end to Canada's missile defence saga was without doubt a significant departure from previous practices of avoiding a decision one way or another. Pearson, Trudeau, Mulroney, and Chrétien had said neither yes nor no to missile defence and, except for Mulroney's partial decision on SDI, had avoided saying much at all. All had been hedged to keep the door partially open and partially closed, as had been the actual case for the SDI decision. Also in contrast to Pearson and Mulroney, the decision of the Martin government took place without a clear understanding of the likely US reaction. In 1967, McNamara provided Canada with the way out of ABM. In 1985, senior US officials fully blessed the Mulroney SDI policy months before the September 7, 1985, announcement. As for Trudeau and Safeguard, Mulroney and GPALS, and Chrétien and NMD, a variety of factors made a non-decision possible if not desirable. All three also had the ABM Treaty safety blanket to live under, which would disappear in December 2001. Moreover, neither Mulroney nor Chrétien faced an operational deployment decision, and, even though Trudeau faced Safeguard's deployment and brief operational life, the technology of the day sufficiently insulated Canada from any negative fallout in the US defence relationship as a function of the Cold War strategic environment, and half-hearted political support in Washington, to say the least.

On all these counts, Martin did not have the luxury of his predecessors with a deployment decision made by a president firmly committed to the program. Of course, he also inherited from the Chrétien government negotiations with the United States, or discussions – the preferred term by the government at the time – on Canadian participation, and these had largely come to a successful conclusion before Martin took office. On the surface at least, the very existence of negotiations, and the public rationale behind them, suggested that Canada would participate – the only issue being the details of participation. However, the Chrétien decision to engage in discussions with the United States was not tantamount to a complete yes. It was as much informed by the need to acquire from the United States the necessary information on the GMD system and possible Canadian roles in order to make an informed decision. It was, in effect, the long-standing Canadian missile defence dilemma that External/Foreign Affairs had always been deeply concerned about – how to say yes to information without giving the United States and domestic opinion the perception of saying yes to participation.

It was not just the negotiations that Martin inherited but also a basic agreement between the United States and Canada on the way forward. As he proceeded to implement the first two of the three-part agreement – the open letter from Defence Minister David Pratt to Secretary of Defense Donald Rumsfeld seeking participation through NORAD (which Rumsfeld publicly replied favourably to), and the amendment to the NORAD Agreement assigning the GMD early warning mission to NORAD – everyone believed that Canada had essentially said yes. Thus, all were surprised, if not shocked, when the prime minister unexpectedly reversed course and said no, even though negotiations on the final part of the agreement concerning the details of Canadian participation had ended a year earlier. The decision was completely unexpected by the United States. Even more, in saying no in effect to future negotiations, the prime minister had also unknowingly said no to a draft memorandum of understanding on Canadian access to missile defence research, development, and testing – an agreement that would have had great value for Canadian strategic interests.

Throughout the entire missile defence saga, the United States had ostensibly left the door open for Canadian participation, providing at best only hints about the form Canadian participation might take, with the high-water mark occurring in 1996 with the US JROC decision to assign command and control for NMD to NORAD if Canada agreed. Since then, the United States had pulled back, emphasizing that NMD was a US-only program that required no Canadian contribution, generally understood as territory for supporting missile defence radars – interceptors were never part of the equation despite fears expressed by missile defence critics in Canada. The Bush administration stuck to the same policy line: GMD was designed to operate without Canada. Moreover, it was obvious, albeit implicit, that Canada would have to take the initiative in seeking to participate in the US program. There would be no overt or covert American pressure on or invitation to Canada to participate. Nonetheless, existential pressure existed, and, once Bush announced that the United States would proceed to deployment in Presidential Directive – 25, three of the four no's of Canada's unofficial missile defence policy evaporated – an architecture existed, a deployment date had been set, and silence on the part of the Canadian government meant no to participation.

Except for the timing of the Bush announcement, the decision to deploy an operational system should have come as no surprise. With the election of Bush and the re-election of a Republican majority in both Houses of Congress, there was little doubt that missile defence would proceed. Missile defence had long been central to Republican defence policy, and George W. Bush, like his father, was an ardent advocate. Republicans had repeatedly criticized Clinton for his half-hearted commitment. A Republican president would certainly view the technology envelope much differently from Clinton. A deployment decision was inevitable the minute after Bush was declared president-elect. The only question was when the decision would be made, and it was highly likely that the decision would be made before the end of his first term. September 11 – 9/11 – simply guaranteed it.

What this meant for Canada, and whether it would include some form of private invitation, would be a different question, especially after the political difficulties in the relationship in the late 1990s. At the same time, domestic opposition began to become more vocal again, but in the absence of a public invitation or government pronouncement it made little impact until the fall of 2004. Even the announced US withdrawal from the ABM Treaty on December 14, 2001, led to little public reaction. Until December 2002, National Defence's four no's managed the problem – no architecture, no deployment decision, no invitation, no decision. It would take a formal deployment operational decision and the war in Iraq to alter the private and public environment.

Like its predecessor, the new administration moved relatively quickly to make missile defence its own. On May 1 in a speech to military officers at the National Defense University, Bush outlined the administration's new security strategy, which included an expansive layered missile defence program.[1] The strategy eliminated the Clinton-era artificial division between NMD and TMD, integrating the research efforts. Along with dropping the term "NMD" and replacing it with a ground-based mid-course defense, "GMD," this integration reflected a subtle yet important shift in US thinking about missile defence. The small, limited system of NMD was to be bolstered by at least one, if not more, layers – specifically the forward-deployed US naval systems of the formerly labelled Aegis cruiser- and destroyer-class theatre-wide program also in the development, test, and evaluation phase. These systems ideally would permit a very early intercept attempt against an East Asian (North Korea) or Middle Eastern (Iran) attack before the GMD system in North America engaged. In addition, the US Army THAAD system, once operational and forward deployed with US forces overseas – in South Korea, for example – might also be able to engage. The US Air Force's airborne laser – the least advanced of the US missile defence efforts and designed to strike at missiles during the boost phase – offered another layer in the defence of North America.[2] Even without the airborne laser, and depending upon quick target identification and cueing of the theatre systems, and their distance from target, such capabilities might also permit a boost-phase intercept – the most vulnerable stage of an ICBM, when the missile rises up from its launcher or silo.

The first steps toward an operational site at Fort Greely were taken in 2001. The administration announced that Fort Greely would serve as the second test site for missile defence, along with Kwajalien Atoll in the South Pacific.[3] Shortly thereafter, BMDO officials began to speak privately of the possible employment of the test site in an emergency operational defence role. Technically, a second test site did not violate the ABM Treaty, but the issue quickly became irrelevant. On December 14, 2001, Bush gave formal notice of withdrawal from the treaty under the provisions of Article 15.[4] To complete these initial steps, on January 4, 2002, Bush elevated the status of the BMDO within the American federal administrative structure by creating the Missile Defense Agency (MDA) in its stead.

Certainly, there was some recognition among Canadian officials that Bush would speed up the process. Art Eggleton, the defence minister, informed cabinet early on that a deployment decision would come much sooner rather than later. But no one in Ottawa or elsewhere in Canada truly expected the Bush administration to destroy the cornerstone of strategic stability, as Lloyd Axworthy, the former foreign minister, had labelled the ABM Treaty.[5] Indeed, opposition voices in Canada continued to raise doubts about whether the administration would proceed. Under the guiding hand of Axworthy as a private citizen, it was still suggested that it was possible to influence the US decision-making process

Terminal (Theater) High Altitude Area Defense intercept launch.
Courtesy US Missile Defense Agency

against deployment leading to at least postponement through the first term of the Bush administration.[6] Certainly, 9/11 created a favourable international climate of cooperation (which would quickly evaporate after the decision to invade Iraq) for Bush's dramatic decision to withdraw from the treaty. Regardless, the writing was on the wall for a revolutionary (or reactionary depending upon one's perspective) jolt to the world of arms control. This administration had very different views from the Democrats in Washington and Liberals, New Democrats, Foreign Affairs, and National Defence in Ottawa.

The Republicans were on record as opponents of ABM and doubters of arms control, at least in the strict formal sense inherited from the Cold War experience. ABM was simply an obstacle to defending America against the growing threat of proliferation, and without saying so a case for withdrawal from the treaty was being laid. Some had previously argued that the ABM Treaty was null and void on legal grounds after the collapse of its signatory, the Soviet Union, in 1991.[7] Republicans had also fully opposed the theatre missile defence demarcation agreements negotiated with the Russians in 1997, which were never submitted by the Clinton administration to the Senate for ratification, knowing full well as it did that demarcation would be rejected by the Republican majority. Even more, Republicans had been very critical of the administration's attempts to play on the legal margins regarding treaty violations, such as debating whether pouring concrete at a site prohibited by the treaty would amount to a violation.[8] Finally, and perhaps most significantly, arms control was seen as a legacy of the Cold War. Turning one of the central pillars of the arms control gospel upside down, the ABM Treaty was portrayed as an agreement that perpetuated mistrust and Cold War attitudes.[9]

Of course, hindsight is twenty-twenty. More realistic in the views of many Canadian officials was an expectation of an amended ABM Treaty that would permit GMD to proceed without any changes to the prohibitions on third parties in Article 9. Even with a Republican in the White House, and Republican control of Congress, the United States would not walk away from ABM. Rhetoric outside office was one thing. Saner heads would prevail once the reality of the presidency and the international environment was faced. During the Clinton years, with the president a reluctant NMD supporter, international opposition – especially from Russia and Europe – could be used, if not welcomed, to buttress the preferences of the administration. Indeed, Clinton's fourth criterion for deployment up until the 1999 NMD Act had been international peace and security, and even after the act passed this criterion was not to be entirely ignored. Even more, one could always rely upon the sanctity of the rule of law in American thinking (which, ironically, would be proven correct on December

14, 2001). For missile defence opponents beyond North America, fervent opposition made sense as long as the Democrats controlled the White House. NMD might be inevitable, but inevitable with a Democrat in office would be a very long time.

This was not the case, however, for the new president and administration. Republicans had a much deeper unilateralist impulse on national security. They were passionate advocates of missile defence and doubters about the efficacy of arms control. Indeed, many believed that arms control was a one-way street providing advantages to adversaries, while undermining US security efforts. As such, not only would fervent international opposition not likely carry much, if any, weight in Washington, it just might push the new administration over the edge. Bush and the Republicans might not hesitate to destroy ABM if the Russians in particular dug in their heels. A more conciliatory, understanding tone might be more productive than outright opposition. Obviously, the Russians would conclude that some legal constraints on US missile defence programs, especially the outer space prohibition, were better than none. In this they could be reassured by the absence of any discussion of the old SDI narrow-versus-broad interpretations. Indeed, following Bush's May 1 speech, Putin responded favourably in welcoming Bush's promise of consultation and his commitment not to proceed unilaterally. Putin also hinted at a possible deal on ABM in exchange for deep cuts to offensive strategic nuclear forces.[10] Similarly, the Europeans would recognize the value of avoiding an alliance rift and the possible spillover of missile defence into their public domains. Above all else, if pushed, the need to modernize the treaty was self-evident.

Technology had changed greatly since 1972. Then a single site could not hope to defend the territory of either the United States or the Soviet Union. At best, a site could defend a relatively small part of each country, which informed the logic of provisions in the ABM Treaty on the distance between a national capital and ICBM ABM site and the 1974 protocol, which limited each party to a single site. Moreover, the five-hundred-mile radius of the long-range Spartan missile of the ABM Safeguard site could not defend one single major American city, unless an incoming warhead was on a track that passed relatively close to the ABM site on its way to Minneapolis. In addition, ABM had no choice but to employ nuclear warheads, and in order to use them effectively without major collateral damage, especially to satellites, they had to be detonated at the upper reaches of the atmosphere.[11] In contrast, the new GMD interceptors would fly further and faster and engage targets in outer space using kinetic-kill technology unavailable in the 1970s. With this new technology, even a single site at the old Safeguard location in North Dakota could provide near-national coverage. If ABM was in part designed to prohibit an actual national defence capability, then technology had largely negated its purposes, and even arms control and ABM advocates might be persuaded to support treaty modernization.

For Canada, the new environment-technology interface did not necessarily mean that the Article 9 security blanket would disappear. There were no indications that the United States wanted to deploy any GMD components or adjuncts on Canadian soil, despite some thinking in Canada and NORAD in the 1990s of the possible value of Canadian radar sites. Moreover, GMD in Alaska appeared to be a relatively minor violation of the treaty at the time – a single site with interceptor numbers well within the treaty's provision and a new X-band radar somewhat distant from the Fort Greely site but still close enough to be considered colocated. The possibility of treaty modernization to permit a limited system also suggested a more passive, cooperative international environment in which Canadian participation might not engender a vocal public debate. As long as the United States and Russia agreed, no one could effectively employ the threat to global and strategic stability arguments. In so doing, the cover of a modernized ABM Treaty could dampen down whatever domestic opposition might exist and be sufficient to mollify internal dissension within the Liberal caucus, especially with the real issue hinging almost exclusively on the possibility of GMD operational command and control being assigned to NORAD. Besides, command and control were outside the treaty, as suggested implicitly during NMD in the late 1990s.

Even more, the position of the European allies started to soften almost immediately after the election of Bush. Many of the European allies had long opposed US strategic defence efforts and NMD, being concerned about negative implications for relations with Russia to the point of dragging out the old ABM and SDI arguments of strategic decoupling. Although unwilling to publicize their own missile defence efforts in and outside NATO for fear of an unwelcome domestic debate, they had identified proliferation as a key security issue in the post-Cold War era and supported the establishment of a Non-Proliferation Centre within NATO.[12] Moreover, the door had been opened to European strategic missile defences in the 1999 new Strategic Concept, which would be followed in 2003 by allied agreement to extend the NATO feasibility study on missile defence into the realm of the protection of national territory and population centres. At the NATO summit on June 13, 2001, Bush made his case for missile defence to the allies, which included his concerns about the ABM straightjacket. Not surprisingly, Chirac, the president of France, was the most critical of Bush's views. Otherwise, criticism was highly constrained.[13] According to the administration, positive responses were received from Great Britain, Italy, Turkey, Spain, Hungary, the Czech Republic, and Poland.

For Canada, the shift in European attitudes was again welcome to movement toward participation. As in the past, being able to stay in step with Europe was a perceptual means to multilateralize the missile defence issue and thus legitimize a bilateral relationship, at least in the mind of the public. Foreign (External) Affairs had spent a great deal of time watching how ABM, SDI, GPALS, and NMD had played out among the NATO allies, and GMD would be no different. This did not mean that the Europeans had had an epiphany on missile defence. More accurately, their open hostility had softened greatly, and their public pronouncements had become much more nuanced.[14] In this regard, the British and Danes stood out. Subsequently, British agreement in 2004 to modernize the Fylingdales radar came as no surprise. The Danes would closely follow with an agreement on the Thule radar. In between, the missile defence question would also take a significant turn in the wake of 9/11 and the beginning of the war on terror.

As the allies' position began to change, at least in terms of public rhetoric, so did that of the Canadian government. Even before the election of President Bush, senior Canadian officials began to display some empathy with the American assessment of the threat. Although not formally agreeing with the view enunciated in early 1999 by Secretary of Defense Cohen, the government began to speak of an understanding of the US threat perception. Reflecting this new nuanced policy direction, in spring 2001, the Canadian Security and Intelligence Service, CSIS, released a public report on missile proliferation. It noted that ballistic missile proliferation would continue to increase, concluding, "the ballistic missile programs of some other states (such as Iran, Israel, North Korea, Syria, and potentially Iraq) are also worrisome because they have acquired, or soon will have, the capability to deliver weapons of mass destruction against neighbouring states and foreign military forces within their respective regions, and even, in some cases, beyond."[15]

Moreover, National Defence officials began to recognize the inevitable. Daniel Bon, the director general of policy planning in National Defence, in an internal departmental memo noted that the US would not need funds, territory, personnel, or technology from Canada to go ahead but "could benefit from our political support with the Russians and in their discussions with other NATO Allies." He added that the value of support would depreciate to nothing the closer the United States came to making a decision: "If we are ever to come out and agree if NMD is a good idea, and especially if we decide we want to participate in it, therefore the time to do so is now – or really soon – while that support is of some real value to the US."[16]

The destruction of the twin towers, the attack on the Pentagon, and the downed airliner in Pennsylvania produced a clear message from the Bush administration to the international community: you are either with us or against us. NATO invoked Article 5 and dispatched its Airborne Warning and Control aircraft to support the NORAD mission to defend the skies of North America.[17] Russia and China fully supported the US effort, not least of all because of their own problems with Islamic terrorism in Chechnya and Xinjiang province respectively. In a decision unthinkable before 9/11, Russia did not object to the deployment of US military forces into bases in the former Soviet Republics of Uzbekistan, Tajikistan, and Kazakhstan for Operation Enduring Freedom in Afghanistan. Canada's immediate role was found in NORAD, with a Canadian officer in command on the day of the attack. Shortly thereafter, the government issued a blank-cheque offer of its military forces to the United States and became one of the first allies to send forces to Afghanistan.

In the wake of 9/11, missile defence disappeared from the public agenda, except for a few critics who suggested that missile defence should be scrapped because it would not have stopped the attacks. However, if missile defence ever truly needed a shot in the arm, 9/11 provided it. The United States was vulnerable, and it would take whatever steps necessary to deal with its vulnerability, and those who stood in the way would be swept aside. And there was little doubt that missile defence would be one of these steps. This did not mean that all of the allies had to come on board, as they had done for the initial stages of the war on terror. But Bush's message had all the hallmarks of a loyalty test, and, like SDI before, a nation's response might be interpreted as such. At a minimum, no one wished to be perceived as an obstacle to the American search for security, and this was especially true for Canada in 2001, as it had been ever since Mackenzie King's reply to the extension of the US defence umbrella over Canada by Roosevelt in 1938. The first element of the test for missile defence would be the removal of the ABM Treaty.

The day before 9/11, senior US defence officials met with their Russian counterparts to discuss, among other things, missile defence. At the meeting, the American officials laid out details of the US missile defence effort and its various aspects that would violate the treaty. Without bluntly saying so, they hinted strongly that ABM was an obstacle whose revision was problematic. Two months later, the first summit between the (new) US and Russian presidents was held at Bush's ranch in Crawford, Texas. In the lead up, administration officials placed the ABM Treaty constraints in their crosshairs. On November 8, just before the summit, Condoleezza Rice, the US national security advisor, forcefully reiterated the president's position on the treaty: "The acquisition of an effective missile defence system for the United States and its allies is one of his highest priorities, that he [Bush] believes the only way to get there is a robust testing and evaluation system, and that he is not prepared to permit the treaty to get in the way of doing that robust testing [sic]."[18] On November 13, the two met to discuss a range of issues, including missile defence and arms control. Although there is no public indication that Bush specifically informed Putin that the United States would withdraw from the ABM Treaty, the public comments of both clearly indicated the possibility that this step would be taken. While adhering to the traditional Russian line on ABM as the cornerstone of strategic stability, Putin publicly recognized US security concerns. Bush, for his part, reiterated that the ABM Treaty was obsolete and ought to be scrapped. At the same time, the two presidents agreed on a new approach to strategic arms control – moving from the highly formal, detailed START process toward a more informal strategic offensive arms reduction (SORT) process. A month later, Bush gave six months' notice of the US intention to withdraw from ABM. In February 2003, the United States and Russia signed the Moscow Treaty (SORT I), committing each to reduce their strategic arsenals to between 2,000 and 2,500 warheads as each saw fit.

Well before SORT reductions put paid to the arms control gospel that missile defence would destroy the possibility of significant cuts in strategic weapons, the response of the world, including Canada, to the US withdrawal from the ABM Treaty was met with deafening silence. Putin simply called the decision a mistake but noted that it would not harm relations with the United States.[19] Pragmatism had replaced the high drama of the Yeltsin years. The Chinese reaction to the withdrawal was also very constrained and distinct from the previous high level of rhetoric against missile defence.[20] It was part of a wider effort to tone down the adversarial rhetoric of both parties, especially in the wake of the E-3 spy plane incident. Regardless, theatre defences were of much more concern to the Chinese, and in April Bush had deferred a decision on providing missile defence to Taiwan. Bush also provided reassurances to Chinese president Zemin in a private call before the announcement.

The cornerstone of strategic stability disappeared with a whisper, and the path was open for movement on all missile defence fronts. Of course, one might have expected a greater response from a government and prime minister who had in the past labelled the ABM Treaty as an indispensable pillar of international security and were on record about the dire consequences of an arms race and the collapse of disarmament prospects if it was eliminated. But, of course, this was not a Democrat in the White House, and 9/11 had altered the political landscape entirely. By then, Axworthy, the most vocal ABM advocate on record, had resigned as minister of foreign affairs and retired from politics. His successor, John Manley, carried no such baggage and was on record supporting improved relations with the United States. Indeed, Axworthy as foreign minister had long been a major critic of US policy across the board, and had he continued as foreign minister with a Republican in office, Canada-US relations were likely to deteriorate greatly. Whether pushed or not, Manley represented a fresh start, and neither he nor anyone else in government was going to stand up and criticize the United States over the ABM Treaty decision, especially after 9/11.

Of course, like most predictions of disaster, the end of ABM did not alter in any significant way the strategic security environment. As noted above, two months later, the United States and Russia signed the Moscow Treaty. Other elements of the arms control/disarmament/non-proliferation treaty regimes continued, with all their attendant problems. No one spoke of negotiating a replacement for ABM, and the idea of negotiating a ban on weapons in outer space remained closed. International peace and security as it concerned strategic forces looked no different years after the decision than it had before. The world had not come to an end, but Canada's safety blanket had.

To the extent that the treaty enabled governments to avoid making a decision, the Canadian government was now cursed with the likelihood of having to decide after decades of indecision. The internal debates on the meaning of participation relative to the treaty provisions evaporated overnight. The way was now open, so some feared, for the United States to pressure Canada to come on board and even offer Canadian territory. If missile defence was a loyalty test, the identification of North Korea, Iran, and Iraq as the axis of evil in Bush's 2002 State of the Union address – the three states threatening North America by seeking nuclear weapons and long-range ballistic missiles – might be translated to a US need to base elements of GMD on Canadian soil to ensure an effective defence. With Iran in particular having already tested the two-stage Shahab-3 medium-range ballistic missile, the track of a future Iranian ICBM launched against the eastern seaboard would fly directly over northeastern Canada. The Fort Greely site was optimized for a North Korean launch.

No pressure appeared, however. Instead, the United States continued on the course laid out before 9/11. But the writing was on the wall. In the process of being freed from ABM Treaty encumbrances, two important steps occurred. On May 15, 2002, Lieutenant General Kadish, the head of the MDA, announced that construction would begin on interceptor silos and radars in Alaska the day after the official end of the treaty on June 13.[21] Arrangements had also begun for direct bilateral consultations with the allies. In the summer of 2002, the MDA dispatched a briefing team to discuss missile defence with all the allies on a bilateral basis. They began in Turkey and, at the request of National Defence, concluded in Ottawa in July. The talks were largely of a technical and industrial nature not dissimilar in some ways from the SDI era. For Canada, one might expect the talks to take a different turn. But this did not occur. Instead, the briefings remained at the general level, and, like all the other allies, possibilities of technical and industrial involvement were focused on. If the visit to Canada was to signal the possibility of an invitation to participate in GMD or to pressure Canada to move forward, neither occurred. Certainly, the meetings held out the carrot of technological and industrial benefits from participation, which always attracted Canadian political interest. But in terms of forward movement, the ball remained in the Canadian court.

Alongside the death of ABM, 9/11 led to a major restructuring of the US military command system, with direct implications for Canada. For decades, the Pentagon was mandated by Congress to undertake biennial reviews of US commands under the Unified Command Plan (UCP). At the time of 9/11, the UCP contained four major operational commands: European Command (EUCOM) with responsibility for Europe and Africa, Central Command (CENTCOM) for the Middle East and Central Asia, Pacific Command (PACOM) for Asia and the Pacific, and Southern Command (SOUTHCOM) for Central and Latin America (Figure 5.1). There was no integrated joint operational command for the United States itself and North America as a whole, with the exception of NORAD – an aerospace command divided into three regional headquarters, with Canada as a single regional command located in Winnipeg. The absence of a joint operational command reflected the belief that the only real threat to North America during the Cold War was an aerospace one, and NORAD thus sufficed. This all changed after 9/11 with the announcement of the creation of US Northern Command (NORTHCOM) – the fifth joint operational command whose headquarters would be colocated with NORAD at Peterson Air Force Base in Colorado Springs.[22]

With no Canadian equivalent to NORTHCOM, there existed no formal institutional mechanism save NORAD for defence cooperation.[23] NORAD, however, had only aerospace terms of reference, with no mandate to expand into the maritime or land sectors, which were part of NORTHCOM's terms of reference. In this context, National Defence proposed that the minister write US Defense Secretary Rumsfeld and suggest a meeting to discuss future enhanced defence and security cooperation. Foreign Affairs vehemently opposed such a move, fearing negative implications from expanded cooperation on Canadian sovereignty, especially if the result was an image of NORAD being subservient to NORTHCOM. National Defence would have none of it. Minister Eggleton and the deputy minister, Jim Judd, agreed to ignore Foreign Affairs and proceed on grounds that this was within the jurisdiction of National Defence. The letter was sent and an initial meeting agreed to. Although Foreign Affairs had opposed the move, department officials did agree to participate. As a result, a fight broke out over the leadership of the Canadian delegation. Foreign Affairs argued that,

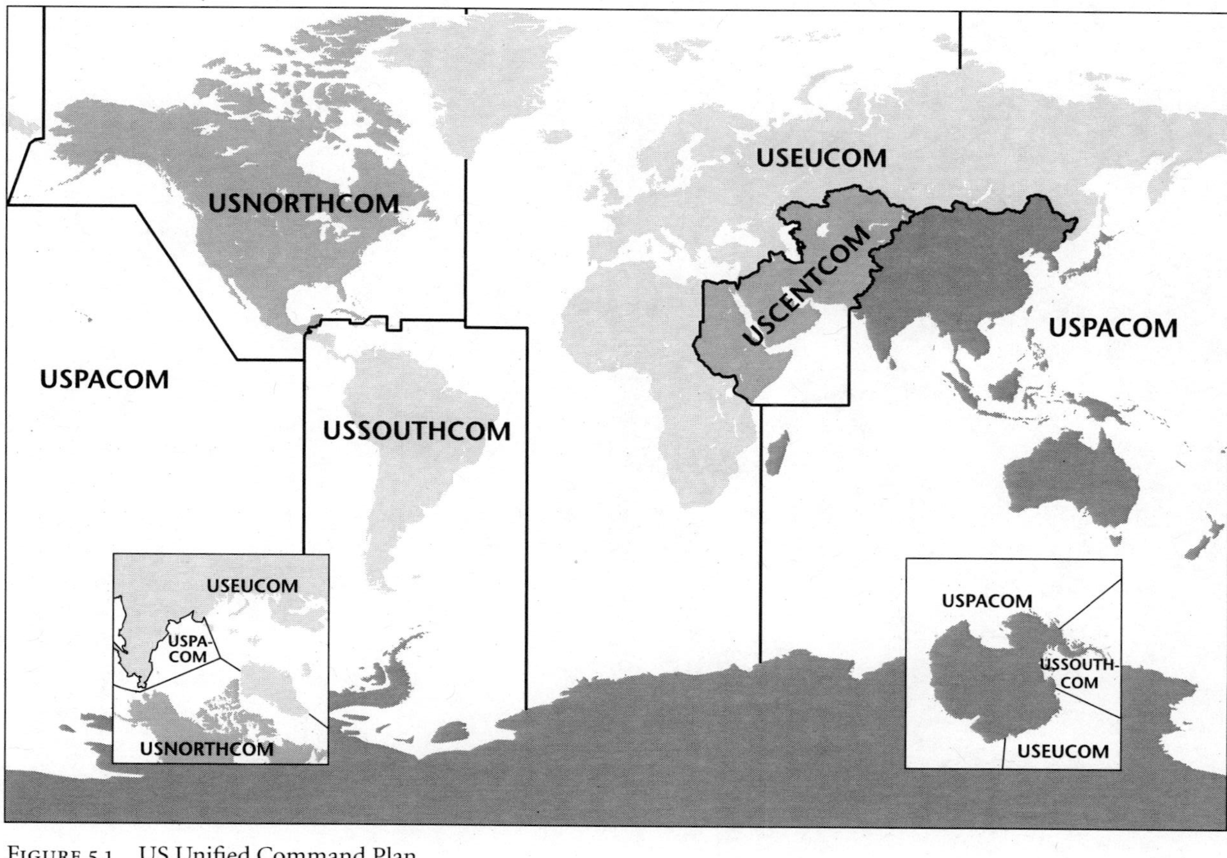

FIGURE 5.1 US Unified Command Plan
Source: US Department of Defense, www.defenselink.mil

NORAD-NORTHCOM Command Centre.
Courtesy NORAD-US Northern Command

as it concerned discussions with a foreign country, it had the lead. National Defence argued that the discussion concerned defence cooperation and was thus within its jurisdictional mandate. Despite the high level of acrimony between the two departments over the issue – an atmosphere as bad as that during the Axworthy years – a joint delegation was finally agreed to, led by Foreign Affairs' deputy minister James Wright and National Defence's assistant deputy minister for policy Ken Calder.

Canada, however, was not alone in trying to grapple with the implications of 9/11 and the establishment of NORTHCOM for North American defence and security. Despite initial suggestions in the Canadian media of NORTHCOM being a US conspiracy to control Canada and the Canadian Forces, the opposite was the case. The United States had caught the Canadian sovereignty bug. Many of its officials were as concerned about the political implications of Canadian military forces crossing the border to help the United States during a crisis as Canadians were concerned about the inverse. The net result of the discussions was a compromise in agreeing to the creation of a binational planning cell, succeeded by the binational planning group in 2003 under the command of the Canadian deputy commander of NORAD and located at Peterson Air Force Base – the home of NORAD. The planning cell and then planning group were tasked to look at other areas of defence cooperation, of which scenarios including freighters carrying weapons of mass destruction (WMD) or containers with WMD being offloaded and transported by rail through Canada to the United States were to be considered. However, at the apparent insistence of the Canadian delegation, missile defence was excluded from their mandate – it would have to be dealt with separately and elsewhere.

Initially, the future of the Canada-US defence relationship appeared headed for an expanded NORAD or perhaps another separate binational command for the maritime sector in particular, as both Canada and the United States shared a common interest and need for an integrated approach, and it would be hard to ignore the highly successful NORAD arrangement as the template. The nature of the 9/11 attacks, NORAD's response that day and subsequently through Operation Noble Eagle, fears of terrorists infiltrating the porous Canada-US border, the threat of freighters and containers, and the emergence of the idea of a North American security perimeter substantially changed the NORAD–missile defence dynamic.

Historically, one of if not the strongest argument for Canadian participation in missile defence had been NORAD. Indeed, ballistic missile defence concerns among Canadian officials had been largely dominated by the future of NORAD, and thus the Canada-US defence relationship, rather than the possibility of defending Canadian cities. Pearson had been concerned about the future of NORAD in considering ABM. NORAD had been sidestepped in the SDI decision only because the issue did not relate to deployment whatsoever. As deployment concerns rose with first GPALS and then NMD, the future of NORAD became dominant. With the end of the Cold War and collapse of the Soviet bomber/air-launched cruise missile threat, there was no direct military threat to North America, and the strategic value of Canada had largely evaporated. As the Cold War faded further and further into memory, many Canadian officials, especially those with links to NORAD, feared that the United States would start to wonder whether its investment was worth it, especially with a declining US defence budget. Indeed, the attempt to downgrade NORAD in the 1993 UCP was a message about the future for NORAD outside missile defence.

NORAD as a window into US defence plans, US strategic developments, and US military space was much more important to Canada than it was to the United States. Some observers in Canada regularly mused about why the United States continued to bother with Canada as a free rider in NORAD. Certainly, the United States gained very little in having access to Canadian thinking, strategic planning, and defence investments. To stand aside from the only significant threat to North American security in American eyes in the 1990s, WMD and ballistic missile proliferation, by being outside NMD would place NORAD on the chopping block. Why fund an aerospace defence institution that would not participate in the only relevant area of defence? Besides, Congress had paid no real attention to NORAD. The agreement had never been submitted to the US Senate for consent and advice. Furthermore, many US officials were surprised, if not shocked, to learn that a Canadian officer was in command of the air defence of the United States on 9/11.

Nonetheless, 9/11 gave fresh life to NORAD and altered the parameters of the missile defence debate in Canada. Integrated binational air defence was significant not only for 9/11-type attacks but also given growing concerns that terrorists might acquire crude cruise missiles, marry them to WMD, and stand off the North American coast and strike at US cities.[24] New NORAD early warning capabilities were needed, and air defence assets had to be redeployed to meet these new threats. As an air defence institution, NORAD's future was secure, at least in the short term, and similar maritime and land concerns even suggested the possible expansion of NORAD. In this situation, even if Canada stood aside from GMD, the future of NORAD appeared not to be at stake. Canada was again strategically relevant by virtue of geography and economic interdependence, and NORAD was the vital institution for cooperation. Like ABM, Canada might sidestep GMD, and perhaps all that was needed was a US message to relax and not worry about GMD, just as McNamara had done for ABM in 1967. This message never appeared, and, of course, Rumsfeld was not McNamara. Certainly, NORAD would survive, but what the NORAD mission might entail with GMD in operation and Canada on the outside was a different question. Moreover, it wasn't just the question of how Canada and NORAD would relate to NORTHCOM that confronted officials in Ottawa, especially with the NORAD agreement not up for formal renewal again until 2006; it was also the implications of the other elements of the new UCP.

As part of the responsibility for all elements of the operational defence of the North American region, the commander of NORTHCOM took command of NORAD. In so doing, the UCP removed the NORAD link to US Space Command, folding it into US Strategic Command, located in Omaha, Nebraska. The NORAD commander also lost USAF Space Command, as it too was moved underneath Strategic Command. The organic link between NORAD and Space Command was severed, even though the operational space element remained just outside Colorado Springs at Schriever Air Force Base, and Strategic Command continued to operate the Space Control Operations Center in Cheyenne

Mountain, feeding early warning data into the NORAD operational command centre.

The aforementioned initial Canada-US binational committee spent no appreciable time discussing the Space Command merger with Strategic Command, and this was also out of bounds for the binational planning group. On the surface, the changes had no appreciable impact on NORAD. It had always been a supported command and would remain so even with the organic link gone. Indeed, the organic link had not prevented the fallout from the SDI decision in which Canadians were frozen out of strategic aerospace defence planning for North America and the Space Control Operations Center closed to Canadian officers. As long as NORAD maintained the early warning mission, the link could not be severed, and whether any significant changes occurred, especially concerning access to US planning, would still likely depend upon Canada's ultimate decision on missile defence. However, the organic link was much more important than most thought. As long as NORAD and Space Command were linked by virtue of a common commander, an atmosphere more conducive to cooperation and information exchange beyond the formal bounds of agreements between Canada and the United States was more likely. Canadian officers might not have the formal need or right to know, but close relationships could lead to a potentially more favourable interpretation of the boundaries between them.

Symbolic of the state of affairs, years after the doors swung open again to Canadians following the Gulf War, the organic link even spread to uniforms. A senior American officer of Cheyenne Mountain, believing that the command looked sloppy in a range of different uniforms, ordered all to dress in a single USAF working uniform. The only thing that would distinguish Canadians from Americans was the flag patch on their shoulders. The image for a Canadian nationalist might be horrifying, but it was a symbol of unity, and with unity came an environment in which all would be treated the same. For Canadians this could only help to produce closer relations and more access.

Unfortunately, closer relations witnessed in the 1990s following the Gulf War slowly began to degrade even before 9/11. Canadian access via NORAD had declined significantly as US officials began to apply strictly the no-foreign-disclosure rules. Of course, many of these areas were outside the formal NORAD-Cheyenne Mountain environment, such as found in the Joint National Test Facility. Here then was the problem of severing the link, amplified with the entrance of Strategic Command into the equation – the most unilateralist of all US commands because of its nuclear strike mission. Severing eliminated the favourable environment that benefited Canadians within the mountain. Assigning Space to Strategic Command meant the involvement of not just an actual operational command but one responsible for US strategic nuclear forces. For Canadians always keen to keep a distance from US strategic nuclear forces, their presence inside the mountain could not help but bring these forces too close, if anyone in Canada was paying attention. For the United States, there was no historical close relationship here with Canada, and the top secret environment of Strategic Command would most likely bring a very narrow interpretation on need and right. With responsibility for US military space, this would likely extend much further into early warning as well. The relatively favourable environment from the old organic link with missile defence assigned to Space Command facilitated at the military level the case for NORAD obtaining the North American missile defence mission. With Strategic Command, this favourable environment evaporated.

NORAD and Canadians might need access to early warning information to complete their mission, but this would likely be limited only to basic data being transmitted into the Ballistic Missile Early Warning Center. Outside of such basic information, Canada had no need or right to know, and Strategic Command officers were more likely to follow this to the letter, erring on the side of caution. NORAD's future might not be at stake, but the space side of its aerospace mandate appeared to be, possibly demanding greater Canadian investment in space in order to defend its mandate. A Canadian contribution of the Space Surveillance Network now was no longer just an asymmetric contribution to missile defence if need be – it now was pressing for other reasons as well. Two years after the UCP revisions, funding was finally released for Project Sapphire, a Canadian military satellite designed to track objects in space from space and contribute to the network.

The entrance of Strategic Command also had questions for command and control and early warning of the future missile defence system itself. The missile defence mission, like ground-based air defence, had long been assigned to the US Army. GMD would be an army mission, and responsibility for the system itself would fall to the army and, in the case of Fort Greely, to the Alaskan National Guard. As NORTHCOM was the operational command of all US forces in North America, command and control of the operational GMD system logically fell to this command. NORTHCOM by virtue of its organic link to NORAD still provided an opening for Canadian participation. However, the odds had dramatically changed. Prior to NORTHCOM, aerospace defence was NORAD, especially with the decision years before to eliminate US Continental Air Defense Command. Then NORAD was the logical home for a GMD-type system, and thus the costs of a Canadian no were likely to be extremely high. Now NORTHCOM provided an alternative to NORAD.

Ironically, the new UCP resolved one missile defence political problem for Canada, even though critics in Canada would portray the issue of Canadian participation almost exclusively as weapons in space. GMD was not weapons in space. However, US missile defence strategy was to develop over time a layered defence, and weapons in space might be a future layer. A layered defence was a strategic mission, which now belonged to Strategic Command, which would coordinate ground, sea, air, and space systems. Previously, this mission would have been assigned to Space Command, with its organic link to NORAD. NORAD in missile defence generated an image of Canada linked to future weapons in space. With the link now severed, Canadian participation through NORAD in GMD would be fully separated from the weapons-in-space question. The United States had unintentionally made Canadian participation easier, even though the exact options open to Canada with NORTHCOM at play were an issue.

Canada would not be alone in being concerned about Strategic Command's role. Certainly, missile defence from space when or if it materialized would fall to Strategic Command. Responsibility for the space-based assets supporting missile defence, which included early warning, was not problematic in providing data to other commands as required. But a single integrated operational command had problems, as forward-deployed assets would come under the jurisdiction of other operational commands, and they would not likely give them up easily. For example, a site in Europe would fall under European Command. Forward-deployed assets on naval vessels off the North Korean coast would be under PACOM. As these missile defence assets could play a role defending forward-deployed forces, regional allies, and North America, the command and control issue was a difficult one. Indeed, the airborne laser system would serve all these defence functions, belong to the air force, and be potentially part of every regional command.

For now, the differential development rates of the various missile defence programs and technological limitations on systems integration worked in favour of disaggregated command and control. NORTHCOM would win for the time being, but exactly how GMD would be integrated as future systems came on line would be open to future UCP reviews. If Canada engaged well before the system was fully operational with either the mission assigned to NORAD or just a positive Canadian response that would facilitate the Canadian right and need to know, Canadian involvement would become a given in the future debate on the broader command and control issue within the United States. It would give Canada a venue for input into US decision making, as had occurred with the 1993 UCP. The Americans would have to proceed from the basis of one missile defence element potentially being binational, linked, and integrated with the other missile defence layers, but independent of them.

If the United States went down the weapons-in-space path despite Canadian objections, Canada could stand aside while keeping a window into these and other strategic developments. Above all else, operating GMD with a substantial Canadian role through NORAD would give Canada a need and right to know while holding the space element at arm's-length – the hidden benefit of the US decision to separate NORAD from Space Command. As the old NORAD arrangement kept US nuclear forces at arm's length, the new one would keep weapons in space, if or when they appeared, at a similar distance.

The key, of course, was timing. The longer Canada waited, the more Strategic Command might become involved, and the more difficult Canadian engagement might become, as it concerned the two primary elements of direct strategic interest to Canada – early warning mission and command and control, which also included possible Canadian input into operational planning decisions. Even if Strategic Command did not come into play, the longer Canada waited, the more the United States would proceed on its own, and the less likely, all things being equal, the United States would be willing to reverse direction to satisfy Canada.

With regard to early warning, on the surface at least, it would be straightforward simply to add GMD through NORTHCOM to the list of NORAD early warning receivers. But like most things, it was not necessarily that straightforward. The nature of the early warning mission had changed since Safeguard. Since the 1960s, data from various US sensors was collected in the Ballistic Missile Early Warning Center to identify potential aerospace threats to North America. This information would then be passed to the Cheyenne Mountain Operations Center for assessment, and the assessment of whether North America was under attack would then be passed to the national command authorities. Assessment from multiple data sources was essential in order to avoid an error, which might lead to an accidental decision to launch US nuclear retaliatory forces to ensure that these forces were not destroyed in their silos or on their runways.[25] With flight times of an ICBM from Russia and China approximately thirty to thirty-five minutes, assessment times for NORAD of somewhere between seven and twelve minutes still provided ample time for the president to make a decision. Only a few minutes were needed from the order to launch to the actual escape of the missiles from their silos or the launch of strategic bombers from their runways.

GMD, however, did not necessarily demand the same level of assessment because an error would not be catastrophic and time was at a premium. The United States might be embarrassed having shot at nothing or space junk if assessors made too quick a decision in error. If it destroyed a civilian satellite by accident, the United States would be liable only for restitution. Even if it struck a single military satellite, it would not spark thermonuclear war. With a kinetic-kill system, there would be no nuclear explosion to damage other satellites on orbit (which would not be the case with the Russia Galosh system defending Moscow), nor would it cause any blast or radiation effects back on earth, though the intercept would likely leave a debris field that could damage other satellites and corrupt an orbital path. Regardless, the quicker one could shoot at a target, the more opportunities the system had to shoot again. More shots increased the probability of a successful intercept, and this was all possible because of longer ranges and faster interceptors.

A more effective defence meant quicker decisions, and this required the compression of the early warning assessment process. With the costs of error low and the benefits of a quick decision high, colocating early warning with command and control made sense. NORAD looked redundant to GMD unless it either took command with Canadian agreement or Canada agreed to participate and the United States decided to keep GMD under US-only command. This is not to suggest that NORAD's early warning mission would disappear entirely even if Canada did not agree. At the end of the day, the United States would still need some mechanism to inform the prime minister of an aerospace attack against North America, and a binational mechanism would be preferable to simply having a US officer call the prime minister to tell him to duck. But with Strategic Command controlling the assets and the retaliatory capabilities, and NORTHCOM responsible for GMD, the scope of NORAD's early warning mission might become very narrow indeed. The future appeared highly constrained for NORAD and Canada outside GMD – a limited early warning mission to serve the purpose of notifying Canadian national command authorities and only a redundant role elsewhere. It summed up to a very marginal aerospace role for NORAD and with it the end of Canada's privileged access to American strategic defence thinking and developments.

Command and control also meant access to US operational planning decisions for GMD employment. As GMD moved forward, decisions would have to be made about defensive priorities. With a limited number of interceptors, a strategy to shoot more than once if necessary at a single target or at numerous potential targets, and the need to make quick decisions, agreement had to be reached beforehand on which cities would be defended and in what order. As the system was being designed to deal with limited missile threats from the North Koreas of the world with relatively crude targeting capabilities, the obvious targets would be major North American cities – very large, high-value targets where missing by miles (a large circular error probability) made little real difference when it came to nuclear weapons. Arguably, not all cities could be defended, and, as evolving early warning technology increasingly made the ability to predict the targets relatively quickly after the launch of the missile, it would not necessarily be the case that incoming warheads would be targeted in order of launch. Even though the decisions would still be made by humans, the software had to be programmed, and programming meant making a priori decisions. For example, an attacker might launch three missiles at North America against six interceptors, and each could be engaged initially. If all three missed, they would not all necessarily be engaged a second time. One might engage only two, holding a third interceptor in reserve in case one or the other, or both, missed again, especially if the target in two cases was Washington and New York and the third open farmland in the west.

For Canadian defence planners, command and control participation in GMD potentially meant a role in the intercept planning process and ensuring Canadian cities were included on the intercept priority list. Perhaps one could not expect that Ottawa would be second after Washington even if Canada was on board. But it would certainly be somewhere on the list and hopefully near the top. This would be one of the key elements for negotiation. Otherwise, Canada would have to rely on US goodwill in deciding how to employ the interceptors in the case of attack. As a US-only system, there was no formal requirement for US planners to consider the defence of Canada. None of this was at play with Safeguard, as the only major Canadian city it could protect was Winnipeg, and defending the Grand Forks ICBM field at Grand Forks meant defending Winnipeg by default.[26] Regardless, at least during the initial Safeguard debate, concerns about the amount of coverage it would provide for Canadian cities were enunciated. In effect, the issue reflected the long-standing belief that the United States would defend Canada – in spite of Vice-Admiral Browne's statement in 2000 that the United States had no obligation to do so if Canada did not participate – and Canada's strategic interest in ensuring that Canada would have input in the manner in which such defence occurred.

The shoot-look-shoot strategy required the ability of the system to assess the success or failure of an initial intercept in order to determine if a second or third shot was necessary. The Clear, Thule, and Fylingdales radars were oriented out from North America. Once the warheads passed overhead, they were lost from view. Other radars in the continental United States at Beale, Cavalier, and Cape Cod could look only so far north. There was a gap between them, which potentially could be filled by Canadian radar – locations originally identified in the 1991 NORAD study as Frobisher Bay, Churchill, and Goose Bay. These could serve the vital role of battle damage assessment in order to determine

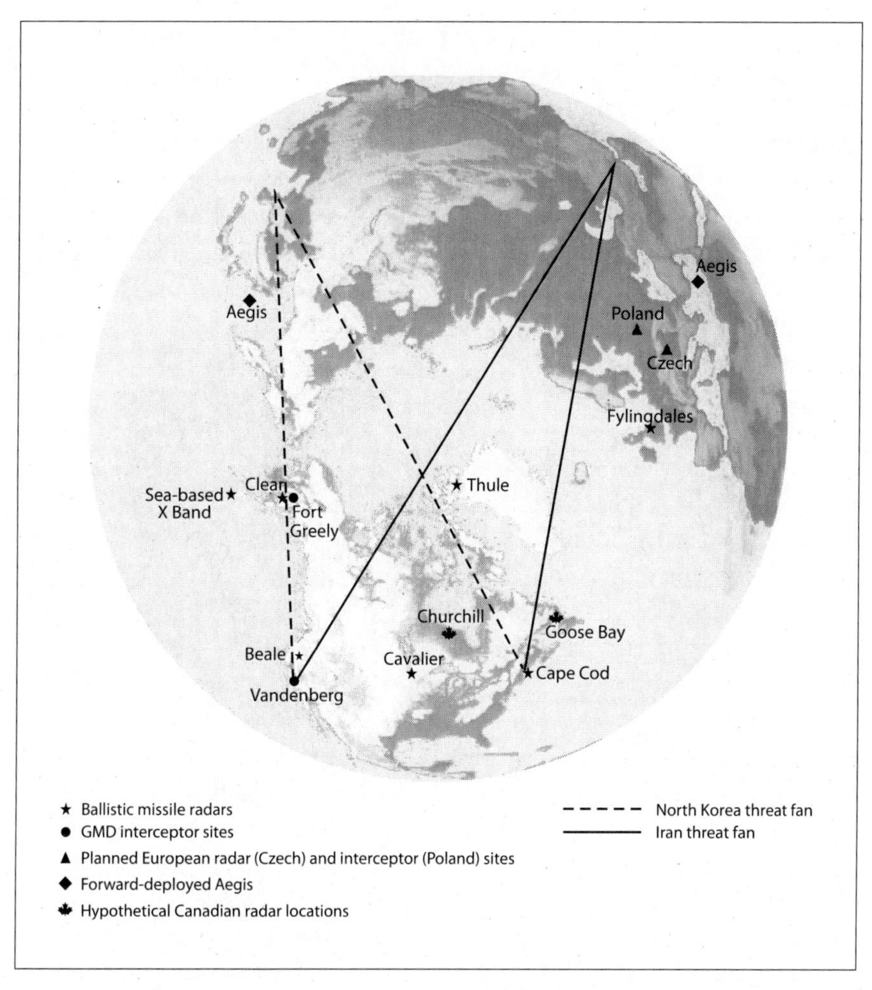

FIGURE 5.2 Polar projection of US missile defence and ballistic missile threat fans from North Korea and Iran

whether to assign another interceptor to the target. This was key Canadian leverage or a possible asymmetric contribution of participation. This was also a danger point because, if Canada said no, the United States might consider the answer as effectively undermining US security, with the wider political implications that might entail.

These technical, strategic, and institutional considerations all came fully into play once Bush announced the decision to deploy. Certainly, there remained numerous questions about the details of GMD architecture, technologies, and operational options that were needed before Canada could make a specific decision, and the answers would influence the options open to Canada. The problem after December 12, 2002, was twofold, however. First, many of the detailed questions had no answers yet, because the system was still in the developmental phase and had never been truly tested as an integrated system. The United States could not answer because it did not have the answers yet. These would unfold over time, and some of them might not be available even once the system went operational in the fall of 2004, if it remained on schedule.[27] But, as they were made, the possibility of Canadian participation would start to decline, or the costs to participate might rise in order to get the United States to reverse course, having made decisions on a US-only basis. In Canada, the fear of having to make a major investment, as recognized by Pearson in 1968, with the demands to recapitalize the Canadian Forces after a decade of darkness, as General Hillier would subsequently put it, was paramount. Second, there was no reason for the United States to answer unless Canada could signal that the endgame was participation. US officials had never wanted to be in a position of asking for something and then not getting it. But Canada could or would not signal the endgame unless it knew the nature of the end. Thus, the Canada-US dance resumed, and someone had to go first. Canada worried about what it might agree to, and the United States worried that any hint of an open door might be counterproductive. In the end, the ball was in Canada's court, and Canada would take the first of many steps on the path to which all observers, even the opponents of missile defence, expected would be full participation.

A month before the Bush deployment announcement, Foreign Affairs began to recognize the need to move forward toward possible participation. National Defence officials were surprised by, if not somewhat suspicious of, the sudden expression of Foreign Affairs' interest in talking to the Americans about missile defence. Both ministers, with the support of the Privy Council Office, agreed on an approach to enter into exploratory discussions with the United States. All concluded that missile defence could no longer be ignored and might have negative repercussions if it was not at least discussed with the United States. Moreover, it was recognized that Rumsfeld was not predisposed to Canada and played hardball. Canada could no longer give no answer.

On January 20, 2003, a joint Foreign Affairs and National Defence group led by Jill Sinclair, the director general of international policy, and Daniel Bon, the director general of policy planning, with Maj. Gen. Pierre Daigle from the chief of defence staff's office in support, arrived at the MDA in Washington to discuss GMD.[28] They were armed with more than fifty questions about the system. The response of the MDA was cold, to say the least. Having heard variants of the questions for years, MDA officials did not take the visit as a signal of any significant movement in Canada's non-committal position. In fact, the MDA expected that the visit was the opening of negotiations for participation. When it quickly became clear that the Canadian delegation had no such mandate, they were informed to return when they were truly interested.

It is difficult to estimate the extent to which personalities and past experiences underpinned the outcome of the meeting. Most of the individuals at the meeting had been around a long time and knew each other. Sinclair, for example, had been at the point of the land mines issue in the late 1990s, and many perceived her as sharing Axworthy's views. Many of the MDA officials had been around from the days of GPALS and NMD. Some were unprepared to answer Canada's questions; others were unwilling to do so until Canada was publicly in. Regardless, both delegations played their cards close to their chests. The Canadians were unwilling to commit to a need to be defended, and the Americans wouldn't provide technical answers to the capacity of the announced GMD system to defend Canada. In effect, the MDA, responsible for the system until handed off to NORTHCOM, had written Canada off by not responding to the Canadian questions and not interpreting the meeting as movement forward. Canada would have to signal that it was truly serious, and this required higher-level intervention. Canada would have to generate a stronger signal, and to do so effectively meant bringing the issue to cabinet. National Defence had long perceived this as a dangerous step, because the outcome could not be predicted. Cabinet members, like most of the actors in the Canadian debate, held fairly rigid views on missile defence, and it would be difficult to persuade those opposed to agree to move forward at least in an exploratory fashion on the basis of complicated, nuanced, and relatively intangible strategic political considerations. Even the position of the prime minister could not be predicted.

By now, domestic politics also loomed large. Pro-American domestic sentiment that had followed 9/11 had slowly begun to dissipate, and, for a government that had treated foreign affairs as secondary to domestic politics, forward movement on missile defence might now be liable to significant domestic political fallout – not because of missile defence itself but because of the crisis over Iraq, which damaged Canada-US relations. As the Iraq issue moved into and through the Security Council, indications were strong that Canada would join the coalition, at least so thought Ambassador Cellucci.[29] By the time the Canadian diplomatic effort to find a compromise between the US and French positions had failed, the government's signals had become vague and ambiguous. When the United States, supported by the United Kingdom and Australia, decided to proceed on the basis of Security Council Resolution 1441 of November, following the report of the Blix committee, Canada stood aside.[30] In so doing, as stated by the prime minister, Canada chose the United Nations over the United States.

Like most of these issues, the actual decision was not as problematic as the manner in which it was communicated and reinforced over time. Public commentators had long concentrated upon the relationship between the decision – either for or against the US position – and the state of relations, and this always ended up in an independence or dependence, ally or satellite, debate. The more Canada agreed, the better the relations. The less Canada agreed, the poorer the relations. But Foreign Affairs and National Defence had long recognized that the impact of any Canadian negative decision significantly depended upon the manner in which it was communicated. Taking a different position from the United States was one thing – proclaiming it in a manner designed to offend publicly the United States was another. Not only did it cross the bounds of proper diplomatic etiquette, but such public affronts were generally reserved for adversaries, not one's closest friend and ally. Whether intentional or not, the manner in which the Iraq decision came out publicly, not least of all because the lack of discipline within the Liberal caucus and the failure of the prime minister to act forcibly in responding to such outbursts, as well as wider public comments, did not help the Canadian cause in Washington.

Of course, the mismanagement of Iraq might lead some in government to suggest that movement elsewhere would limit the impact of the Iraq decision. Indeed, shortly after the Iraq decision, the government appeared to reverse itself, committing troops to Afghanistan even after it had said that none was available and then hinted that the no on Iraq was because of commitments elsewhere. On the surface at least, this decision looked like an attempt to recover from the damage of the Iraq one. A favourable decision or forward movement on the president's prized missile defence program might prove salutary as well, even though missile defence had apparently not been raised in cabinet discussions on Iraq. Regardless, the possibility of using missile defence as a means to improve relations with Washington confronted growing public disapproval of President Bush personally, especially after evidence began to mount that Iraq had no weapons of mass destruction. The dilemma was clear: missile defence might mollify the president and improve relations, but missile defence might be met by resounding domestic opposition because it was identified with a disliked president. No matter how the bureaucracy framed the issue, moving forward would have to be done very cautiously, especially with growing public disapproval of the president.

Even before the endgame of Iraq, in the wake of the January meeting with the MDA, and much to the surprise of National Defence, Foreign Affairs pushed for a quick memorandum to cabinet for beginning discussions on Canadian participation. The department appeared to have done an about-face from ardent opposition to qualified proponent, or perhaps it just felt that the time had come for a decision. Regardless, the interdepartmental process began to construct a memorandum on missile defence based on several considerations. The US deployment decision meant that Canada could no longer ignore the issue, because one way or another Canada would be in or out. Missile defence was a reality whether one liked it or not and whether the technology really worked or not. There were many questions that the United States would answer only if Canada indicated forward movement. But indicating forward movement had to be carefully managed in order not to raise US expectations of a Canadian yes, which would raise the costs if the ultimate decision was no. Both departments agreed that the talks would be in reality negotiations. Instructions from both ministers under orders from the prime minister had made it clear that for political reasons they were not to call them such, as it committed the government too much.

The memorandum raised the issue of the defence of Canada and recognized the future ballistic missile threat from proliferators and the importance of moving quickly rather than waiting for the threat to become manifest. In some ways, Canada had been involved for years in missile defence as a function of NORAD's role in the detection and assessment of missile launches. In addition, industrial access issues were identified, as were the question of avenues of influence on the weapons-in-space issue and possible burden-sharing concerns. The memorandum also identified the different forms participation might take and their associated costs and benefits. Moreover, it was stressed that, because of geography and physics, Canada was involved whether it was inside or outside the tent. Finally, and in some ways above all else, there was the NORAD question.

The most pressing question was whether Canada should seek to have the early warning mission assigned to NORAD, which would keep the door open for a future decision to seek the assignment of the GMD mission itself to NORAD, if the United States agreed. Even if Canada remained silent, one might expect the Americans to raise the early warning question, and in this situation the government would have to decide, and it could decide only having entered into discussions with the United States. Alternatively, no action might simply lead the United States to act unilaterally, thereby marginalizing, if not gutting, NORAD's aerospace early warning mission.[31] Furthermore, early indications hinted that the United States preferred that the entire scope of GMD operation, from early warning to intercept, be brought together on a single screen, and that screen should be NORTHCOM. Canada needed to be in at a minimum in order to ensure that NORAD could at least see the screen. The key political goal, then, was a small forward step to keep the option open as the United States progressed and to get more answers without formally or publicly committing or trapping the government one way or the other. The answer was cabinet agreement to seek discussions with the United States.

Prior to the memorandum and the cabinet decision to enter into discussions, Paul Martin, the prime minister in waiting, appeared to come out in favour of Canadian participation after his departure from cabinet. On April 30, 2003, in a speech to the Canadian Newspaper Association, Martin emphasized the importance of improving relations with the United States and the need for Canadian involvement in a continental missile defence system: "If a missile is going over Canadian airspace, I want to know. I want to be at the table before that happens."[32] It may not have been a ringing endorsement, but it certainly indicated future movement forward. Moreover, Martin commanded significant support within the Liberal caucus and had come to control the Liberal Party administrative hierarchy such that his viewpoint would significantly affect the government's calculus. Chrétien's days were numbered, having announced several months earlier his decision to retire. Movement could be expected at the latest in the fall after the Liberal leadership convention. Whether this motivated the prime minister and cabinet to move forward is unclear. Regardless, Chrétien decided to move.

On May 29, the new defence minister, John McCallum, announced that the "government has decided to enter into *discussions* with the United States on Canada's participation in ballistic missile defence." Strangely, the minister then followed the sentence with a reference to negotiations: "No final decision will be taken before returning to cabinet after these *negotiations*."[33] Subsequently, government spokespersons would reiterate repeatedly that these were discussions, not negotiations. The latter implied that the government had already made a decision to participate, and negotiations would be about the nature of participation. Formal discussions, instead, could be understood as obtaining detailed information as sought in the January meeting with the MDA, upon which a decision could then be taken whether to negotiate. In effect, it was the long-standing missile defence dilemma: how to get information from the United States without committing, even though getting the information from the United States required some form of commitment.

Regardless, the rationale provided by the minister for formal discussions strongly implied at a minimum that cabinet was leaning heavily toward participation – an essential signal to the United States. In making the announcement, McCallum identified the rationales for the discussions or, in effect, participation. The rationales were the protection of Canadians and influence on the weaponization of space. Indeed, he stated, "if we [Canada] are not inside the tent our ability to influence the U.S. decisions in these areas [weaponization of space] is likely to be precisely zero."[34] The announcement was as bold a signal as one could expect from a divided government faced with the inevitable, and certainly it was viewed as a strong signal in the United States of forward movement.

Bolstering this forward move, the government had obtained some indication of where the House stood, and thus the Liberal Party, on Canadian participation. Before the announcement, the Canadian Alliance Party introduced a motion in the House calling for NORAD to take responsibility for the command and control of a continental missile defence system. The non-binding motion passed by a vote of 156 to 73, with 38 Liberals (most from Quebec), the Bloc Québécois, and the NDP opposed. Martin skipped the vote. Public opinion polls also spoke in favour of forward movement. In the fall of 2002, one poll found 62 percent of Canadians believing that a missile launch against Canada was a possibility – up from 36 percent in a poll two years earlier.[35]

While the government may have spoken publicly about discussions, the bureaucracy walked a fine line between discussions and actual negotiations with its American counterparts. The first was a high-level meeting in Washington on June 18 led by Wright (Foreign Affairs) and Calder (National Defence), along with the former deputy commander of NORAD and current vice-chief of the defence staff, Lt. Gen. George Macdonald.[36] At the meeting, the Canadians presented the key elements of the Canadian position – the defence of Canadian cities, the role for NORAD, industrial access, the non-weaponization of space, and costs.[37] Wright emphasized that Canada would not be able to provide any direct monetary contribution. Upon conclusion, one senior Canadian military official stated, "there was no concrete proposal that was put on the table by the American side ... and we will not be responding definitely until there is one."[38]

Following this initial meeting, formal discussions began on July 22. Distinct from the public debate, which fixated upon the threat of missile defence to international peace and security and the weaponization of space, the dominant issue for National Defence was the capacity of the system and willingness of the United States to provide protection for Canadian cities. From the onset, the US delegation had difficulty with making any such commitment. It noted that the specific target became known only very late in the warhead's flight, and it was thus difficult for the United States to give a guarantee about the unknown, especially in terms of making any form of a legal commitment to defend Canadian cities. Canada was welcome to be part of the missile defence effort, but there could be no specific guarantees. In response, the Canadians stressed that, without some form of commitment to defend Canadian cities, it would be difficult, if not impossible, for the Canadian government to see any benefits or sense in participation.

For the Canadian officials, the means to ensuring some degree of protection for Canadian cities was to obtain a role in the missile defence intercept decision-making process, and this naturally meant a role for NORAD. For the entire missile defence saga, this role had been understood as assigning NORAD command and control of any ground-based missile defence system for the defence of North America if Canada was on board. Such a decision in 2003 required the United States to reverse course, having already assigned command and control to NORTHCOM. In addition, and alongside technical and legal concerns about a formal commitment, the US team had concerns about foreign involvement in a US defence system. In around 2000, US officials had begun to apply strictly the no-foreign-disclosure rules to Canadians in NORAD. Immediately after 9/11, many senior US officials were shocked and dismayed to find out that US air defence was in the hands of a foreigner, Canadian Lieutenant General Finlay, on the morning of the attack.[39] However, a distinction could be made in formal command and control – the agent delegated with release authority – and input into the operational software that determined interceptor targeting. Regardless, some form of decision-making involvement would be make-or-break for Canada, and any such involvement still meant NORAD.

The Canadians needed to impress upon the US delegation that this involvement was essential, and this led to the scheduling of a NORAD visit for educational purposes. At a minimum, the visit would serve to impress upon the Americans the role NORAD already played as a function of its early warning mission. Whether the visit served its purpose is not clear. However, to the surprise and consternation of the Canadian team, it found an implicit separation of the United States and Canada at NORAD headquarters. NORAD organization charts identified the US Element NORAD (USELM-NORAD) within the traditional structure. It appeared that the US intent was to structure the organization to work around Canada for ballistic missile defence events, without any Canadian involvement. The military feared that, unless Canada moved quickly, the United States would start to wonder whether Canada was needed at all even in the early warning role, and the position of command director would become closed to Canadians. If all this occurred, Canada-NORAD would be marginalized, and one could see the end of the binational command.

Following the visit to NORAD Headquarters in Colorado Springs, the two teams met again in Washington (September 25) and Ottawa (September 30). At the conclusion of the two meetings, progress toward an agreement had been made. The delegations agreed that a public ministerial statement of intent to participate was required, and the US provided a basic text for Canadian consideration, which would be subsequently revised to reflect distinct Canadian views. Both delegations agreed that the text of the Canadian letter and US response would be mutually agreed upon before public release. And both agreed that the venue for Canadian participation would be NORAD, which would require some amendments to the NORAD Agreement. At this time, the amendments would formally give NORAD the early warning mission for GMD and permit NORAD to have some access to the developmental stage of GMD. As for a Canadian contribution, the United States remained vague. The delegation did not raise any specifics, including any reference to a possible radar site on Canadian soil or the Sapphire space-based optical space surveillance satellite, and neither did the Canadian delegation. The US lead, J.D. Crouch, the assistant secretary of defense for international security policy, stated that an in-kind offer of NORAD as Canada's contribution would likely be sufficient.

Thus, by October, the outline of an agreement was in place. The process leading to Canadian participation would consist of two phases. The first would be the exchange of letters in which Canada formally requested negotiations on participation through NORAD. The second would be an amendment to the NORAD Agreement regarding early warning for GMD and NORAD involvement in GMD development. Exactly what NORAD involvement would entail was left for final negotiations following the exchange of letters. Nonetheless, the US delegation had clearly signalled increased US flexibility for Canada to have a role in planning and decision making on technical defence questions.

In the process leading to the exchange of letters in January 2004, two important events occurred that raised concerns in Washington. First, press reports in the late fall suggested that Canada was about to initiate formal negotiations with the United States. In response, the United States sent a letter to Foreign Affairs emphasizing that in its view the ongoing talks were formal in nature. Second, the agreed Canadian letter to Secretary of Defense Rumsfeld, approved by National Defence Minister David Pratt and Foreign Affairs Minister Bill Graham, was held up in the Prime Minister's Office before Christmas. When approved, the text had been slightly amended. Whereas the original text had been emphatic in declaring what the two parties were going to do in light of the growing ballistic missile threat, the text now spoke of the two parties exploring what they should do (discussed below). This change was viewed in Washington as a possible step backward by Canada.

The outcome of the discussions might be seen by some as a bureaucratic conspiracy to bind the hands of the government. In reality, it simply reflected standard diplomatic interaction between sovereign states – no different perhaps from bureaucrats meeting to agree upon a summit communiqué before the summit is held, as had occurred, for example, in negotiations on the communiqué for the Shamrock Summit in March 1985. Moreover, the entire process of discussions, rather than formal negotiations, reflected the political sensitivity of the issue. Canada needed information on which to base its ultimate decision, but the sensitivity of the information required Canada to signal its support beforehand. The United States wanted a yes before it agreed to the meaning of the yes. Canada wanted the details before it said yes. This dilemma had been in many ways of Canada's own making. The previous government nearly a decade earlier made it clear that it did not want to discuss the issue. The United States responded by emphasizing that the program was US-only and designed to be US-only. In other words, Canada had said no as far as the United States was concerned, and this certainly was the view in the MDA. If Canada had changed its mind, it had to say so, and the United States would not accept simple private assurances. In effect, the entire process had been prudent and pragmatic. Nonetheless, as Fowler had said eighteen years earlier, the time had come for the Canadian government publicly to fish or cut bait.

As the initial discussions concluded, National Defence officials in particular proceeded on the belief that the new prime minister was pro-ballistic missile defence. Even if Martin's rhetoric of repairing Canada-US relations as a priority of the new government was designed only to set himself apart from his predecessor, missile defence offered an immediate avenue to do either or both. Moreover, Martin appointed David Pratt, known for his pro-US position and on record as a proponent of missile defence, as the minister of national defence. Pratt had been chair of the Standing Committee on National Defence and Veterans Affairs and had overseen its report that had advocated Canadian engagement within certain parameters. In addition, the new foreign minister, Bill Graham, had been convinced of the necessity of Canadian participation.

However, in reality, Martin's views on missile defence were not very far from Chrétien's and those of many within the Liberal Party. He doubted that the technology would work and whether it was worth the investment. For him, missile defence looked like a significant misallocation of defence dollars. Even so, he recognized that this was a US decision and that, even if he disagreed, Canada needed, if possible, to be supportive. This did not mean Canada had to say yes to missile defence. Rather, if the final decision was no, it should not be accompanied by a condemnation of missile defence – a view that would be followed to the letter in the Pettigrew statement to the House.

In late January 2004, the Pratt letter, dated January 15, and Rumsfeld's response were posted on the National Defence website for all to see. Identifying a shared vision, Pratt wrote: "In light of the growing threat involving the proliferation of ballistic missiles and weapons of mass destruction, we *should explore* extending this partnership to include cooperation in missile defence, as an appropriate response to these new efforts and as a useful complement to the missile defence mission." In implicit reference to the early warning question, he wrote, "we believe that our two nations should move on an expedited basis to amend the NORAD agreement to take into account NORAD's contribution to the missile defence mission," adding, "it is our intent to negotiate in the coming months a Missile Defence Framework Memorandum of Understanding with the United States with the objective of including Canada as a participant in the current US missile defence program and expanding and enhancing information exchange." NORAD would be the centrepiece of cooperation, and he noted that the United States was "prepared to consult with Canada on operational planning areas."[40] Rumsfeld replied with his endorsement of the process in totality.

With the public release of the two letters, the mixed discussions/negotiations now would become formal negotiations when the delegations next met in February. If, however, the Canadian delegation expected a relatively quick resolution based upon US flexibility from the previous meeting, they were seriously disappointed. The leader of the US delegation, J.D. Crouch, had resigned in October, replaced by John Rood, the deputy assistant secretary of defense for forces policy. Whether a function of the Canadian change to the text of the letter, a new personality in the lead, or a change in instructions from the secretary of defense's office, the United States began to pull back from Crouch's undertaking regarding the protection of Canada and Canadian input into the developmental process. In particular, the United States now wanted a clear, unambiguous Canadian commitment before dealing with the command and control question and protection. Even more, the United States again raised the problems of making any legal commitment to defend Canada.

From the onset of discussions in June, it had been clear to the Canadians that the US team was divided on the question of Canadian participation. This had played out initially around the question of Canada's contribution. Those interested in Canadian involvement sought little in the way of a Canadian contribution – NORAD in kind sufficed. Those uninterested pushed for a significant contribution as a means to get Canada to say no. In February, this division was reflected in whether Canada should have any role in command and control and whether any commitment should be made to defend Canada.

For the Canadians in February, there was recognition of US concerns, especially in making a formal commitment to the defence of Canada. However, Canadian officials emphasized interest only in being factored into the US intercept process to obtain some reasonable level of protection. If a missile was going to strike Canada but not the United States, the United States should commit to protect Canada. Moreover, if several missiles were targeted at low-priority targets, such as rural areas in the west, and one at a Canadian city, the city would get priority. In addition, all Canada sought was for its officers to be formally involved in the information and consultation process and, if Canada was targeted, that the prime minister be involved when the president made a decision. Canada did not expect nor request a command role. In so doing, the delegation stepped back from the 1998 NORAD concept of operations proposal.[41] The actual command would remain US-only. All Canada sought was that its concerns be factored in, while at least obtaining some access to the US decision-making process.

The Canadian attempt came to naught. As far as the United States was concerned, the flexibility demonstrated in September had gone too far. It would be the last meeting of the two delegations but not the end of the dialogue. Instead, the missile defence question for Canada bifurcated into two components – the question of NORAD and the early warning mission for GMD, and an earlier draft memorandum of understanding (MOU) from the United States providing for Canadian participation in the areas of industrial and technological research and development. The latter, of course, was reminiscent of the SDI question in 1985. At the same time, the mood of the minority Liberal government began to change significantly with the possibility of an election and the Quebec advertising scandal, which would lead to the establishment of the Gomery Inquiry, in play.

Even though GMD built upon the NMD research and development process (as NMD built upon GPALS and GPALS on the basic research of SDI), and the Bush administration had announced early on that Fort Greely would be developed as a missile defence test bed long before the December 2002 operational deployment decision, the Defense Department and the MDA had many major decisions to face in making GMD operational by the fall of 2004. Among these decisions was the issue of early warning for the basic cueing of the dedicated missile defence radars. NORAD, of course, was responsible for ballistic missile early warning, but this did not extend to missile defence within NORAD's terms of reference. Getting Canada to agree to the inclusion of GMD in the NORAD early warning mission was a means to at least resolve quickly and easily one of the many decisions facing the United States on the path to an operational capability. Otherwise, the United States would have to invest in a separate early warning system for GMD or seek a change to eliminate NORAD entirely from ballistic missile early warning.

The NORAD early warning question was Canada's only real card in the negotiations. When and how to play the card comprised an issue of disagreement among the Canadians. Some felt that GMD could simply be included as just another recipient of NORAD's early warning assessment. Whether recognized or not, this would replicate NORAD's role during the brief operational life of Safeguard in 1975, which had not required any formal change to the agreement.

For the military in particular, this position would signal Canada's commitment to missile defence participation. Moreover, if Canada blocked early warning or made life difficult, the United States would turn to Strategic Command, which had responsibility for the early warning assets that supported the NORAD early warning mission. If this occurred, NORAD, at least in terms of its ballistic missile function, would become redundant. With NORAD renewal looming on the distant horizon (2006), its future could be at stake. Furthermore, the US military had made the very case to Secretary of Defense Rumsfeld as a significant potential Canadian contribution.

However, the lead Canadian defence official on the team opposed this view on three grounds. First, the NORAD Agreement was not to be necessarily interpreted to allow for the simple addition of GMD into the early warning mission. Second, such a decision, once public, would be viewed by some observers as an attempt to fool everyone, and the government stood to be accused of duplicity by participating through the back door – a view reminiscent of the department's position on SDI guidelines. Finally, Canada would give up its only bargaining card for nothing. The United States would get the only thing Canada could contribute, and if Canada wanted more the United States would be in a position to demand more from Canada. Instead, the Canadian interests, as adopted in the discussions with the United States, were to have early warning as a package deal, including greater Canadian and NORAD access as a GMD participant. However, with the failure of the February discussions to reach an agreement, the early warning issue was put off until July.

Although the United States was unwilling to meet Canadian interests for at least some input into the GMD intercept planning process, there remained the commitment by both parties to negotiate a MOU. Shortly after the February meeting, the United States forwarded a draft memorandum for consideration. It was not new, however. Its roots predated the fall 2003 discussions and the December 12, 2002, Bush announcement. Following the 2002 summer bilaterals between the MDA and the allies on missile defence cooperation, initial discussions were held between the MDA and defence officials from the assistant deputy minister for material's office on potential areas of research and development cooperation. For reasons of optics, these discussions moved from the MDA to the Office of the Deputy Under Secretary of Defense for Research, Development, Testing and Evaluation. On January 10, 2003, just before the January 20 meeting with the MDA, the deputy under-secretary, Pete Aldridge, wrote to the ADM (Material), proposing the establishment of a missile defence research, development, testing, and evaluation (RDT&E) MOU. It received a favourable response, but little more could be done until the government moved forward, as occurred in May. By October, this part of the discussions on the US side had now shifted to the Office of the Deputy Under Secretary of Defense for Acquisitions, Technology and Logistics and on the Canadian side to the director general for international and industrial programs to establish a framework MOU. At the time, it was still recognized that a framework RDT&E MOU would have to potentially take into account NORAD and Canadian operational roles in GMD.

The draft MOU was to establish a framework for RDT&E cooperation to facilitate Canadian access, contributions, and support. The United States proposed the establishment of an executive steering committee and an equitable cost-sharing arrangement based upon each individual project, with additional costs related to unique national requirements being borne by that nation. The MOU was to be twenty-five years in duration but wouldn't preclude any other arrangements, and, of course, each party could end the MOU with notice. Among the initial projects identified for consideration were kinetic energy and electromagnetic pulse effects, research on a gas gun (reminiscent of the 1950s Project Helmut), and research on target discrimination and decoy identification employing small and micro-satellites. The only contentious part of the draft MOU was the US initial desire for a formal clause indicating that Canada had agreed to participate in the development and deployment of missile defence capabilities, without specific mention of GMD, a clause the United States quickly backed away from.

The framework MOU was largely consistent with long-standing Canadian policy on the meaning and interpretation of participation. It was to be the official government part of the research and development side of the equation that had been left out of Mulroney's 1985 SDI decision. More importantly, it was entirely consistent with the policy framework on missile defence laid out in the 1994 White Paper, with the exception that the ABM narrow interpretation element was now null and void. Its research and technology focus was further consistent with Canadian efforts that can be traced back to ABM and clearly identified in the 1994 White Paper as well. Agreement would open the door to Canadian firms and, above all else, promote Canadian access to US technical specifications on missile defence and establish an environment in which Canada would have a need and right to know. Even if command and control for GMD remained US-only, broader Canadian and NORAD access and potentially input might be secured in the future. Canada would be on board without the government sacrificing its weaponization of space policy and without a requirement for large Canadian investments. Like the Joint Strike Fighter arrangement, Canada could reap significant economic and technological benefits with relatively few direct costs.[42]

In the end, Canada did not respond to the draft MOU. In early March, orders were received from the government for the civil service to stop all work on missile defence and refrain from making any public statements. As far as the government was concerned, negotiations had come to an end, even though the prime minister and the Prime Minister's Office were largely unaware of their actual state. With a federal election looming, missile defence had come to be seen as a political liability. In Martin's first major defence speech at Gagetown, New Brunswick, on April 11, all references to missile defence from the initial draft were removed. In his April visit to Washington, missile defence was quietly removed from the agenda, and neither Bush nor Martin raised the issue. All that Martin would say when questioned in a news conference during the visit was that a decision had not been made and that the weaponization-of-space prohibition remained in place.[43] As Canada entered into a federal election, with polls indicating the possibility of the government's defeat over the ad scandal, Martin turned to a strategy to portray Stephen Harper and the Conservatives as representing values alien to Canada – code for American values. Painting his opponents as bringing the Bush Republican agenda to Canada, he successfully reversed the trend in the polls and squeezed out a minority government. Subsequently dependent in the House upon two parties firmly opposed to missile defence, the Bloc Québécois and the NDP, Martin was unlikely to move forward on missile defence, even if a decision could be made by cabinet alone through an order-in-council or a favourable vote in the House on a binding or non-binding resolution through the combined votes of Liberals and Conservatives.[44] Indeed, on February 24, 2004, the combination had defeated a Bloc Québécois motion calling for the government to oppose missile defence and "cease all discussions with the Bush administration on possible Canadian participation" by a vote of 155 to 71.[45]

Regardless of the political calculations, the question of early warning remained to be resolved. With the operational date for GMD roughly two months away, the government had no choice but to act. For Martin, the missile defence issue had become strictly about the preservation of NORAD, which meant agreeing to the inclusion of GMD in the early warning loop. In July, discussions were held on the pressing issue of NORAD and early warning for GMD. There was, however, some disagreement among the Canadians on what this legally entailed. The Americans pushed for a formal amendment to the NORAD agreement. Foreign Affairs, however, believed that no formal amendment was needed. With GMD command assigned to NORTHCOM, it was simply a technical matter to add it into the end-users group of early warning assessment, which had always included US-only commands, such as Strategic Command. As a technical issue, it was ostensibly a matter for the Canada-US Military Cooperation Committee. Besides, the 1996 NORAD Agreement had added a clause for just such a contingency, and, for all intents and purposes, missile defence had been the reason for the clause. If someone had remembered, precedent was on the side of Foreign Affairs. NORAD provided early warning to Safeguard during its brief operational life, with no amendment or public fanfare. Of course, there was also the fear of duplicity if the government agreed privately, as it would eventually leak to the public.

National Defence was ambiguous, but the Americans were insistent. On the one hand, ensuring the letter of the law had become increasingly important in the relationship. Canadian NORAD loss of access had been explained as a clamping down on illegal activities – no foreign disclosure meant no foreign disclosure, and Canadian NORAD officers were foreigners, even if they had been treated as otherwise in the past. The same phenomenon had been evident in the ITAR issue, where Canadians had also been given a free ride. Of course, the legal question masked political considerations. Central to the US position was ensuring that the Canadian government publicly committed itself. If missile defence was a loyalty test, the loyalty had to be expressed publicly one way or another. Canada acquiesced.

On August 5, Bill Graham, the new defence minister following the defeat of David Pratt in the June election, announced that Canada and the United States had agreed to amend the NORAD Agreement in order to assign the early warning mission for missile defence to NORAD, as allowed under the consultation clause inserted into the 1996 NORAD Agreement.[46] According to the minister, the "decision [was] made in the interests of preserving NORAD because it was clear that the United States intended to construct a parallel system if they couldn't use this connection with their anti-missile defence ... and that would have rendered NORAD obsolete."[47] In addition, he believed that the decision was necessary to keep the option open for participation in the US missile defence program. But Graham went to great lengths to distance this agreement from the issue of participation, saying, "this decision does not affect or in any way determine the ultimate decision as to whether Canada will participate in missile defence."[48]

Following the August announcement, the government again went into a period of silence on missile defence. Martin had committed to placing any future agreement on Canadian participation before the House of Commons for debate and vote. He could expect that some portion of his caucus might oppose participation in a vote, as had occurred earlier in votes on non-binding Alliance and Bloc motions. This had been a minority of the caucus then. But the political dynamics of missile defence had changed. Concern about the caucus split led Martin to task Graham with monitoring the mood of the caucus. Regardless, Martin still could rely upon significant cabinet support.[49] If necessary, he could evoke party discipline and, either way, rely upon the Conservatives to ensure passage.

Despite the attempt to separate the early warning agreement from the issue of participation, those opposed to Canadian participation saw the agreement as tantamount to participation. Indeed, from June 2003 onward, opponents acted upon the belief that the government was intent on participating – period. Their response was to mobilize anti-missile defence and anti-American/Bush forces to begin a public campaign to lobby the government to reverse course before it actually signed a formal MOU with the United States. In so doing, the opposition did not entirely ignore the old arguments about the arms race, international stability, cost, and technological feasibility. But this time the centrepiece of opposition was the weaponization of space.[50] Weaponization had long been part of the opposition's quiver and like the other arguments, Cold War in origin, drew from the purported plans of SDI days – the old Brilliant Pebbles idea. Non-weaponization was also a sacrosanct policy in Canada that cut across party lines and had widespread public support, not least of all because the Canadian public, like all Western publics, is intuitively predisposed to oppose policies that speak of killing and war. Finally, the government in May 2003 had raised the issue of weaponization of space and thus was very sensitive to it. All one needed was to push the linkage between missile defence and the weaponization of space to create a climate that would lead the government to reconsider. It didn't matter whether there was a linkage relative to the issue at hand – GMD, a ground-based system. All that mattered was making the linkage, and, consciously or not, the polemics rarely spoke about the actual system. Instead, opposition spoke in general terms about US weapons-in-space programs and missile defence.

There was a lot of grist for the mill for the critics to draw upon, and central to the linkage was evidence of US intent primarily in the form of US Space Command's Long Range Plan (LRP). According to the critics, the LRP clearly laid out the American intent to weaponize space.[51] That the LRP stated that the technology lay twenty years into the future and the decision whether to proceed was "not in our [US Space Command's] lane" didn't matter.[52] All that mattered was evidence of intent to proceed with the complicated arguments for and against, regardless of the real issue at hand.[53] Into this basic environment, one other announcement fed the critics' case. In its 2004-05 budget submission, the MDA identified in its plans for 2008 the intent to "initiate a space based Test Bed development to determine the feasibility of exploiting the inherent advantages of intercepting threat missiles from space. We will begin developing a space-based kinetic energy interceptor in FY04 with initial, on-orbit testing to commence with three to five satellites in Block 2008."[54] Here was clear evidence of the linkage.

The proponents of Canadian participation sought to counter the weaponization of space on several grounds. In so doing, they fell back on the relatively unfertile ground of Canadian national interest and national security. Reciting the importance of Canadian input into operational planning in order to enhance the defence of Canadian cities and the value of Canadian access to US plans and thinking, proponents argued that GMD had nothing to do with weapons in space. It was a ground-based system, and committing to participation did not equate whatsoever to participation in weapons in space. In fact, it was typical Canadian hubris to think that the United States wanted to trap Canada into joining with them in weaponizing space. Indeed, if the critics of US unilateralism were to be believed, why would anyone think the United States would want to share space? The critics themselves had emphasized the US desire to dominate and control space to the exclusion of all others. It would be strange for the Americans suddenly to make an exception for Canada. Moreover, by merging Space Command with Strategic Command and assigning GMD to NORTHCOM, the United States had intentionally or not placed a wall between space weapons and GMD.

It is difficult to estimate the impact of the debate on the public at large, and in many ways the opponents and proponents simply fought among themselves. But the key, as it had always been, was perceptions of the public, and the public, having turned against George Bush for Iraq, was not well disposed to support his pet project, missile defence. If Bush lied about WMD in Iraq, he would lie about weapons in space, or at least one could hypothesize such a linkage underlying the shift in public opinion. By 2004, public opinion had begun to move against missile defence and Canadian participation. In February, an internal Liberal Party poll found 70 percent of Canadians in favour of participation.[55] In a November 2004 poll, 56 percent of Canadians were opposed to Canadian participation, with the strongest opposition in Quebec.[56] It didn't matter to a government campaign in support of participation that Canadian public attitudes were soft and potentially malleable. Any government, especially the Martin government elected in June 2004, would be sensitive to public views, given its minority position.

After August and the subsequent entry into operation of GMD, the government continued its policy of silence. It basically left the public debate to the war between the attentive publics, and with no formal negotiations under way with the United States there was no appetite to take on the issue, despite indications that the government might be about to move forward. Reflecting the strange political environment attendant to the first minority government in twenty-five years and a government shrouded in scandal, even the Conservatives took a somewhat bizarre position on missile defence. After years of full support for participation, Harper announced that the Conservative Party would make up its mind when the details of participation were placed before the House.[57] On the surface, the position made some sense, as it would for any government not to agree to something it did not know the details about. Perhaps, then, it was just a signal that a future Conservative government would be prudent and responsible and not simply walk blindly in step with the United States. Stung by being labelled a party that represented alien American values, missile defence was an opportunity to demonstrate the opposite. Yet everyone knew that the Conservatives had always fully supported missile defence, and unless the Martin government gave the farm away the Conservatives would back the government on participation.

Even stranger, missile defence appeared as politically dangerous for all concerned, as if the future of the minority government and the outcome of the next election were at stake. It was reminiscent of the lead up to the 1963 federal election in which the Liberals led by Lester Pearson defeated the minority government of John Diefenbaker partially on a defence issue. Diefenbaker had prevaricated on acquiring nuclear warheads for the Bomarc-B surface-to-air air defence missile. The issue tore cabinet apart and led to the resignation of the defence minister, Douglas Harkness. Pearson, never a proponent of nuclear weapons, argued that it was Canada's duty to share the burden with its allies, and the nation had to fulfill its commitment, which his minority government subsequently did.

The 1963 election was not strictly decided on defence, but it did have elements in common with the current missile defence environment – anti-Americanism, a divided government, and perceptions of an indecisive prime minister. In 1963, Pearson demonstrated leadership in coming to grips with the nuclear issue. In 2004, neither Martin nor the government provided any real national leadership. Except for the Bloc Québécois and the NDP, which took every opportunity to condemn missile defence as the most evil plan ever conceived, the government, the bureaucracy, cabinet, the prime minister, the Liberal Party, and the Conservatives remained more or less mute. For the media, with their penchant for taking a position of opposition to the government (and one still being seen as ultimately committed to participation), this simply fed the strange environment surrounding the politics of missile defence – overwhelming negative press reports on the front pages combined with significant editorial support.

The missile defence issue came to a head with President Bush's first official visit to Canada – a strange occurrence in itself, as presidents usually make a point of visiting Canada early on in their first term, not at the beginning of their second. Nonetheless, the relatively poisoned environment gave cause to the prime minister not to invite Bush to address the House, fearing inappropriate insulting behaviour by some of its members.[58] Instead, the Bush visit included a private meeting with the prime minister and the leaders of the opposition parties in Ottawa and a public event in Halifax to formally thank all Canadians for opening their doors to air travellers on 9/11 after US airspace was closed and many international flights in the air were diverted to Canadian airports.

Prior to the visit, Bush had been advised by the US Embassy and State Department officials not to raise ballistic missile defence. The prime minister was informed that missile defence was not on the agenda and would not be raised. However, Bush ignored the advice and the agenda. Not only did he raise the issue in their private meeting, but it was the first one to come up. In response to Bush's expressed hope that Canada would participate, Martin replied that there were several outstanding questions needing to be answered before a decision was made and that yet he was getting no answers. Bush did not reply, and the discussion moved on to the next issue. Later that day, Bush raised missile defence in his meeting with Harper. Harper's response echoed Martin's in the sense that Canada's answer depended upon the exact meaning and nature of participation.

The impact of the Bush visit on the missile defence question could have been negligible had the issue remained private. Instead, Bush at the press conference in Ottawa stated, "we talked about the future of NORAD and how the organization can best meet emerging threats and safeguard our continent against attack from ballistic missiles."[59] The next day in Halifax, Bush directly spoke of his hope that "we'll also move forward on ballistic missile defense cooperation to protect the next generation of Canadians and Americans from the threats we know will arise."[60]

All things considered, Bush's remarks in Ottawa and Halifax were relatively innocuous. He expressed hope and desire. Nonetheless, they set off a near firestorm, and Martin's discomfort was palatable for all to see. Whether Bush realized it or not, Martin's response was a tentative signal that Canada might say no. Moreover, the prime minister sensed that, by raising the issue publicly, he had unleashed political pressure in Canada that demanded a decision. Any hope that Martin might follow his predecessors in avoiding a formal public decision was, to Martin, problematic. Instead, from the Bush decision to the February announcement of the decision, missile defence was one of the prime targets of the Opposition. Throughout January and February, the government was bombarded with missile defence questions inside and outside the House.

Following the Bush visit, Martin tried one final time to obtain answers from senior officials to what he perceived as the two most pressing questions, cost and protection. Shortly after the Bush visit, he met in early December with senior defence and foreign affairs officials.[61] Martin, reminiscent of Pearson in 1968, sought an iron-clad guarantee that a Canadian yes would not result in any demand for Canadian defence dollars, then or in the future. He also wanted formal assurances that Canadian cities would not be second after US cities in intercept priority. In effect, he wanted to dictate US intercept strategy. Warheads should be intercepted on a first come, first served basis. None of the officials could provide either assurance. More importantly, the military and defence experts who might have been able to provide detailed answers were not invited to the meeting. In the absence of such expertise and Martin's fears of being sandbagged by the bureaucracy – ironically, a fear that Mulroney also felt after years of Liberal rule under Trudeau – the prime minister came to the conclusion that Canada would not be able to proceed any further. He had not yet decided to say no but was leaning in that direction. Regardless, this would be the last time Martin met with officials to be briefed on missile defence. The decision was now in the hands of the Prime Minister's Office and the prime minister.

Subsequently, Martin in his Christmas news conferences reiterated Canadian policy – no weapons in space, no interceptors on Canadian soil, no Canadian money to missile defence, and the right and importance of Canada to have a seat at the table and a say with interceptions taking place in Canadian airspace (even though the interceptions would take place in outer space, and by international law no nation can possess outer space). To most observers, even the opponents, Martin's statements seemed to have confirmed the inevitability of participation. GMD did not include weapons in space, and even if the United States moved in that direction, Canada could avoid being tarnished or trapped because outer space was outside NORAD and part of Strategic Command. The United States had not hinted at, never mind requested, interceptor sites in Canada. In fact, the United States had begun initial discussions with Poland, the Czech Republic, and Hungary about locating a second site on their territory. Certainly, one might fear that the United States would demand a second site in Canada in return for participation at no cost, but this would be unlikely, as the president, in publicly welcoming if not formally inviting Canadian participation, was well aware of the limits involved. The same could be said about costs, and whatever additional costs to NORAD would be funded through the traditional arrangements in which the United States paid all capital costs in the

United States and shared costs in Canada. Moreover, whatever additional costs Canada might face would likely exist anyway because of the early warning mission. Indeed, in the first meeting in Washington in June 2003, the Foreign Affairs lead had made it clear that no Canadian capital investments were likely to follow with participation.

Martin concluded that there was no table for Canada to sit at that justified a yes. Even worse, he feared that a yes might lead to future US demands on Canada, including funding, even though the actual MOU he would implicitly and unknowingly saying no to and which he was never formally briefed upon was some measure of assurance. Nonetheless, it would be difficult to say no to a specific US request on missile defence cooperation after having said yes for all intents and purposes in January. Much would depend, of course, on the reasons that would be attached to the yes, and saying yes to a generic research and development MOU seemed minimal and justifiable only on economic grounds. Politically, however, it was a lose-lose outcome. The anti-missile defence forces would call it a sellout to Bush and Imperial America. The pro-missile defence forces would see it as a disaster, having achieved very little.

As the new year began in an atmosphere of a government believed to be headed toward participation, the general political mood pointed to a short-lived minority government, especially with near-daily revelations of corruption in the Quebec wing of the Liberal Party, which led Martin to establish the Gomery Inquiry on February 10. In reality, missile defence had become a relatively minor issue by February 2005, with a federal budget alongside Gomery dominating the public agenda. On the foreign policy and defence front, the long-awaited White Paper, which would emerge as the International Policy Statement (IPS) in April 2005, continued to languish, with rumours of new authors being brought in to rewrite the paper. As the National Security Policy Statement of April 2004 had sidestepped missile defence, the same could be expected of the IPS. As for the defence component, it had already been rewritten by the new chief of defence staff, Rick Hillier, and parts of it were being briefed to military officials. These were almost exclusively about transforming the Canadian Forces for overseas missions in light of present and future Afghanistans. Anyway, the 1994 White Paper's open door and statements by the government over the past year and a half, including the Pratt letter, sufficed to deal with missile defence, if need be.

In effect, there was also no need to make a decision on missile defence. There were no formal negotiations or discussions under way, and there had not been any since the conclusion of early warning negotiations in the summer of 2004. Nor would the United States seek to push a decision or negotiations. GMD as an operational system was still in a premature state, and US decision makers were well aware of the dynamics of Canadian politics. Unfortunately, it was those very dynamics that led to a complete about-face and the announcement by the foreign minister in the House on February 24 that Canada would not participate, at least "not at this time."[62] With no formal briefing or memorandum to cabinet and the bureaucracy shut out of the process, the prime minister decided the time had come to put the issue to bed.

To most observers at the time, the decision was the triumph of short-term domestic politics. First, Liberal Party fortunes had fallen significantly in Quebec as a result of the ad scandal, and Quebec appeared key to the government's survival in the next election. Quebecers were the most opposed to missile defence and US policy, especially over Iraq. The Quebec wing of the Liberal Party, particularly the youth element, along with the Youth Wing as a whole, were planning to push a resolution at the forthcoming national convention in March opposing Canadian participation, which might pass or at least embarrass the leadership. Missile defence might also drive the soft centre-left vote into the hands of the NDP and Bloc. Facing a united right, the Liberal Party's future could well depend on ensuring that this vote stayed Liberal. Finally, Martin's image had taken a battering as reports leaked out of Ottawa of his indecisiveness and impulsiveness, combined with the *Economist*'s labelling of the prime minister as "Mr. Dithers."[63]

On the surface, the solution was to say no to missile defence publicly, especially with the possibility that the government might fall sooner rather than later. Deep down, however, it was troubling, if not shallow, to say the least. Defence had never played a very significant role in federal elections, and to think that Quebec voters and other voters would ignore the ad scandal and Gomery in light of a no on missile defence was strange thinking indeed. If Martin could not control the Liberal convention and its agenda and avoid embarrassment, he was in much greater trouble than anyone thought. Finally, a no on missile defence in the absence of any pressing reason for one, and in light of the prime minister's repeated declarations on the importance of Canada having a seat at the table with missiles being intercepted overhead, would also be strange.

In the end, ironically, Martin felt that a public no was essential to clear the growing hostile anti-American environment and open the way for better relations in order to deal with much more pressing and important economic issues in the relationship. According to Martin, anti-Americanism was being fed by the failure to put the missile defence question to bed. Missile defence was becoming a referendum on Bush and the entire Canada-US relationship. In dealing with the issue once and for all, anti-American sentiment might abate, enabling the government to deal with the United States more effectively on issues such as mad cow, Devils Lake, and softwood lumber. The occasion for the no also appeared with the forthcoming NATO summit on February 22 in Brussels – two days before the formal announcement in the House.

Without any discussion in cabinet, Martin asked Pettigrew to inform the US secretary of state, Condoleezza Rice, of Canada's no to missile defence at the summit. No time had been set aside for a bilateral meeting with Bush, which would have provided the prime minister with an opportunity to personally inform the president – an oversight that proved somewhat embarrassing after the fact. Indeed, to add to the situation, on the day Pettigrew informed Rice, Frank McKenna, Canada's new ambassador to Washington, following his testimony before the House Committee on Foreign Affairs and International Trade, publicly stated that Canada was already a participant in the US missile defence program. In so doing, he simply confused the NORAD early warning agreement of August 2004 with actual participation – a distinction that had been central to Canadian policy for over three decades.

Prior to making his public statement, McKenna had met with Martin to discuss Canada-US relations, as is the norm for any new ambassador. Strangely, missile defence was not one of the issues discussed, even though Martin felt that the issue was having a negative impact domestically on his ability to deal with more pressing and important issues. The prime minister did not inform his ambassador that the decision to say no had been made. Subsequently, McKenna was briefed by the Privy Council Office, with officials apparently making the distinction about participation clear. Whether McKenna misunderstood or acted out of a different agenda remains open to speculation. Regardless, his statement generated confusion about the government's position and a round of questions in the House in the following days leading up to Pettigrew's statement.

McKenna's statement did not drive the Martin decision to make a public statement in the House rejecting Canadian participation in ballistic missile defence. Martin had to make a public statement sometime once the United States had been informed. It would and could not have been kept secret. At best, then, it drove the timing of the announcement to put an end to the debate inside and outside the House. McKenna was not alone in being unaware of the Martin decision. The bureaucracy had also not been informed beforehand, and Martin made no attempt to obtain senior bureaucratic advice before the final decision or about the mechanics of the private and public announcement. He had given up on the bureaucracy months earlier. There is little doubt that, if the bureaucracy had been informed, senior officials would have counselled a cautious public announcement perhaps structured around the specific MOU that had been on the table a year earlier to avoid possible embarrassment and confusion. Indeed, with a little imagination, such an announcement could have said no without actually saying no. But this was policy made by the prime minister and the Prime Minister's Office, and Martin's actions were little different from Mulroney's failure to inform the bureaucracy of his early decision on SDI.

In the end, image is everything in politics, and observers could not help but be surprised and puzzled by Pettigrew's statement and even more. Not only did the prime minister shortly thereafter reiterate his view that Canada should have a seat at the missile defence table, but the foreign minister suggested that Canada would still be able to reap economic benefits from missile defence. Whatever benefits might have resulted from either saying yes or nothing evaporated overnight not necessarily because the United States would close the door (though this was likely simply because Canadian firms would not have access to the secret information necessary to bid on missile defence contracts) but because Canadian firms would simply not bother. Canadian firms had long believed that, in the absence of official Canadian government participation, they would not succeed and thus had no incentive to invest.

The Martin decision did not completely close the missile defence file, even though it had no appreciable impact on the political drama surrounding the minority government in May or on the election that followed a little less than a year later. Shortly after the announcement, one of the prime US contractors for missile defence, Raytheon, publicly raised the issue of an X-band radar site on Canadian soil at Goose Bay, Labrador.[64] The question for Raytheon was whether the no applied to the possibility of a tracking, cueing, and battle-damage assessment radar, which would prove useful, as NORAD had suggested in 1991, especially if the United States proceeded to deploy interceptors in the northeastern United States. The age-old issue of the meaning of participation thus resurfaced, if only briefly. One could argue that this radar would be part of the early warning mission and also contribute to the Space Surveillance Network and civil space. Moreover, such a radar would give purpose to Goose Bay, which had long been an expensive redundant military base. Regardless, Martin having said no to participation was unlikely even to consider such a request. To do so would smack of duplicity and raise fears still held by some in the public that more was going on with missile defence than met the eye – that McKenna had in fact gotten it right.

Anyway, there was no need for the government to tackle the question, as no formal request from Raytheon or the US government resulted. With many technical and operational hurdles still to overcome at Fort Greely, time again was available, and the value of Canadian territory might offer a chance to revisit GMD when the environment was right and the United States was in need. Besides, negotiations with Poland (interceptor site) and the Czech Republic (radar site) were nearing completion. This also put any urgency for a Canadian site much further down the road. Besides, the United States had not raised the issue of such a potential Canadian contribution during the 2003 discussions. Nonetheless, this future possibility as hinted at by Raytheon was dangerous. If the United States truly believed that such a radar was essential to its security, another Canadian no might be disastrous for the relationship. It is one thing when the no is said in a diplomatic manner and, hopefully, has no strategic significance to the defence of the United States; it is another thing when the no is strategically significant – a situation that every government had been very sensitive to.

Following the announcement in the House, the prime minister formally called Bush, to no effect. Later on and for the last time, Martin raised the decision with Bush at the trilateral Canada-US-Mexico summit in Texas. He told the president that, because he couldn't get answers to his two key questions, he felt he had to say no in order to get on with more important bilateral issues. Bush simply replied "fine," and the issue was dropped. At least in dealing with Martin, Bush was not a man of lengthy discussions. Things were different, however, on the defence front. Even though a phone call from Graham to Rumsfeld, received by Deputy Secretary of Defense Paul Wolfowitz, did not indicate any US displeasure, the reality was the opposite.[65] Rumsfeld was furious with the decision, noting that the United States had not asked for anything and there were no negotiations under way. At a subsequent meeting with the United States, the senior US defence official starkly complained about the manner in which the decision was made and announced. He directly linked it to the modality the previous government had employed when saying no to Iraq in 2003. Furthermore, the United States felt insulted that it was being used as a crass tool to gain votes in Canada.

The apparent triumph of short-term domestic politics around an embattled government is perhaps the deserving end to the decades-long missile defence question. Pearson, Trudeau, Mulroney, and Chrétien all succumbed, albeit in different ways, to the vagaries of domestic politics on missile defence caught by latent public anti-Americanism. For Pearson and Trudeau, it was the era of US imperialism and the war in Vietnam. For Mulroney, it was Reagan the global gunslinger. For Chrétien, it was misguided US unilateralism. For Martin, it was all of the above combined with a government under siege – arguably the worst time for a government to make any significant strategic decision, particularly on a national security issue. Like his predecessors, Martin could have avoided making a decision, with at least the tacit understanding of Washington, especially with so little on the table. Indeed, he could have reiterated existing government policy going back to the 1994 White Paper and developments among the allies and NATO to justify moving forward with a research and development agreement. Arguably, such an MOU could also be used to sidestep the commitment to take the missile defence issue to the House for a vote. As in the past, it all depended upon one's definition of participation, and previous governments had implicitly attempted to define it in very narrow terms around the command and control–NORAD dimension.

In this sense, Martin's no may well be transitory, though the current Conservative minority government has shown no indication of seeking to reverse its predecessor. For Harper, like Martin, missile defence appears too dangerous domestically. For a government seeking to eliminate any images of it being too radical and pro-American/Bush, missile defence is not worth the possibility of alienating the soccer mom vote, as one official put it. Besides, the United States, always sensitive to political realities in Canada, continues to leave the door open. Nonetheless, like ABM, SDI, GPALS, and NMD before it, GMD remains a work in progress. The wait-and-see policy that has been the hallmark of the past still remains partially viable. For now, however, one aspect of the sovereign defence of Canada has been ceded entirely to the United States.

Epilogue
Forward to the Past (2006 and Beyond)

It is not difficult to imagine that sometime in the not too distant future, say 2020, an American president or secretary of defense will announce plans to develop the next layer of US missile defence – a constellation of boost-phase kinetic-kill interceptor satellites designed to defend North America, as well as essential military, civil, and commercial satellites, from a limited attack. The president would likely note in the announcement that the final layer of this integrated, multidimensional ballistic missile defence system would be used to defend not just North America but the entire world from attack and call upon all nations to join in the effort, with a promise to share US technology. The announcement, of course, would not come as a complete surprise. Sometime before the announcement, a space-based kinetic-kill test bed would have been deployed and successfully tested against a missile launch. In addition, the long-delayed Space Tracking and Surveillance System (STSS – formerly Brilliant Eyes and SBIRS-Low) would also have become operational.

If Canada's missile defence past is any indication, the Canadian script will have already been written. Naturally, Canadian officials would have been well aware for some time that the United States might move forward, and much would hinge on the outcome of the 2016 presidential election. Not wishing to be seen as favouring either candidate, despite calls from proponents of and opponents to space-based defences that a Canadian signal could influence the US debate, senior officials in National Defence and Foreign Affairs would have argued that Canada should not commit itself one way or the other. Besides, it remained unclear how the Canadian government might decide to proceed on the issue. A divided cabinet and caucus and significant opposition in Parliament could make the government's way forward difficult to predict. A decision one way or the other might prove disastrous on several fronts. Instead, it was best to keep the issue off the table for as long as possible, especially with significant disagreement between National Defence and Foreign Affairs. Even were the government inclined to make a decision, it remained to be seen how far and how fast the United States would go, owing to domestic economic issues in the United States, significant opposition to the weaponization of space, and ongoing international opposition in Europe and Asia, which might also slow down

Delta II launch of the STSS Advanced Technology Risk Reduction Satellite.
Courtesy US Missile Defense Agency

development or force the administration to reconsider. In any event, the United States had privately signalled that the system would not require any Canadian resources and that Canada would be welcomed to participate in a manner that it defined and initiated within the broad scope of US global security interests. Nonetheless, one could expect that notions of US pressure on Canada to participate would continue to circulate.

Whether the United States proceeds with a formal space-based kinetic-kill or alternative exotic-technology missile defence layer remains to be seen. If the past record of the ebb and flow of US missile defence efforts is any indication, any attempt to speculate on the date of such an announcement, assuming that the United States takes this next step, is extremely problematic. Moreover, even if such an announcement were to occur, the history of the US missile defence effort serves as a cautionary tale on whether set deadlines can be achieved or are even achievable. Missile defence may have seemed inevitable once ballistic missiles emerged on the strategic scene, but the combination of technology, costs, and internal and international politics made the "inevitable" a project that has taken more than four decades. These same factors will naturally determine future developments, which may eventually take missile defence to space.

Since the Martin decision in February 2005, US missile defence capabilities have continued to grow at a relatively gradual pace as informed primarily by technological developments. From an initial alert capability in 2004, on June 20, 2006, the system consisting of nine interceptors at Fort Greely and two at Vandenberg was declared fully operational. By the end of the Bush administration, twenty-one interceptors were in place at Fort Greely, and three at Vandenberg, with plans to increase them to forty-four. Three of the four intercept tests since 2005 had been successful by the spring of 2009.[1] In addition, the US naval and ground theatre systems became operational. The naval Aegis theatre system witnessed thirteen out of sixteen successful tests since 2005, including the intercept of two warheads in flight simultaneously. Reflecting the missile defence-ASAT linkage, on February 21, 2008, the Aegis missile defence system successfully destroyed a failed deorbiting US intelligence satellite. In March 2009, the US Navy announced the creation of the Navy Air and Missile Defense Command.[2] Currently, sixteen Aegis missile defence ships have been deployed with Pacific Command and two with Atlantic Command, with plans to convert another three.[3] With six successful tests since July 2006, the first battery of the US Army's Terminal (formerly Theater) High Altitude Area Defense (THAAD) was activated on May 28, 2008, and expectations were of an initial operating capability sometime in 2009. The remaining US missile defence development

US Airborne Laser.
Courtesy US Missile Defense Agency

program, the boost-phase Airborne Laser, saw the integration of the system on its platform, a Boeing 747, and a successful ground test firing in 2008. The first airborne test against a ballistic missile has been currently planned for 2009.[4]

The proliferation of weapons of mass destruction and ballistic missiles remain the primary driver behind the multilayered US missile defence effort, North Korea and Iran being the two primary states of concern. In 1998, the first long-range North Korean missile test had legitimized the findings of the Rumsfeld Commission. In July 2006, the second North Korean test failed on the launch pad. At the time, it was rumoured that forward-deployed Aegis systems were prepared to attempt an intercept. In April 2009, a third North Korean test supposedly designed to place a satellite in orbit also failed.[5] By June, following a successful North Korean nuclear test, bellicose North Korean pronouncements in response to international condemnation and Security Council Resolution 1874, and reports that North Korea was preparing for another long-range missile test for July 4, the United States increased its missile defence capabilities around Hawaii, including the deployment of a THAAD battery as a precautionary measure.[6]

While Iran has not been as bold as North Korea to date, the international community has failed in its efforts to block its nuclear energy development program, which at least allows Iran to retain a credible nuclear weapons option. As far as its ballistic missile program is concerned, on February 3, 2009, Iran launched its first satellite into orbit, demonstrating a fledgling long-range ballistic missile capability.[7] Even if one were to set aside Iran's anti-Israeli rhetoric and the tensions it has created throughout the region, Iran's growing capabilities threaten Israel and pose a threat to the United States (whose GMD system is optimized for North Korea) and Europe. There have been no indications that the United States will move to create a second GMD site in the northeastern United States, with its potential implications for Canada, and the brief US interest in Goose Bay supporting radar as proposed by Raytheon in 2005 came to naught. Instead, in July 2008, the United States signed an agreement with the Czech Republic to base an X-band radar there, and in August 2008, after a year and a half of negotiations, the United States and Poland signed an agreement to deploy ten US GMD interceptors in Poland in return for Poland acquiring the Patriot PAC-3.[8] Moreover, five months earlier at the NATO Bucharest Summit, the final communiqué recognized "the substantial contribution to the protection of Allies from long-range ballistic missiles to be provided by the planned deployment of European-based United States missile defence assets. We are exploring ways to link this capability with current NATO missile defence efforts as a way to ensure that it would be an integral part of any future NATO-wide missile defence architecture."[9]

Not surprisingly, NATO developments and particularly the US-Polish agreement met with significant Russian opposition. From the onset of US-Polish negotiations, President Putin and senior Russian officials argued that a forward-deployed missile defence battery in Poland represented a threat to the Russian strategic nuclear deterrent. Officials hinted that, if the deployment occurred, Russia might withdraw from the 1987 Intermediate-range Nuclear Forces (INF) Treaty, which prohibits the deployment of intermediate-range ballistic missiles – at that time primarily the Russian SS-20 and US Pershing II missiles.[10] At the same time, Putin at the G-8 summit on June 7, 2007, offered the outdated Russian early warning radar in Azerbaijan as an alternative to the Czech site, which the United States rejected on technical grounds because it is too close to Iran and would be unable to provide in-flight tracking once a missile overflew the radar.[11] Nonetheless, in response to Russian objections and concerns, President Bush proposed increased cooperation with Russia on missile defence and transparency measures for the future missile defence site in Poland. This was echoed by NATO at the Bucharest Summit: "We are committed to maximum transparency and reciprocal confidence building measures to allay any [Russian] concerns. We encourage the Russian Federation to take advantage of United

States missile defence cooperation proposals and we are ready to explore the potential for linking United States, NATO and Russian missile defence systems at an appropriate time."[12]

The future pace and scope of US missile defence will naturally hinge upon the level of commitment from the new Obama administration and a Democrat-controlled Congress. In the lead up to the 2008 election, the two candidates' platforms reflected the traditional differences between Republicans and Democrats on missile defence. Senator McCain, echoing Reagan, called for an America that "should never again have to live in the shadow of missile and nuclear attack."[13] He sought the continued development of "effective" defences against rogue states and "potential strategic competitors like China and Russia." In contrast, the Obama platform, echoing the policy of the Clinton administration, called for a pragmatic, cost-effective missile defence effort that would not divert resources from other priorities until the technology is proven.[14]

Not surprisingly, after Obama's election in November, the new administration applied the brakes to plans to increase the number of GMD interceptors and cut spending on missile defence by $1.2 billion.[15] The spending cuts targeted the GMD system (no increase in interceptors), new hit-to-kill technology, ABL development, and the planned site in Poland. Although rumours emerged of a secret deal with Russia in which the United States would scrap the Polish site in return for increased Russian support on Iran, the administration initially has kept the option open to deploy sometime in the future.[16] The administration's more cautious approach was reflected in the pronouncement on missile defence at the most recent NATO summit in Strasbourg and Keil: "We recognize that additional work is still required. In this context, a future United States' contribution of important architectural elements could enhance NATO elaboration of this Alliance [missile defence] effort."[17]

In a dramatic and somewhat surprising step on September 17, 2009, the president cancelled the Bush administration's plans for a fixed X-Band radar in the Czech Republic and the deployment of GMD ground-based interceptors in Poland.[18] In their place, the president announced a "phased, adaptive approach," which is to consist of forward-deployed Aegis missile defences and a transportable radar by 2011. This is to be followed around 2015 by additional sea-based Aegis defences and a ground-based variant of the naval Standard Missile-3 (Block 1B variant), although the location of these ground-based interceptors has not been specified. Echoing the Clinton administration, the president argued that the Iranian threat is evolving slowly, and the priority should be based upon the greater Iranian missile threat to deployed US military forces and Middle Eastern and European friends and allies.

In addition, Obama proposed a return to formal traditional arms control negotiations with Russia, in contrast to the relatively informal approach of the Bush administration that led to the SORT agreement in 2002, though to date there has been no suggestion of revisiting an ABM-like treaty. There are also indications that the administration may reverse course on the question of negotiating an international treaty prohibiting the weaponization of space.[19]

The election of Obama has significantly changed the political atmosphere of Canada-US relations. As in the past, the election of a Democrat to the White House has psychologically eliminated the existential pressure on Canada to participate in the US missile defence effort. Ironically, Obama's popularity in Canada could pave the way for Canadian participation in terms of public support, in the same sense that public opposition to Bush made missile defence participation untenable for the Martin government in 2004-05. Regardless, from the moment of Martin's announcement, missile defence disappeared from the Canadian political agenda and, on the surface at least, did not appear to have any impact upon either NORAD or the Canadian-US defence relationship. Indeed, in 2006, Canada and the United States agreed to the indefinite renewal of NORAD.[20]

Beneath the surface, however, all has not been well, and the problems engendered by the 2005 decision were compounded by the creation of Canada Command, announced in the May 2005 Defence Policy Statement and stood up on July 1, 2005.[21] Reflecting the fallout of the decision, Rumsfeld and then commander of NORTHCOM and NORAD Admiral Keating placed the binational command directly in their sights. In so doing, Keating, acting as a combatant commander similar to the other UCP combatant commanders, began the process of downgrading NORAD by locating it as the air component command alongside the sea and land components within the NORTHCOM command structure. For all intents and purposes, NORAD was becoming subordinate to the US-only NORTHCOM. Concurrently, Canada Command assumed responsibility for all aspects of the defence of Canada – land, sea, and air. In so doing, the decision in mirroring the United States implicitly signalled a more nationalist approach to continental defence in Canada. Moreover, the relationship between Canada Command and NORAD emerged as a new problem, with both under the command of lieutenant generals (NORAD's Canadian deputy commander). Who was actually in charge of the air defence of Canada was unclear, a problem further compounded by the existence of the dual-hatted commander of Canada-NORAD Region and the 1st Canadian Air Division, colocated in Winnipeg, and reporting to both commands. This translated into the larger problem of sorting out the tri-command relationship among Canada Command, NORAD, and NORTHCOM.

Both Foreign Affairs and National Defence officials recognized from the start the problems created by the Martin decision. The decision put paid to any notion of a Canadian right or need to know about US missile defence technology, planning, and operational strategy, in contrast to the 1985 Mulroney decision where senior US and Canadian officials at least grappled with the exact meaning of a Canadian no to SDI. Moreover, the 2005 decision suggested that the real future of NORAD would be restricted to an operational air defence function, along with a role in information collection. Even then, NORAD's early warning mission appeared at risk of becoming little more than a redundancy given its reliance on US early warning assets and the inherent logic of sidestepping NORAD in favour of a direct link between US Strategic Command, which owns the assets, and NORTHCOM, which has North America as its UCP-assigned area of responsibility. Furthermore, the interrelationship between missile defence and space strongly suggested that access to developments in US military space planning might well be compromised. In effect, the no implied not only the marginalization of NORAD but also the strategic marginalization of Canada. Above all else, the Canadian no had raised serious doubts in the United States about Canada's commitment to the defence of North America, and such doubts could only serve to increase US proclivities toward unilateralist responses to the defence of North America and jeopardize the historical US preference to see North American defence in binational terms.

NORAD's status improved at least rhetorically with the resignation of Rumsfeld and the appointment of Keating's successor, air force lieutenant general Renuart in March 2007. The new secretary of defense, Robert Gates, and Renuart largely removed NORAD from its subordinate role, joining in on the fiftieth-anniversary celebrations. Roughly at the same time, General Hillier, the chief of the defence staff, and General Pace, the chairman of the Joint Chiefs of Staff, agreed to establish a bilateral tri-command study to sort out the command question. As was traditional in the relationship, this meant a military-led functional, tactical approach. The only strategic guidance provided by both the minister of defence and secretary of defense was to do no harm to NORAD. Exactly what this means beyond the symbolic value of the institution remains to be seen.

Having recognized the problem of the missile defence no, the issue for some National Defence officials was how to get the Harper minority government to reverse course at least symbolically. Although National Defence officials believed there was sentiment within the government to do so, political realities as seen through the eyes of the Prime Minister's Office had made missile defence too dangerous. Consideration was given to re-engage the United States by dusting off the defunct research and development memorandum of understanding

(MOU) on missile defence that had been on the table in 2004, with the possibility of integrating the US MOU into a broader multilateral set of research and development MOUs with key European allies and the NATO missile defence effort. In so doing, officials recognized the irony here: if NATO developed and deployed a strategic missile defence capability integrated with a US European site or naval systems, Canada would stand out as an anomaly within the alliance – a non-participant in North America yet engaged in Europe through NATO, with the possibility of Canadian funding of the NATO effort through the NATO common fund. Members of Pearson's cabinet had wondered about the domestic impact of the United States possessing a defence with Canada on the outside, and in the future the Harper or successor government might do the same. Moreover, engaged in the NATO effort, at least on the basis of Canadian agreement to the missile defence, paragraphs of the NATO Bucharest Summit communiqué smacked of potential duplicity – if the government was not participating in the US missile defence program, why was it participating in the NATO effort? Regardless, the effort came to naught, and, with the downplaying of missile defence in the most recent communiqué in light of current US policy under Obama, any pressing need for the government to revisit the missile defence question has subsided for the time being.

Nonetheless, the United States has quietly left the door open for a Canadian initiative. All that may be needed is a majority Conservative government. Ironically, there may be little to support this expectation. Certainly, there has been strong support within the party for participation, especially among the old Alliance/Reform element, and Conservatives have long been proponents of defence, at least rhetorically. Yet Mulroney, working with two large majorities, faced divisions within his cabinet and caucus and sought to finesse the SDI question. In the political arena, even a majority government may well wonder what is truly to be gained from a policy reversal, even though a quick decision to reverse will ensure that the issue is either long forgotten or lost in an election dominated by the traditional set of socio-economic concerns, especially after the 2008 economic meltdown. But then a quick decision to reverse in the spring of 2006 after the Harper victory would have also likely seen the issue disappear. Indeed, it would have been as strange for the Harper government to fall over the relatively innocuous MOU on research and development, with a subsequent election fought over the issue, as it would have been if Martin had chosen a different course. Indeed, the entire missile defence saga prompts one to question what appears to be a strange reality: if defence, including missile defence, doesn't matter domestically, why do politicians act as if it does?

More importantly, the 2003-04 negotiations indicate that the United States is unlikely to give Canada privileged access to and input into the missile defence of North America. It is possible that an initial research and development MOU may lead in the future to a greater role for Canada by cementing its bona fides in being committed to North American defence. Nonetheless, Canada missed repeated opportunities to ensure a prominent role in North American missile defence. Indeed, the story of Canada and ballistic missile defence is not only déjà vu but also repeated lost opportunities to protect Canadian strategic interests.

At the onset of ABM in the 1960s, the opportunity existed for Canada to ensure access to an operational US missile defence system for the defence of North America and thus the defence of Canadian cities. The primary means was to signal Canada's interest in, and support for, assigning command and control of the system to NORAD. Even though it would take another forty years before an operational system was actually deployed, a commitment early on would have established a policy precedent and, possibly, have put the missile defence question to rest decades ago. Certainly, each Canadian government feared that such a commitment, despite domestic political considerations, might generate pressures from the United States to place interceptors on Canadian soil and to get Canada to invest significant amounts of money.[22] However, in the early days, not least because of US political dynamics on ABM (the half-hearted commitments of McNamara and Nixon), the possibility existed to set the foundation for a firm long-term policy – yes to research and development, early warning, and command and control but no to interceptors on Canadian soil and significant money outside the traditional funding model of NORAD. In other words, the opportunity to define the meaning of participation according to Canadian interests was lost – not once but repeatedly, from ABM to GMD. Even with the ABM Treaty, Canada still had the opportunity to define the meaning of participation. Indeed, the treaty's real safety blanket was not keeping missile defence at a distance but providing a mechanism for Canada to participate without participating by removing the option or possibility of US pressure on deploying interceptors on Canadian soil and limiting demand for Canada to invest.

By the time of the Bush decision in December 2002 to deploy GMD, the United States had unilaterally decided the meaning of participation, and this did not include a Canadian role in the command and control of the system. Even so, a yes would still have provided some strategic value in obtaining at least some form of access to US missile defence developments and operational plans, with the door open to the possibility of influence and a greater role in the future. At least Canadians might have been able figuratively to stay in the missile defence room. Even more, a Canadian yes would have had significant psychological value in demonstrating Canada's commitment to North American defence.

Even though successive prime ministers, ministers, cabinets, senior Defence and External/Foreign Affairs officials, including the military, saw little value and utility in defending against ballistic missiles, doubted that the technology would ever work, and feared the impact of missile defences on international stability, their US counterparts to a lesser or greater degree saw missile defence as significant for various strategic and political reasons. In avoiding a decision for decades – what became by 2000 tantamount to a no and then eventually saying no to missile defence – Canada actually said no to something much greater. It was a no to the primacy of the defence and security in US thinking.

In many ways, Canada's dithering and then its subsequent no demonstrated a divorce from reality. As a small, weak, middle, or secondary power – take your choice – Canada could do little but react to the defence initiatives of its superpower partner. Despite Canadian hubris that its decision on missile defence mattered internationally – the idea that a Canada by saying yes or no would grant legitimacy or illegitimacy to the US program – the reality was the opposite. The no, except for its impact on short-term Canadian politics and long-term Canadian strategic interests, was simply irrelevant.[23] Certainly, many suggested over the decades that the Canadian government publicly engage in US defence and security debates in an attempt to influence the United States. Paradoxically, this suggestion implies that Canada can influence the outcome of these debates, even though the idea that Canada can influence the United States through such institutions as NORAD has been regularly discounted. Regardless, Canadian governments have been understandably reluctant to engage the United States publicly, especially during US presidential elections, fearing possible repercussions if the nation found itself on the wrong side and recognizing that Canadians deeply resent any US engagement in Canadian debates.

In the end, the primary issue for Canada should not have been cast as principally an international question but a national one. With the United States slowly but certainly moving forward over time – which should have led to the conclusion that missile defence was inevitable – the question was simply assessing the implications of being in or out. Even if critics were right that missile defence will generate an arms race, destroy all prospects for nuclear disarmament, undermine Western non-proliferation goals, and create an unstable international system, the unavoidable fact is that this would have occurred whether Canada is on board or not. Faced with the inevitability of a US missile defence system for North America, the strategic logic that has implicitly or unconsciously guided Canadian defence policy since the end of the Second World War dictated Canadian participation in some form or guise.

This strategic logic can be summarized in terms of five principles based upon the reality of geography that dictates that the United States must defend Canada in some way in order to defend itself. First, given the long-standing reluctance of Canadian governments to provide sufficient funds to defend the nation on its own, Canada has long sought to expropriate US defence resources for its own interests. Second, the process of expropriating US defence resources raises issues of Canada's sovereignty – the solution being to manage these through close cooperation with the United States, ensuring that Canadian sovereignty concerns are voiced and accounted for in the US planning process. Third, close cooperation provides key points of access to US defence planning not just for North America but globally – information that otherwise would be either unobtainable or completely dictated by the United States.[24] Fourth, access gives opportunities for Canada to insert its defence interests into the US process, with the potential to influence US decisions. Finally, to close the loop, Canadian cooperative engagement reassures the United States of Canada's commitment to the defence of North America, thereby reducing the inherent internal pressures for the United States to act unilaterally.

Canadian indecision on missile defence has consistently harmed Canadian interests. Canada is unable to expropriate US missile defence resources relative to its own defence interests. Canadian sovereignty is undermined by ceding the defence of Canadian cities to the United States. Canada has at best limited access to US missile defence operational planning and US research and development. There are few, if any, means for Canada to ensure its interests are injected into the planning process and obtain opportunities to influence the operational system. Finally, Canada has fed US unilateral proclivities by undermining its commitment to the joint defence of North America.

The net result is an ongoing transformation of the defence relationship with the United States and arguably the collapse of the fundamental strategic principles that have guided Canadian defence and foreign policy since the end of the Second World War. This does not mean that the Canada-US defence relationship will end. Indeed, on the surface, both the United States and Canada will continue to pay homage to their close relationship. Besides, neither Canada nor the United States, for reasons of geography, economic interdependence, shared interests, and values, and the close working relationship between the military services, can completely break with the past. These reasons, in turn, feed a Canadian complacency and attitude that all is fine, and the missile defence decision did little, if anything, of substance to the relationship.

Certainly, missile defence in the greater scheme of everyday politics was not a very important issue and more often than not took a backseat on the national political agenda. It was a political irritant that successive governments would have preferred to ignore. However, this preference ensured that the missile defence file would be mishandled. Despite the best attempts of National Defence to frame the issue in terms of Canadian strategic interests, the message never truly resonated among political elites. Instead, the issue simply became framed around the future of an institution, NORAD – as if saving the institution ensured Canadian strategic interests were protected. NORAD became conceptualized as an end in itself rather than a means to an end.

Even the fundamental responsibility of the government to defend its citizens in an uncertain and dangerous world – when the opportunity arose at little cost to Canada – was largely forgotten or ignored throughout most of the missile defence saga. Instead, political elites unconsciously accepted Trudeau's view – "I do not want, in other words, to negotiate with the United States for a little more protection for Canada, if I feel that that protection is such that it might entail more danger for the whole world."[25] The Canadian foreign policy principle, reiterated by successive governments, that Canada's security is identical to international peace and security was more than just rhetoric. It was, for good or ill, a view deeply imprinted, like Konrad Lorenz and his baby geese, into the Canadian elite psyche. The dominance of international considerations in the Canadian attentive public debate came at the expense of Canadian strategic and national security interests.

At the end of the day, the familiar notion that a week is a long time in politics was always near the surface. When combined with no strong sense of Canadian strategic interests at play, viewing the issue in terms of preserving NORAD as an institution, the belief that Canadian interests were synonymous with the interests of everyone else in the international system, and living in an environment of latent anti-Americanism masquerading as Canadian nationalism, it is not surprising that time and time again Canadian governments failed to provide leadership on the missile defence file. This result has been part and parcel of an overall drift toward the marginalization of Canada in North American defence. Of course, if the nationalist proclivities of the current government, as most clearly demonstrated by the current obsession with Arctic sovereignty, are met with sustained investments, perhaps then Canada will be able to make a significant national contribution to North American defence and generate conditions for the United States to take Canada seriously. This would potentially reinvigorate the relationship, albeit on different grounds than in the past. Unfortunately, Canadian history provides little evidence of such rhetoric being translated into practice in the long term. With the economic downturn that began in 2008 likely to produce a sharp decline in government revenues, it is more likely that the current government, like its predecessors, will turn to defence cuts, knowing full well that these will incur few domestic political costs.

The fundamental problem for Canadians at the end of the day is the attitude expressed in the 1920s by Senator Dandurand, which continues to resonate within the Canadian mindset – Canada is a "fire-proof house far away from inflammable materials" – along with the notion that the United States will defend Canada regardless.[26] Despite their self-image and rhetoric, Canadians remain more or less isolationist. Canadian overseas engagement will continue to be tolerated as long as the costs remain relatively low. Governments will continue to trumpet Canadian internationalism in some form or guise, and Canada's unique contributions and place on the world stage, despite the reality of being a bit player. The services will continue to privilege overseas operations to the neglect of North American and national missions, while, from an operational perspective, drawing ever closer to their counterparts in the United States. The bureaucratic struggles between Foreign Affairs and National Defence will continue, especially in the absence of clear government leadership, and both will seek to define and defend Canadian interests in somewhat of a policy vacuum. An overwhelmingly uninterested public will remain the rhetorical whipping boy being used to justify the preferences and predilections of elites, scared by the "punditocracy," who generally get it wrong.

These elements, unless shattered by some unpredictable cataclysmic event, will continue to underpin a Canadian policy of avoidance and inaction. But by the next decade, Canada will face an international environment in which offensive and defensive strategic systems coexist.[27] Sometime in the future, Canada will confront the most significant strategic development of the twenty-first century: the defence and security issues surrounding the weaponization of outer space as the technology to exploit space cost-effectively evolves. The weaponization of space, both for defence against ballistic missile attack and protection of vital military, commercial, and civilian satellite services, has the same air of inevitability that missile defence had the day after the first German V-2 attacks. The political pressure to deal with it already exists, regardless of the technology. Canada's response will be déjà vu all over again. The future is known – it is just a matter of waiting for the details to be filled in.

Notes

Prologue

1 Douhet is generally seen as the first theorist of airpower, despite H.G. Wells' 1908 visionary novel of future warfare, *The War in the Air* (London: W. Collins and Sons, 1921). He is thus representative of the new vision. Many others came to similar conclusions around the same time without having read Douhet. For a basic overview, see David MacIssac, "Voices from the Central Blue: The Air Power Theorists," in *Makers of Modern Strategy from Machiavelli to the Nuclear Age*, ed. Peter Paret (Princeton, NJ: Princeton University Press, 1986), 624-47.

2 German air forces did strike at Warsaw and Rotterdam, but this was in the context of Germany's invasion of both countries. For a discussion of interwar thinking, see George Questor, *Deterrence before Hiroshima* (New York: Transaction, 1986).

3 Some analysts point out that German production in particular rose significantly until the final stages of the allied strategic bombing campaign. However, Hitler had purposely decided to provide guns and butter to the population out of concern that domestic support might evaporate, as it had done in the First World War. The strategic bombing campaign began roughly at the same time as the full mobilization of the German economy, and it is difficult to estimate how much the economy might have been able to produce in the absence of the allied bombing campaign.

4 Richard Overy, *The Air War 1939-1945* (London: Stein and Day, 1981).

5 Brodie stated: "Thus far the chief purpose of a military establishment has been to win wars. From now on its chief purpose must be to avert them. It can have no other purpose." See Bernard Brodie, *The Absolute Weapon: Atomic Power and World Order* (New York: Harcourt, Brace, 1946), 76.

6 The US Airborne Laser program began in the mid-1990s. It consists of a chemical oxidize laser to be deployed on a Boeing 747. It is estimated to have the range of around five hundred kilometres. By focusing the laser beam on one of the three booster stages, the integrity of the outer missile shell would be compromised, resulting in the missile and its warhead falling back to earth. The first airborne test of the laser in the air is scheduled for 2010.

7 During the 1991 Gulf War, an unintended penetration aid for the Iraqi modified mid-range Scuds was their poor engineering. The missiles broke up upon re-entry in the terminal phase, providing too many targets for the first generation Patriot-2 interceptors.

8 In the Trudeau-commissioned foreign and defence policy review, the Standing Committee on External Affairs and National Defence (SCEAND) received several briefs and testimony on the option of neutrality and non-alignment, which would be rejected in the committee's final report to the House. See Library and Archives Canada, Ottawa, House of Commons SCEAND, Minutes of Proceedings and Evidence, no. 27 (February 1969); no. 28 (February 18, 1969); and no. 35 (March 26, 1969).

9 The standard history of Canada-US air defence cooperation leading to the creation of NORAD remains Joseph Jockel, *No Boundary Upstairs: Canada, the United States, and the Origins of North American Air Defence, 1945-1958* (Vancouver: UBC Press, 1987).

10 Peyton Lyon, "Defence and Foreign Policy," *Special Studies Prepared for the Special Committee of the House of Commons on Matters Relating to Defence*, Supplement, Canada, House of Commons (Ottawa: Queen's Printer, 1965).

Act 1: Anti-Ballistic Missiles

1 The actual location was just outside Cavalier at Nekoma, North Dakota. The phased array radar that remains is named for Cavalier.

2 Department of National Defence, *General Air Staff Requirement for a Defence System to Counter the Intercontinental Ballistic Missile,* November 10, 1954. Library and Archives Canada. RG #24.

3 Department of National Defence; Report on Ballistic Missile Defence to the Chiefs of Staff Committee; Defence Research Board; September 1958. Library and Archives Canada. RG #24.

4 Department of National Defence; Project Helmut – Defence against Ballistic Missiles: Memorandum to Acting Chief of the Naval Staff; January 12, 1961. Library and Archives Canada. RG #24.

5 Department of National Defence; Report on Ballistic Missile Defence to the Chiefs of Staff Committee; Defence Research Board; September 1958. Library and Archives Canada. RG #24.

6 Department of National Defence; *Defence of North America against ICBM Attack in Time Period 1960-1969;* Defence Research Board; February 1959. Library and Archives Canada. RG #24.

7 This ratio of defensive to offensive costs became prominent for a brief period during the debate on the Strategic Defense Initiative (SDI). It was raised by Paul Nitze as one of four criteria for SDI deployment and known as cost effectiveness at the margins. It also underpinned the long-standing argument that missile defences produced arms races. If offence was always cheaper than defence, then any deployment of defences was logically met with more offence. See Act 2.

8 The other radars were located at Cape Cod, Beale (California), and Cavalier (after the closure of Safeguard). As the three outer radars were fixed, these radars were designed to track objects after they are flown over the outer radars and also provide warning of submarine-launched missiles. For example, Cavalier was to track submarine launches from the Arctic up to Hudson Bay.

9 Defence Research Board; Report on Ballistic Missile Defence to the Chiefs of Staff Committee; September 1958.

10 Department of National Defence; Memorandum to JPC; Chief of the Air Staff; November 3, 1958. Library and Archives Canada. RG #24.

11 Department of National Defence; Brief to the Chairman, Chiefs of Staff, on JBMDS; Joint Ballistic Missile Defence Staff; January 15, 1960. Library and Archives Canada. RG #24.

12 See Andrew Richter, *Avoiding Armageddon: Canadian Military Strategy and Nuclear Weapons, 1950-1963* (Vancouver: UBC Press, 2002).

13 Department of National Defence, "Armaments and Modern Weapons," *Special Studies Prepared for the Special Committee of the House of Commons on Matters Relating to Defence*, Supplement (Ottawa: Queen's Printer, 1965). Library and Archives Canada. RG #24.

14 See John Clearwater, *Canadian Nuclear Weapons* (Toronto: Dundurn Press, 1998).

15 On Canada and the Cuban Missile Crisis, see Peter Haydon, *The Cuban Missile Crisis Reconsidered: Canadian Involvement Reconsidered* (Toronto: Canadian Institute for Strategic Studies, 1993); on Diefenbaker's relationship with Kennedy, see Lawrence Martin, *Presidents and Prime Ministers: Washington and Ottawa Face to Face: The Myth of Bilateral*

Bliss (Toronto: Doubleday, 1982). See also Knowlton Nash, *Kennedy and Diefenbaker: The Feud that Helped Topple a Government* (Toronto: McClelland and Stewart, 1990).

16 Both gaps were in fact myth. Estimates of the size of the Soviet bomber fleet were way off the mark. In 1960, Kennedy exploited the missile gap in the presidential election campaign against Nixon. Nixon was well aware of the real situation of a marked US superiority, but as this information was top secret he did not divulge it during the campaign. It was not until the end of the decade that the gap between US and Soviet strategic forces had narrowed to a point of relative parity after the US decision to freeze the growth of its forces paving the way for the beginning of arms control talks.

17 Estimates of the accuracy of missile warheads are expressed as circular error probability, and these estimates partially determine how many missiles and warheads are needed. Circular error probability is the estimate of the distance or radius between the actual target and the point at which 50 percent of the warheads are likely to land.

18 According to Herken, this decision was also motivated by a lack of USAF interest in continental air defence. Gregory Herken, *The Counsels of War* (New York: Oxford University Press, 1987), 187.

19 Paul B. Stares, *The Militarization of Space: US Policy, 1945-1984* (Ithaca, NY: Cornell University Press, 1985), 108.

20 Although little attention was paid to civil defence measures, the logic opposing ABM naturally implicated civil defence as undermining MAD. Whereas the Soviet Union continued extensive civil defence efforts, the United States and the West basically quit the effort mostly on the ground that it was of little utility in the case of a thermonuclear war and thus not worth the investment for the population as a whole.

21 Wohlstetter was the author of the Rand Corporation's major study on Strategic Air Command basing in the early 1950s. He would also become a major supporter of ABM in the 1960s and 1970s. Gregory Herken, *The Counsels of War* (New York: Oxford University Press, 1987), 88-94.

22 In addition to Herken, a valuable overview of the intellectual debates is found in Fred Kaplan, *The Wizards of Armageddon* (New York: Simon and Schuster, 1987). It is difficult to trace the origins of the "whiz kids" label, but it came to academic prominence after the release of David Halberstam's *The Best and the Brightest* (New York: Random House, 1972).

23 Henry S. Rowen, "The Evolution of Strategic Nuclear Doctrine," in *Strategic Thought in the Nuclear Age*, ed. Laurence Martin (Baltimore: Johns Hopkins University Press, 1979), 148.

24 McNamara's assured destruction criteria consisted of the capability to destroy 20 to 25 percent of the Soviet population and 50 percent of its industrial capacity on each leg of the strategic triad. Ibid., 146.

25 Among the thinkers on the possibility of limited nuclear war were Henry Kissinger and Thomas Schelling. See Henry Kissinger, *Nuclear Weapons and Foreign Policy* (New York: Doubleday, 1958), and Thomas Schelling, *The Strategy of Conflict* (New York: Oxford University Press, 1963).

26 The Lisbon Force goals were premised on an estimate that effective conventional defence and deterrence required the allies to deploy ninety-six divisions, of which roughly half would be combat ready. These goals were simply unrealistic and far beyond the capabilities of the Western Europeans, who were still recovering from the devastation of the Second World War. The failure to meet the Lisbon goals propelled the allies to rehabilitate and rearm West Germany, which became a member of the alliance in 1955. Jane E. Stromseth, *The Origins of Flexible Response: NATO's Debate over Strategy in the 1960s* (New York: St. Martin's Press, 1988), 11.

27 NORAD was operationally established in the fall of 1957 but formally consummated by the political leadership with the exchange of notes in May 1958.

28 In a 1965 study for the House of Commons Special Committee, National Defence in making the case for arming Canada's air defence force with nuclear weapons noted that the Germans in the Second World War destroyed one bomber for every 3,343 heavy anti-aircraft shells fired, and the kill probability for the Bomarcs and Voodoo interceptors without nuclear weapons was not much better. Department of National Defence, "Armaments and Modern Weapons," *Special Studies Prepared for the Special Committee of the House of Commons on Matters Relating to Defence,* Supplement (Ottawa: House of Commons, April 1965), 173.

29 On the evolution of Flexible Response, see Jane E. Stromseth, *The Origins of Flexible Response: NATO's Debate over Strategy in the 1960s* (New York: St. Martin's Press, 1988).

30 Gregory Herken, *The Counsels of War* (New York: Oxford University Press, 1987), 189.

31 The groundbreaking conference on arms control was held in Boston in the summer of 1960 and sponsored by the American Academy of Arts and Sciences. The proceedings were published in the academy's fall 1960 issue of its journal, *Daedalus,* and in an edited volume by Donald Brennon, *Arms Control, Disarmament, and National Security* (New York: Braziller, 1961).

32 Colin Grey, *Modern Strategy* (London: Oxford University Press, 1999), 234.

33 Arms control negotiations took off in the mid-1960s, leading to the Outer Space Treaty (1967), the Non-Proliferation Treaty (1967), the Agreement on Measures to Reduce the Risk of Outbreak of Nuclear War (1971), Strategic Arms Limitation Treaty (SALT I 1972), and the ABM Treaty (1972), with the last three considered as constituent components of the SALT basket. The other centrepiece was the normalization of relations between West and East begun with West German abandonment of the Hallstein Doctrine, which had forbidden any formal diplomatic relations with any country recognizing East Germany as a sovereign state, and culminating with the negotiation of the Helsinki Accords in 1975.

34 North Atlantic Treaty Organization, *The Future Tasks of the Alliance,* Brussels, December 14, 1967.

35 Alexi Kosygin quoted by Dr. Sherman in testimony to the Standing Committee on External Affairs and National Defence, Minutes of Proceedings and Evidence, no. 48, June 10, 1969, p. 1719.

36 At the time, 1973 was identified as the initial date of a Chinese operational ICBM capability. Allen S. Whiting, "The Chinese Nuclear Threat," in *ABM: An Evaluation of the Decision to Deploy an Antiballistic Missile System,* Abram Chayes and Jerome Wiesner, eds. (New York: Signet Books, 1969), 162.

37 Department of External Affairs; Memorandum to the Secretary of State for External Affairs; Under-Secretary of State for External Affairs; February 4, 1964.

38 Department of External Affairs; Letter to the Minister of National Defence; Secretary of State for External Affairs; March 9, 1964.

39 Canada, *White Paper on Defence* (Ottawa: Queen's Printer, 1964), 14.

40 Ibid.

41 Department of External Affairs; Internal Memorandum; Defence Liaison Division; July 10, 1965.

42 Department of National Defence; Talking Paper; Division of Continental Policy; November 5, 1965.

43 In the margin of the paper, an unidentified handwritten note recognizes that this didn't bother anyone, as a result of existing deployments of the Nike-Hercules surface-to-air

missiles (SAM) around cities and similar actions by Canada. The latter likely refers to the Bomarc-B SAM equipped with a nuclear warhead.

44 Operation Rolling Thunder, the limited-controlled bombing campaign of North Vietnam, was still in its initial stages. Two years later, there were half a million US personnel prosecuting the war, and public opposition in Canada and the United States had become a significant political force. Although significant anti-American forces on the left in Canada had existed for many years, from the lament of George Grant to the branch-plant economy perspective of the NDP, their coalescence into a major political force that the Liberal government could not easily ignore would be intimately tied to the war in Vietnam. The early 1960s consensus image of the United States as the defender of democracy and freedom came to be contested by the late 1960s image of the United States as an imperialist warmonger.

45 As far as the Chinese part of the equation was concerned, the Europeans were fully aware of the administration's ABM proposal to Moscow (and implicitly linked with Harmel). China was America's problem.

46 Department of External Affairs; Talking Paper; Division of Continental Policy; November 5, 1965; p. 3.

47 Ibid.

48 The navy also began research on the Early Spring ASAT, employing a modified Polaris SLBM. For an extended discussion, see Paul B. Stares, *The Militarization of Space: US Policy, 1945-1984* (Ithaca, NY: Cornell University Press, 1985), 106-31.

49 The Cold Lake camera was deployed in 1961 and retired in 1981. St. Margaret's Bay became operational in 1976 and closed in 1993. Since then, Canada has provided no hardware contribution to the US Space Surveillance Network, although it plans to deploy a space-based optical sensor under Project Sapphire. See Act 5.

50 Privy Council Office; "NORAD," Cabinet Conclusions; May 22, 1965; p. 10.

51 Department of External Affairs; Note to Washington and NATO-Paris; May 28, 1965.

52 Department of External Affairs; Internal Memorandum; Defence Liaison Division; May 28, 1965.

53 Ibid., July 10, 1965.

54 Department of External Affairs; Aide-Mémoiré on NORAD; July 26, 1966.

55 Department of External Affairs; Memorandum to the Secretary of State for External Affairs; Under-Secretary of State for External Affairs; May 2, 1966.

56 Department of External Affairs; Internal Memorandum; September 23, 1966.

57 Department of External Affairs; Memorandum to the Prime Minister; Secretary of State for External Affairs; September 30, 1966.

58 Department of External Affairs; Confidential Memorandum to Defence Liaison Division; Secretary of State for External Affairs; October 3, 1966.

59 Department of External Affairs; PJBD Meeting Memorandum to the Secretary of State for External Affairs; October 14, 1966.

60 The original memorandum to the prime minister had been sent on January 30. Department of External Affairs; Memorandum to Defence Liaison Division; Under-Secretary of State for External Affairs; February 10, 1967.

61 Privy Council Office; "Discussion of NORAD Agreement between Canadian and U.S. Officials," Cabinet Conclusions; February 7, 1967; p. 11.

62 Department of External Affairs; Internal Memorandum; February 22, 1967.

63 Department of External Affairs; Outline for Memorandum on the Implications for Canada if the US Deploys an ABM System; Division of Defence Liaison; February 23, 1967.

64 Unless the United States contemplated an intercept in outer space (exo-atmospheric), there was no value to moving interceptor bases farther north of likely targets. The problem with an exo-atmospheric intercept largely was the collateral damage to satellites from a nuclear explosion in outer space – damage the United States had discovered during its earlier nuclear testing envelope and that underpinned the logic of the 1963 Partial Nuclear-Test-Ban Treaty.

65 For a full discussion of Canada's nuclear role, see John Clearwater, *Canadian Nuclear Weapons* (Toronto: Dundurn Press, 1998).

66 See Albert Legault and Michel Fortmann, *A Diplomacy of Hope: Canada and Disarmament 1945-1988* (Kingston, ON: McGill-Queen's University Press, 1992).

67 The aborted Multilateral Force proposal by the United States was a product of issues surrounding the management of nuclear forces assigned to the defence of Europe and the attempt to bring French independent nuclear forces under alliance (read US) control. Charles de Gaulle, the president of France, opposed the proposal, seeing it as undermining French strategic independence, as well as for the role West Germany would obtain in the nuclear realm. The proposal was withdrawn in 1965, and the NATO Nuclear Planning Group (NPG) was established to manage the nuclear issue in the alliance. On the NPG and its roots, see Paul Buteux, *The Politics of Nuclear Consultation in NATO, 1965-1980* (Cambridge, UK: Cambridge University Press, 1983).

68 The negotiations of the Outer Space Treaty drew directly from the Law of the Sea in defining space in terms similar to international waters, with vessels (satellites) as national property. M.J. Peterson, "The Use of Analogies in Developing Space Law," *International Organization* 51 (Spring 1997): 245-74.

69 On Canada's role in drafting the NPT, see Albert Legault and Michel Fortmann, *A Diplomacy of Hope: Canada and Disarmament 1945-1988* (Kingston, ON: McGill-Queen's University Press, 1992), 259-67.

70 Department of External Affairs and Department of National Defence; NORAD: Memorandum to Cabinet; May 31, 1967.

71 Canada, *House of Commons Debates,* Vol. 14, 1st Session, 27th Parliament (April 4, 1967), p. 14487. In his memoirs, Hellyer's only comment on the meeting was McNamara's provision of information on good progress being made on an agreement "not to proceed with counter-missiles." He then added, in his only direct note on ABM itself, that "Canada had consistently declined the US invitation to participate ... and we were never convinced of the advisability of the Americans' proceeding with such a massive program." Paul Hellyer, *Damn the Torpedoes: My Fight to Unify Canada's Armed Forces* (Toronto: McClelland and Stewart, 1990), 223.

72 Canada, *House of Commons Debates,* Vol. 14, 1st Session, 27th Parliament (April 7, 1967), p. 14644.

73 Ibid. (April 10, 1967), p. 14703.

74 Ibid. (April 11, 1967), p. 14791.

75 Department of National Defence; *North American Aerospace Defence Plan for 1975: Report on DoD Prepared Plans for Aerospace Defence of USA Coming into Effect in 1975;* January 31, 1968.

76 McNamara also informed the delegation that to protect 70 percent of the population would cost an additional US$1.2 billion, and 90 percent would cost $2.5 billion. One may assume that this would not have required the deployment of interceptors on Canadian soil.

77 Privy Council Office; "Future of NORAD," Cabinet Conclusions; June 1, 1967; p. 5.

78 Department of External Affairs; Report on PJBD Meeting June 5-9; June 14, 1967.

79 Department of External Affairs; Memorandum to the Prime Minister; Canadian Chair, PJBD; July 14, 1967.
80 Department of External Affairs; Report on Meeting of the Defence and External Affairs Committee; Defence Liaison Division; July 14, 1967.
81 Department of External Affairs and Department of National Defence; NORAD: Memorandum to Cabinet; September 5, 1967; p. 1.
82 Ibid., p. 4.
83 Privy Council Office; "NORAD," Cabinet Conclusions; September 12, 1967; pp. 9-11.
84 Ibid., p. 10.
85 Ibid., p. 11.
86 Privy Council Office; "Statement on Anti-Ballistic Missiles," Cabinet Conclusions; September 18, 1967.
87 Cited by Paul Martin Sr.; Standing Committee on External Affairs; Minutes of Proceedings and Evidence; no. 14, March 7, 1968; p. 327. Emphasis mine.
88 Canada, *House of Commons Debates,* Vol. 3, 2nd Session, 27th Parliament (September 25, 1967), p. 2427.
89 Privy Council Office; Cabinet Conclusions; February 8, 1968.
90 Although not specified by the minister, he is likely referring to the deployment of the Defense Support Program infrared sensor satellites in geosynchronous orbit, capable of detecting missile launches. The airborne detection system is likely the Airborne Warning and Control System, which became operational in the 1970s.
91 There is no evidence in Privy Council Office and External Affairs documents that this briefing was ever provided, either written or verbally.
92 Privy Council Office; Cabinet Conclusions; February 27, 1968.
93 Ibid.
94 The minutes do not specify the dissenting voice. Given previous cabinet discussions and the nature of the arguments, it is highly possible the speaker was Pierre Trudeau, the justice minister.
95 There is no evidence that the speaker was cognizant of the possible role of NORAD in the ABM system through its early warning function and/or the logical question of command and control as distinct from the debate that had centred upon the actual ABM system of interceptors and radars.
96 The separate External Affairs and National Defence House committees had been informally meeting together through much of 1967 and would subsequently merge to create the Standing Committee on External Affairs and National Defence (SCEAND).
97 Standing Committee on External Affairs; Minutes of Proceedings and Evidence; no. 14, March 7, 1968; p. 339.
98 Ibid., p. 340.
99 In other words, Canada would be an accidental target. Neither Sentinel not its successor Safeguard would be adequate if Canada was on the Soviet Union's target list. Overall, External Affairs tended to view Canada separate from the United States, whereas National Defence viewed North America as a single target set. Department of External Affairs; Memorandum on BMD C^2; Defence Liaison Division; January 22, 1969.
100 As the September memorandum to cabinet noted of air defence, "plans are jointly drawn and approved by both Canadian and American authorities ... United States influence in planning is no doubt preponderant, but it is not exclusive." Department of External Affairs and Department of National Defence; NORAD: Memorandum to Cabinet; Privy Council Office; September 5, 1967; p. 6.
101 Department of External Affairs; Memorandum to the Minister on Visit of CINC; August 21, 1968.

102 Department of External Affairs; Memorandum on CINC Reeves Visit to the Minister; Defence Liaison Division; August 22, 1968.
103 Department of External Affairs; Letter to Under-Secretary of State; Chairman of the Chiefs of Staff; August 9, 1963.
104 Department of External Affairs; Memorandum to the Defence Liaison Division; United Nations' Division; September 4, 1963.
105 Department of External Affairs; Memorandum to the Defence Liaison Division; Legal Division; September 6, 1963.
106 Department of National Defence; Letter to the Chiefs of Staff; Acting Chairman, US Joint Chiefs of Staff; December 11, 1963.
107 The Military Cooperation Committee underneath the PJBD was established in 1946 and consists of senior military representatives from the United States and Canada. It meets twice yearly, rotates locations between the two countries, and is a forum to discuss direct technical military issues of concern in the relationship.

The ITW/AA or early warning mission refers to NORAD's core mission to identify and assess initially air and subsequently space threats to North America. Early warning of an air attack by the Soviet Union was provided primarily by the three radar lines established in Canada in the 1950s – the Mid-Canada, Pinetree, and Distant Early Warning (DEW) lines. The development and employment of space-based infrared satellites by the US Air Force in 1970 in geosynchronous orbit provided initial warning of a ballistic missile attack and cued ground-based radars. The data from the radar and space assets were funnelled into the Cheyenne Mountain Operations Center, where the command director assesses the information to determine if North America is under attack. If so, the director notifies the National Command Authorities.

108 Department of External Affairs; Memorandum on US Public Announcement on Air Defence; Defence Liaison Division; January 12, 1968.
109 Department of External Affairs; Memorandum to the Chief of the Defence Staff; Under-Secretary of State for External Affairs; November 4, 1968.
110 Department of External Affairs; Memorandum to Under-Secretary of State for External Affairs; Defence Liaison Division; June 27, 1968.
111 Ibid.
112 Canada, *House of Commons Debates*, Vol. 2, 1st Session, 28th Parliament (October 21, 1968), p. 1596.
113 Ibid., Vol. 3, 1st Session, 28th Parliament (November 19, 1968), pp. 2904-05.
114 Safeguard was initially to consist of several interceptor sites. After the signing of the ABM Treaty, it was downsized to a single site to defend the Grand Forks ICBM field.
115 According to one study, the actual defensive capabilities of Sentinel and Safeguard for the continental United States were nearly identical. Abram Chayes, Jerome B. Wiesner, George W. Rathjens, and Steven Weinberg, "Overview," in *ABM: An Evaluation of the Decision to Deploy an Antiballistic Missile System*, Abram Chayes and Jerome Wiesner, eds. (New York: Signet Books, 1969), 43, 45.
116 This point was reiterated by Mitchell Sharp, the secretary of state for external affairs, in November, when he wrote in the context of SALT negotiations: "In the circumstances, a formal statement by the Canadian government at an early date on ABM's is unlikely to be helpful and could well be misinterpreted." Department of External Affairs; Memorandum for the Prime Minister; Secretary of State for External Affairs; November 6, 1969.
117 Department of External Affairs; Memorandum for the Prime Minister; September 26, 1969.
118 Department of External Affairs; Memorandum for Politico-Military Affairs; Canadian Embassy in Washington; May 6, 1969.

119 A month after Cadieux's meeting with Laird, the director of research and engineering in the US Department of Defense, John Foster, met with senior External Affairs and National Defence officials in Ottawa and provided a thorough technical briefing on Safeguard. Department of External Affairs; North American Defence Briefing by Dr. John Foster Memorandum; Office of Politico-Military Affairs; June 12, 1969.
120 Department of External Affairs and Department of National Defence; "Implications for Canada of Anti-Ballistic Missile Defence Systems," Annex D, North American Air Defence: Memorandum to Cabinet; Privy Council Office; July 23, 1969.
121 In the context of the drafting of a future memorandum to cabinet, some External Affairs officials perceived that National Defence was attempting to provide comfort and support for the US ABM program. Department of External Affairs; Letter to Political-Military Affairs; Disarmament Division; January 21, 1970.
122 Department of External Affairs; Letter to the Prime Minister with Attached Memorandum; Secretary of State for External Affairs; May 21, 1969.
123 Department of External Affairs and Department of National Defence; North American Air Defence: Memorandum to Cabinet; Privy Council Office; July 23, 1969; p. 7.
124 Detroit would also be included in the initial scaled-down Safeguard proposal. Within a year, however, Safeguard would be scaled back further to two initial sites, Great Falls, Montana, and Grand Forks, North Dakota, for point defence of two US ICBM fields. Shortly thereafter, the ABM question would disappear from Canadian purview.
125 Canada, *House of Commons Debates,* Vol. 8, 1st Session, 28th Parliament (March 17, 1969), p. 6683.
126 Department of External Affairs; Memorandum to the Prime Minister; Secretary of State for External Affairs; November 6, 1969.
127 At this meeting, cabinet agreed to proceed with a white paper on defence policy. Privy Council Office; Cabinet Conclusions; July 22, 1970.
128 Department of National Defence; North American Defence Policy in the Seventies: Memorandum to Cabinet; Privy Council Office; June 1, 1970.
129 This technical point regarding the value of Canadian territory for a mid-course phase intercept was raised directly by George Lindsey in his testimony to SCEAND.
130 Standing Committee on External Affairs and National Defence, Minutes of Proceedings and Evidence, no. 41, May 6, 1969, p. 1381.
131 Sharp's testimony was raised the next day in the House by Tommy Douglas, who sought assurances that bringing ABM under NORAD would not proceed without discussion in the House. Canada, *House of Commons Debates,* Vol. 11, 1st Session, 28th Parliament (May 7, 1969), p. 8426.
132 Standing Committee on External Affairs and National Defence, Minutes of Proceedings and Evidence, no. 41, May 6, 1969, p. 1387.
133 Standing Committee on External Affairs and National Defence, "Strategic Weapons Systems, Stability, and Possible Contributions by Canada: Part One – Strategic Weapon Systems, Stability, and the Prevention of Nuclear War; Part Two – Possible Future Developments to North American Air Defence," Minutes of Proceedings and Evidence, Appendix ZZZ, no. 43, May 13, 1969; "Part Three – The Implications for Canada of Ballistic Missile Defence," Minutes of Proceedings and Evidence, Appendix CCC, no. 46, May 21-22, 1969.
134 Ibid., p. 1643. Nuclear blackout is the detonation of a nuclear warhead to blind and thus defeat the defence as a function of electromagnetic pulse destroying the electronic signals emitted by the radars.
135 Ibid., p. 1645.

136 Standing Committee on External Affairs and National Defence, Minutes of Proceedings and Evidence, no. 48, June 10, 1969.
137 Ibid., p. 1719.
138 Ibid., p. 1710.
139 Ibid., pp. 49-51.
140 Canada, *Defence in the 70s: White Paper on Defence* (Ottawa: Queen's Printer, 1971), 26.

Act 2: The Strategic Defense Initiative

1 Reagan's motives for SDI remain clouded in history. Many have suggested it was a product of his science fiction experience as an actor in Hollywood, others of a combination of early briefings on missile defence when he was governor of California or of his first visit to NORAD headquarters in 1979. For a brief discussion, see Lou Cannon, "Reagan's Big Idea: How the Gipper Conceived Star Wars," *National Review,* February 22, 1999. For the most thorough discussion, see Frances Fitzgerald, *Way Out There in the Blue: Reagan, Star Wars, and the End of the Cold War* (New York: Simon and Schuster, 2000).
2 The agreement, which expired in 1980, was initially extended without change for one year on the request of the Trudeau government to allow for parliamentary input following the defeat of the Clark minority government in the fall of 1979.
3 "Treaty between the United States of America and the Union of Soviet Socialist Republics on the Limitation of Anti-Ballistic Missile Systems," Appendix A in Matthew Bunn, *Foundation for the Future: The ABM Treaty and National Security* (Washington, DC: Arms Control Association, 1990).
4 Unfortunately, large sections of the early warning discussion were deleted upon release of the document. It is possible, if not likely, that early warning for Safeguard was at least raised. Department of National Defence; North American Defence Policy in the Seventies: Memorandum to Cabinet; Privy Council Office; June 1, 1970.
5 NORAD was renewed in 1973 for two years and in 1975 for five years. Cabinet documents, despite deleted passages, contain no reference to the Safeguard early warning issue. One memorandum to the prime minister was directly concerned with the defensive use of nuclear weapons by NORAD forces, but large sections are deleted, and the text available appears to concern consultations for nuclear air defence weapons, wholly consistent with the preceding memorandum to cabinet. See Department of External Affairs and the Department of National Defence; NORAD and the Employment of Defensive Nuclear Weapons: Memorandum to the Prime Minister; Privy Council Office; March 9, 1973; Department of External Affairs and Department of National Defence; NORAD Agreement: Interim Extension Memorandum to Cabinet; Privy Council Office; February 12, 1973; Department of External Affairs and Department of National Defence; Renewal of NORAD Agreement: Memorandum to Cabinet; Privy Council Office; February 25, 1975.
6 The treaty limited both parties to early warning radars on their territorial periphery oriented outward. The two US sites in the United Kingdom and Greenland, however, were grandfathered. As discussed below, in the 1980s as part of the Reagan rationale for SDI and suspicion of arms control, the United States charged the Soviet Union with violating the treaty through the deployment of a large radar at Krasnoyarsk, in the interior of the Soviet Union. With the end of the Cold War, Russian officials agreed that the radar was in direct violation. For the formal text, see Articles 3 and 6, Appendix A.
7 Brian Mandel, *The United States Strategic Defense Initiative: Implications for Deterrence, Arms Control, Adversaries, and Allies,* Operational Research and Analysis Establishment (ORAE) Report (Ottawa: Department of National Defence, 1984).

8 These pressures remained largely behind closed doors such that they did not generate the public political forces in the United States, as had occurred with ABM. Joel Sokolsky, "Changing Strategies, Technologies, and Organization: The Continuing Debate on NORAD and the Strategic Defense Initiative," *Canadian Journal of Political Science* 19, 4 (December 1986): 751-74.

9 Initial testing began in 1968, followed by an initial operating capability in 1971. Tests ceased in 1972 and were resumed in 1976. Matthew Mowthorpe, *The Militarization and Weaponization of Space* (Lanham, MD: Lexington Books, 2004), 117-25.

10 The Homing Overlay Experiment consisted of four tests employing a kinetic-kill vehicle to intercept a target in space. The first test occurred in February 1983. The Soviets claimed that, in using a Minuteman launcher, the United States had violated the "testing in an ABM mode" prohibition of the treaty. Matthew Bunn, *Foundation for the Future: The ABM Treaty and National Security* (Washington, DC: Arms Control Association, 1990), 98.

11 On March 10, President Carter eliminated the restriction on space-based testing, providing the green light for an operational ASAT unless negotiations with the Soviet Union progressed. United States, "Arms Control for Anti-Satellite Systems: Presidential Directive-33," Washington, DC, March 10, 1978.

12 Electromagnetic pulse produced by a nuclear detonation destroys all electrical systems. If it could be harnessed without a nuclear explosion, the weapon could destroy the arming mechanism of any warhead.

13 Privy Council Office; Defence Policy – Coordination and Communication Memorandum for the Prime Minister; February 5, 1985; p. 5. Sokolsky in 1986 cites one source who links the decision to a desire by Canadian defence officials to keep the door open without any advance knowledge of what was to come. Joel Sokolsky, "Changing Strategies, Technologies, and Organization: The Continuing Debate on NORAD and the Strategic Defense Initiative," *Canadian Journal of Political Science* 19, 4 (December 1986): 762.

14 With the establishment of US Space Command in 1984, the Space Defense Operations Center would transition into the Space Control Operations Center.

15 Department of National Defence; Internal Memorandum; Directorate of Continental Policy; October 1979.

16 The Outer Space Treaty does not explicitly define where outer space begins. By default, it is generally held that space begins where an object is able to complete a single orbit – roughly 150 kilometres above the earth. This definition follows from the Article 4 prohibition on weapons of mass destruction in space, which states that "State Parties to the Treaty undertake not to place in orbit around the earth any objects carrying nuclear weapons or any other kinds of weapons of mass destruction" and recognizes that the Soviet fractional orbital bombardment system (FOB) was outside the treaty. Above all else, both the United States and the Soviet Union would have only agreed to the treaty if ballistic missiles were excluded. Long-range ballistic missiles ascend to a higher altitude than some orbiting satellites but lack the speed necessary to obtain orbit (roughly Mach 25). "Treaty on Principles Governing the Activities of States in the Exploration and Use of Outer Space, Including the Moon and Other Celestial Bodies," January 27, 1967, *United Nations Treaty Series.*

17 Department of External Affairs, International Defence Arms Control, *Chronology of 1984 ASAT Proposal,* September 26, 1986.

18 Naturally, the development of the next generation of strategic forces began long before Carter and Reagan, given the long lead times involved. Nonetheless, Carter had made the decisions to proceed with modernization, including the next generation of ICBMs (MX), SLBMs (Trident), and bombers (B-1), among other programs.

19 A heavier missile simply refers to its lift capacity enabling it to carry a larger payload or number of warheads. Throw weight measures the weight a missile can carry into space or on a ballistic trajectory to its terrestrial target. The largest rocket payload was the US Saturn V, developed for the US-manned moon project.

20 Gregory Herken, *The Counsels of War* (New York: Oxford University Press, 1987), 274. Paul Nitze had a long and distinguished career in the US government. He had been in charge of the report on the effects of the atomic blasts on Hiroshima and Nagasaki, the author of the influential Cold War National Security Council Report 68, secretary of the navy during the Johnson years, a member of the SALT negotiating team under Nixon, and head of the US Intermediate Nuclear Weapons negotiating team resulting in the 1987 INF Treaty. Eugene Rostow would be appointed director of the US Arms Control and Disarmament Agency by President Reagan. See also Strobe Talbot, *The Master of the Game: Paul Nitze and the Nuclear Peace* (New York: Alfred A. Knopf, 1988).

21 Strategic stability, as understood in the Cold War, is a condition whereby neither the Soviet Union nor the United States acquired the capacity to threaten the other's second-strike forces, which would generate pre-emptive first-strike incentives. For a basic discussion of this concept, see James Fergusson, *Strategic Stability Reconsidered* (Ottawa: International Security Research and Outreach Programme, Department of Foreign Affairs and International Trade, September 2001).

22 William Burr, "The Nixon Administration, the SIOP, and the Search for Limited Nuclear Options," National Security Archive Electronic Briefing Book no. 173, November 23, 2005.

23 United States, "Nuclear Weapons Employment Policy: Presidential Directive-59," Washington, DC: White House, July 25, 1980. Emphasis mine.

24 *Galosh* was actually the code name for the initial Soviet interceptor with characteristics similar to the US Spartan missile. On Chevaline, see John Baylis and Kristan Stoddart, "Britain and the Chevaline Project: The Hidden Nuclear Programme, 1967-1982," *Journal of Strategic Studies* 20, 4 (December 2003): 124-55.

25 Before an international treaty can become legally binding upon the United States, the Senate must give its advice and consent based upon a two-thirds majority vote. Reflecting the reality of strategic arms control largely codifying existing plans and preferences, the United States would continue to adhere to the SALT II limits until 1986, when it deployed its 131st air-launched cruise missiles. Institute for Defense and Disarmament Studies, *Arms Control Reporter: A Chronicle of Treaties, Negotiations, Proposals, Weapons, and Policy 1986* (Brookline, MA: Institute for Defense and Disarmament Studies, 1986), 607.B.111.

26 The neutron bomb was a high-yield, short-life-span radiation weapon that would kill people but not destroy buildings. Publicly, the weapon was met with great horror. To the consternation of the Carter administration, the Europeans portrayed the decision as one being imposed upon them by Washington. To the shock of the Europeans, Carter refused to proceed, and the neutron bomb was never added to the NATO arsenal. Carter in moving forward with INF wanted to avoid a repeat of such a public image.

27 In the end, a global double zero, designed to eliminate the possibility of the Soviet Union deploying intermediate-range missiles outside Europe – east of the Urals – would be basically adopted in 1986 at the Reykjavik Summit, with the formal INF Treaty signed a year later. It would become one of the symbols of the end of the Cold War.

28 In reality, détente had died in the last years of the Carter administration, and this administration set in motion the modernization of US military forces. Nonetheless, Carter's move to the right became lost in public memory and Reagan's hard-line rhetoric.

29 The agreement was originally for five years but was subsequently renewed and remains operative.

30 For a discussion of the media and public opposition to ALCM testing, see Ann Denholm Crosby, "The Print Media's Shaping of the Security Discourse: Cruise Missile Testing," *International Journal* 52 (Winter 1996-97): 1.
31 Joel Sokolsky, "Changing Strategies, Technologies, and Organization: The Continuing Debate on NORAD and the Strategic Defense Initiative," *Canadian Journal of Political Science* 19, 4 (December 1986): 751-74.
32 Force enhancement refers to the role satellites play in improving the effectiveness and efficiency of terrestrial military operations, as discussed in Act 3. Besides secure global communications, the US Global Positioning System, for example, enabled troops in the field to identify their positions, eliminating the "lost lieutenant" problem. NORAD provided early warning of Iraqi Scud attacks to Israel and coalition forces. Remote sensing or earth observation satellites provided accurate near-real-time pictures of friend and foe military dispositions in the field.
33 In 1983, the US Navy established its own Space Command.
34 After two years of dealing with numerous issues, including its relationship with NORAD, US Space Command was formally established in August 1985. Joseph T. Jockel, *Canada in NORAD 1957-2007: A History* (Kingston, ON: McGill-Queen's University Press, 2007).
35 Department of External Affairs; Internal Memorandum; February 7, 1984. Also, no focal point for SDI existed until the establishment of the Strategic Defense Initiative Organization in early 1984.
36 In the House and media, SDI discussions remained conspicuous by their absence. One exception followed a presentation by George Rathjen, a long-time missile defence advocate, to a conference at York University on April 30, 1984.
37 Department of External Affairs; Report on SDI Briefing of Minister of National Defence; International Defence Relations; October 30, 1984.
38 Canada, *House of Commons Debates,* Vol. 1, 1st Session, 33rd Parliament (November 19, 1984), p. 341.
39 Ibid. (December 18, 1984), p. 1339.
40 Ibid. (December 20, 1984), p. 1405.
41 Ibid. (December 21, 1984), p. 1442.
42 Canada, *House of Commons Debates,* Vol. 2, 1st Session, 33rd Parliament (January 21, 1985), p. 1502.
43 This ad hoc group simply represented the officials from External Affairs, National Defence, and PCO assigned to the SDI issue by their respective senior officials. There appears to be no formal records or minutes of their meetings, and available documents provide no explicit reference to a working group. Instead, depending upon the originator of the meeting, these officials would be tasked to attend and subsequently report back to their superiors, as evident in, for example, the report on the meeting between Minister Coates and US Under-Secretary of Defense Iklé by the External Affairs representative. Such a process is a long-standing tradition in which formal interdepartmental working groups are rare. Nonetheless, such ad hoc groups, if such a term is accurate, are essential, as both External Affairs and National Defence and their ministers desire to come to an agreement before proceeding to cabinet (i.e., forwarding a joint memorandum to cabinet).
44 Jack Best, "Clark and Coates Disagree over 'Star Wars,'" *Ottawa Citizen,* January 3, 1985. In his memoirs, Brian Mulroney notes the divide between Clark and Coates but, interestingly, sides with Clark. Brian Mulroney, *Memoirs 1939-1993* (Toronto: Douglas Gibson Books, 2007).
45 Department of External Affairs; Briefing Note to the Minister of External Affairs; January 11, 1985.

46 Department of External Affairs; Briefing Note to the Minister of External Affairs; January 15, 1985.
47 Most of the memoranda for Mulroney available for this study were returned with no comment from the former prime minister.
48 The reference of full support being potentially unwelcome in Washington likely refers to its impact on the internal American debate. Privy Council Office; Strategic Defense Initiative (SDI) – Star Wars: Memorandum for the Prime Minister; Foreign and Defence Policy Secretariat; January 7, 1985.
49 Ibid., p. 5.
50 Whether the phrase "currently constituted," which is language used in Article 2 of the ABM Treaty and hotly contested a year or so later in the debate over the *broad* versus *narrow* interpretation, was purposeful in this regard is difficult to know. Certainly, neither Clark nor External Affairs officials who would have drafted the statement would have been aware of the future interpretation debate. See Act 3.
51 Canada, *House of Commons Debates,* Vol. 2, 1st Session, 33rd Parliament (January 21, 1985), pp. 1502-06.
52 Ibid. (January 30, 1985), pp. 1830, 1838.
53 Ibid. (January 31, 1985), p. 1860.
54 Ibid. (February 13, 1985), p. 2301.
55 Ibid.
56 The concept of a rule-based international order is a recent one but does reflect the vital importance of the principle of international law in Canadian policy as enunciated in Louis St. Laurent's famous Gray Lecture in 1947. See Louis St. Laurent, *The Foundations of Canadian Policy in World Affairs* (Toronto: University of Toronto Press, 1947). The phrase is now by and large used as a code to condemn US unilateralism.
57 Privy Council Office; Defence Policy – Coordination and Communication Memorandum for the Prime Minister; February 5, 1985; p. 3.
58 The first part of Broadbent's non-confidence motion concerned acid rain, and the second Star Wars, which called for the government to make it clear that Canada would not participate in SDI. In his speech, Broadbent also referred to statements by US Secretary of Defence Weinberger and Under-Secretary of State for Political-Military Affairs Richard Burt's linkage of Star Wars to a modernized NWS. Canada, *House of Commons Debates,* Vol. 3, 1st Session, 33rd Parliament (March 19, 1985), pp. 3145-46.
59 The purpose of Nitze's visit was to discuss the following week's nuclear and space arms negotiations.
60 Canadian Press, "PC's Deny Link between Star Wars, Modernization of Northern Radar," *Ottawa Citizen,* March 7, 1985.
61 For example, the day after the February 11 Weinberger speech in London, the Associated Press reported that West Germany had received a formal invitation. Associated Press, "Bonn Offered Role in Star Wars," *Globe and Mail,* February 12, 1985.
62 Illness prevented Weinberger from delivering the speech.
63 Department of External Affairs; Memorandum to Ottawa; Canadian Embassy in Washington; March 5, 1985.
64 Department of External Affairs; Cable to the Washington Embassy; March 9, 1985.
65 Department of External Affairs; Memorandum to the Minister; March 8, 1985.
66 Department of External Affairs; Cable to Ottawa; Canadian Embassy in Washington; March 10, 1985.
67 Importantly, actual testing was years, if not decades, away. In fact, US missile defence tests violated ABM only in July 2002, when a US Aegis-class cruiser's radar was employed in

an ABM mode, though by then the ABM Treaty was defunct. In the context of Fitzgerald's study, in which she barely mentions the allies and Canada and ignores entirely the Weinberger March invitation, the United States desire to eliminate the last clause stemmed from domestic political considerations and funding from Congress. Moreover, the great debate on the legality of SDI relative to the ABM Treaty was a year away. Frances Fitzgerald, *Way Out There in the Blue: Reagan, Star Wars, and the End of the Cold War* (New York: Simon and Schuster, 2000).

68 Department of External Affairs; Cable to Ottawa; Canadian Embassy in Washington; March 29, 1985.

69 Charlotte Montgomery, "Star Wars Invitation Extended by US," *Globe and Mail,* March 26, 1985.

70 Charlotte Montgomery, "Clark Denies Receipt of Invitation by US: Unaware of Star Wars Move," *Globe and Mail,* March 27, 1985.

71 Department of External Affairs; Invitation by the US to Participate in SDI Research: Memorandum to Robert Fowler; Gordon Smith, Deputy Minister for Political Affairs; March 27, 1985. Later that year, Gordon Smith was sent off to NATO as the permanent representative (ambassador) of Canada.

72 Privy Council Office; Invitation by USA to NATO Allies to Participate in SDI Research: Memorandum for the Prime Minister; Robert Fowler, Foreign and Defence Policy Secretariat; March 27, 1985.

73 United States, Department of Defense; Letter to the Honourable Erik Nielsen, Minister of National Defence; Caspar W. Weinberger, Secretary of Defense; March 26, 1985.

74 "Agreed Statements, Common Understandings, and Unilateral Statements Regarding the Treaty between the United States of America and the Union of Soviet Socialist Republics on the Limitation of Anti-Ballistic Missile Systems," Appendix A in Matthew Bunn, *Foundation for the Future: The ABM Treaty and National Security* (Washington, DC: Arms Control Association, 1990), 164.

75 Although no formal evidence from PCO files indicates that the attached draft letter was ever sent, Nielsen concluded the memorandum by stating that he would proceed unless the prime minister indicated otherwise on the memorandum when returned to the defence minister. There are no notations on the copy. Other material from PCO files and elsewhere provides no reference to the letter having been sent. The memorandum contains no specific date, but the reference to the Weinberger invitation likely indicates very early in April. Department of National Defence; SDI Related Research: Exploratory Discussions; Memorandum to the Prime Minister; Minister of National Defence; April 1985.

76 Privy Council Office; Update on Strategic Defense Initiative (SDI): Memorandum for the Prime Minister; Robert Fowler, Assistant Secretary, Foreign and Defence Policy Secretariat; April 4, 1985.

77 Department of External Affairs; Letter to the Minister of External Affairs; March 29, 1985.

78 This primarily refers to the Defence Production and Development Sharing Arrangements, with the former dating back to 1957. These were reinforced by the evolution of the corporate relationships over many decades and the recent creation of the North American Defence Industrial Base Organization.

79 Privy Council Office; Star Wars: Memorandum for Mr. Fowler, Assistant Secretary to the Cabinet, Communications; April 12, 1985.

80 Department of National Defence; SDI Related Research: Exploratory Discussions; Memorandum to the Prime Minister; Minister of National Defence; April 1985.

81 Department of External Affairs; Memorandum to the Minister of Industry; April 1, 1985.

82 In response, the French were successful in creating a European research and development coordination organization – EUREKA – within the European Community in 1985.

83 Department of Regional and Industrial Expansion; Economic Implications of Canadian Participation in Strategic Defense Initiative: Memorandum for Arthur Kroeger; Electronics and Aerospace Branch, Research and Development; May 17, 1985.
84 Ibid., p. 7.
85 Department of External Affairs; Internal Memorandum; March 4, 1985.
86 Privy Council Office; Update on Strategic Defense Initiative (SDI): Memorandum for the Prime Minister; Robert Fowler, Assistant Secretary, Foreign and Defence Policy Secretariat; April 4, 1985.
87 "The Objective of Arms Control," lecture transcript, Alistair Buchan Memorial Lecture series (Washington, DC: State Department, March 28, 1985), 9-10.
88 Department of External Affairs; Memorandum on Canada-US Dimension [of SDI]; April 3, 1985.
89 No such consensus emerged. At the summit, President Mitterrand provided a veiled and ambiguous no.
90 Privy Council Office; "Box Score" of Allied Positions on SDI Research, Update on Strategic Defense Initiative (SDI): Memorandum for the Prime Minister; Robert Fowler, Assistant Secretary, Foreign and Defence Policy Secretariat; April 4, 1985.
91 Department of External Affairs; Memorandum to Ottawa; Canadian Embassy in Paris; April 3, 1985.
92 Privy Council Office; Discussion on SDI at Priorities and Planning Committee: Note for the Prime Minister; Robert Fowler, Assistant Secretary, Foreign and Defence Policy Secretariat; April 16, 1985.
93 Department of External Affairs; Letter to Brian Mulroney; President Ronald Reagan; June 24, 1985.
94 Department of National Defence; Memorandum to the Minister of External Affairs; Minister of National Defence; April 16, 1985.
95 There are apparently no copies available of the Kroeger Report at either the Department of Foreign Affairs or the Department of National Defence; rather, there are only references to the report as a PCO document. As such, the report remains classified likely up to the thirty-year deadline and perhaps beyond. Access to information provided mainly blank pages.
96 Shortly after Reagan's re-election in November 1984, Chernenko proposed negotiations over nuclear and outer space weapons, thereby creating a linkage and possible trade-off. Subsequent discussions in January between Soviet Foreign Minister Gromyko and Secretary of State Schultz led to a resumption of strategic arms control discussions and the creation of defence and space talks, which in the end went nowhere. See Frances Fitzgerald, *Way Out There in the Blue: Reagan, Star Wars, and the End of the Cold War* (New York: Simon and Schuster, 2000).
97 Department of External Affairs; Letter to Arthur Kroeger; May 16, 1985.
98 Department of External Affairs; Internal Memorandum; May 31, 1985.
99 Ironically, as noted below, the majority of Canadians actually supported Canadian participation.
100 Privy Council Office; SDI and External Affairs: Memorandum for Gordon Osbaldeston; Robert Fowler, Assistant Secretary, Foreign and Defence Policy Secretariat; June 14, 1985.
101 Department of National Defence; Letter to Arthur Kroeger; Minister of National Defence; May 13, 1985.
102 Department of National Defence; Attachment to Letter to Arthur Kroeger; Col. William Weston, Planning Guidance Team; May 29, 1985.
103 Agreed Statement G in the ABM Treaty prohibited the transfer of blueprints.

104 Privy Council Office; Review of the US Strategic Defense Initiative: Memorandum for the Prime Minister; Arthur Kroeger; May 7, 1985.
105 Department of External Affairs; Memorandum to Arthur Kroeger; May 13, 1985.
106 Department of External Affairs; Memorandum to Ottawa; Canadian NATO Delegation; June 27, 1985; Department of External Affairs; Memorandum to Ottawa; Canadian Embassy in Washington; July 1, 1985.
107 Privy Council Office; SDI: Notes for Kroeger; Robert Fowler, Assistant Secretary, Foreign and Defence Policy Secretariat; June 5, 1985.
108 Ibid.
109 Privy Council Office; Kroeger Paper on Strategic Defense Initiative: Memorandum to the Prime Minister; Robert Fowler, Assistant Secretary, Foreign and Defence Policy Secretariat; June 11, 1985.
110 This conclusion is based upon an evaluation of all available documents and confidential interviews. The majority of the actual report released by PCO, including the identification and assessment of options, is whited-out. Privy Council Office; *The U.S. Strategic Defence Initiative and Canada's Role;* Arthur Kroeger; June 7, 1985.
111 In August, Kroeger provided an update to his report, which was endorsed by the former clerk of the PCO, who had left office at the beginning of that month. Privy Council Office; Special Joint Committee on Canada's International Relations: Memorandum for the Prime Minister; G.F. Osbaldeston, Former Clerk of the Privy Council; August 20, 1985.
112 The joint committee's overall mandate was to examine the 1984 Green Paper (a thought piece rather than a formal White Paper policy commitment) entitled *Competitiveness and Security: Directions for Canada's International Relations.*
113 Privy Council Office; Special Joint Committee on Canada's International Relations: Memorandum to the Rt. Hon. Brian Mulroney; August 20, 1985. The undated Hockin memorandum was attached to this memo.
114 Privy Council Office; Special Joint Committee on Canada's International Relations: Memorandum for the Prime Minister; G.F. Osbaldeston; August 20, 1985.
115 Cited in Douglas Ross, "SDI and Canadian-American Relations," in *America's Alliances and Canada-American Relations,* ed. Lauren McKinsey and Kim Richard Nossal (Toronto: Summerhill Press, 1988), 150. Ross also identifies a Southam poll in August that found only 40 percent in favour and 42 percent opposed.
116 Department of External Affairs; Internal Memorandum; August 30, 1985. It is likely that these views were eventually passed on to Clark in some form, as it references the forthcoming meeting of the Cabinet Committee on External Affairs and National Defence on 4 September. Whether they were presented is unknown.
117 Ronald Reagan, *The Reagan Diaries*, ed. Douglas Brinkley (New York: Harper Perennial, 2007), 351.

Act 3: Global Protection against Limited Strikes

1 For example, the January 29 national edition of the *Globe and Mail* relegated the State of the Union address to page 6 and made no mention of GPALS whatsoever.
2 The *Globe and Mail* ran several stories on the Scud-Patriot exchanges but never looked to their potential wider meaning, even after the GPALS announcement. For example, see Geoffrey Rowan, "Patriot Missile Rising to Occasion," *Globe and Mail,* January 22, 1991.
3 The *Globe and Mail* report on the summit did note the GPS agreement and stated, "Canada and other allies may be ultimately asked to join as well." But no link to SDI was made, and in the weeks to follow no editorial piece raised or discussed missile defence, GPALS, or GPS. Colin McKenzie, "Russia-US Reach Deal on Arms Cuts," *Globe and Mail,* June 17, 1992.

4 Canada, *Challenge and Commitment: A Defence Policy for Canada* (Ottawa: Department of National Defence, 1987), 19.
5 SBIRS-Low/STSS is to be a constellation of roughly twenty-four satellites in low earth orbit. The program would include SBIRS-High, designed to replace the existing Defense Support Program infrared early warning satellites in geostationary orbit that became operational in 1970. Both Low and High have faced significant development problems. See Act 4 and Act 5 for a full discussion.
6 The Soviets also accepted laboratory research. In 1972, immediately after the signing of the treaty, Marshal Andrei Grechko stated that the ABM Treaty "does not place limitations of any kind on the conduct of research and experimental works directed at the solution of the problem of the defense of the country against rocket nuclear strikes." David Yost, *Soviet Ballistic Missile Defence and the Western Alliance* (Cambridge, MA: Harvard University Press, 1988).
7 According to Fitzgerald, the process leading to the debate began in January 1985 and culminated with Sofaer's memorandum sent to Schultz and Nitze on October 3. Three days after the delivery of the memorandum, the first public outline of the broad interpretation was made by Robert McFarlane, Reagan's national security advisor, on October 6, on ABC's *Meet the Press.* Francis Fitzgerald, *Way Out There in the Blue: Reagan, Star Wars, and the End of the Cold War* (New York: Simon and Schuster, 2000), 290-96. For the original arguments, see Abram Chayes and Antonia Handler Chayes, "Testing and Development of 'Exotic' Systems under the ABM Treaty: The Great Reinterpretation Caper," *Harvard Law Review* 99, 8 (June 1986): 1956-71.
8 Article 4, "Treaty between the United States of America and the Union of Soviet Socialist Republics on the Limitation of Anti-Ballistic Missile Systems," Appendix A in Matthew Bunn, *Foundation for the Future: The ABM Treaty and National Security* (Washington, DC: Arms Control Association, 1990), 162.
9 Agreed Statement D, "Agreed Statements, Common Understandings, and Unilateral Statements Regarding the Treaty between the United States of America and the Union of Soviet Socialist Republics on the Limitation of Anti-Ballistic Missile Systems," in ibid., 164. Article 13 refers to the role of the Standing Consultative Commission of Soviet and American officials for treaty compliance discussions, and Article 14 allows for amendments to the treaty and five-year reviews. See above.
10 Department of External Affairs; Memorandum to the Under-Secretary of State; November 25, 1986.
11 Privy Council Office; Strategic Defense Initiative (SDI) – Star Wars: Memorandum for the Prime Minister; Foreign and Defence Policy Secretariat; January 7, 1985.
12 Special Joint Committee of the Senate and the House of Commons on Canada's International Relations, *Independence and Internationalism* (Ottawa: Queen's Printer, 1986).
13 Canada, *House of Commons Debates,* Vol. 1, 2nd Session, 33rd Parliament (October 20, 1986), p. 522.
14 Ibid. (October 21, 1986), p. 533.
15 Clark's statement directly references the communication of this position in a letter to Secretary of State Schultz "indicating the grave importance Canada places on the US continuing to adhere to a restrictive [narrow] interpretation of the Treaty." Canada, *House of Commons Debates,* Vol. 3, 2nd Session, 33rd Parliament (February 13, 1987), p. 3087.
16 Ibid. (March 6, 1987), p. 3908.
17 Department of External Affairs; Memorandum to Ottawa; Canadian Embassy in Washington; February 5, 1987.
18 In the brief arms control section, no reference to ABM was made whatsoever. Canada, *Challenge and Commitment: A Defence Policy for Canada* (Ottawa: Queen's Printer, 1987), 19.

19 Canada, *House of Commons Debates*, Vol. 5, 1st Session, 33rd Parliament (September 9, 1985), p. 6395.
20 Department of National Defence; Guidelines for DND Personnel Memorandum; June 1987.
21 NATO's "extended air defence" was the term applied to the alliance program to provide anti-tactical ballistic missile defence (ATBM).
22 Department of External Affairs; Memorandum to the Director-General Current Policy, Department of National Defence; International Security and Arms Control Bureau; July 13, 1987.
23 In actuality there were two shuttle programs employing the same space shuttles – one led by NASA and the other by the USAF, reflecting the different interests of NASA and the Pentagon. Both, of course, would employ Canada's remote manipulator arm – the Canadarm. See below.
24 Canada, *House of Commons Debates*, Vol. 10, 1st Session, 33rd Parliament (June 16, 1986), p. 14447.
25 Ibid. (June 17, 1986), p. 14529.
26 Privy Council Office; Strategic Defense Initiative: Memorandum for the Prime Minister; Arthur Kroeger; August 1985.
27 Department of National Defence; Memorandum to the Secretary of State for External Affairs; Minister of National Defence; November 24, 1986.
28 According to the Defence Production Sharing Agreements, Canadian companies were to be treated as American in bidding for US defence contracts, thereby creating a single integrated defence market. For a full discussion of the range of issues, see Alastair Edgar and David Haglund, *The Canadian Defence Industry in the New Global Environment* (Kingston, ON: McGill-Queen's University Press, 1995).
29 Department of External Affairs; Memorandum to Assistant Deputy Minister (Policy), Department of National Defence; Political and International Security Affairs; June 23, 1987.
30 Department of National Defence; Memorandum to Assistant Deputy Minister Political and International Security Affairs, Department of External Affairs; Assistant Deputy Minister (Policy); August 21, 1987. A year earlier, the Aerospace Industries Association of Canada and the minister of national defence had written Clark, the minister of external affairs, to lobby for the public distribution of guidelines.
31 Greg Ing and Anne McIlroy, "Atomic Energy Lab Working on U.S. Star Wars Research," *Ottawa Citizen*, July 13, 1987.
32 United States General Accounting Office, *Strategic Defense Initiative Program: Extent of Foreign Participation* (Washington, DC: United States General Accounting Office, 1990).
33 Cited in Douglas Ross, "SDI and Canadian-American Relations," in *America's Alliances and Canada-American Relations*, ed. Lauren McKinsey and Kim Richard Nossal (Toronto: Summerhill Press, 1988), 158.
34 For their report prepared for the director general of International and Industry Programs, see James Fergusson, James Mackintosh, Patrick McDonnell, and Shane Levesque, *Missile Defence: Options for Canadian Industry* (Ottawa: Department of National Defence, 2001).
35 Ronald Reagan, *The Reagan Diaries*, ed. Douglas Brinkley (New York: Harper Perennial, 2007), 576.
36 For the most recent assessment, see Stephané Roussel, *The North American Democratic Peace: Absence of War and Security Institution-Building in Canada-US Relations* (Kingston, ON: McGill-Queen's University Press, 2004).
37 Ironically, this underpinned the position the NDP held, even though its argument naively assumed that Canada could really be neutral during the Cold War and would be seen that way by the Soviet Union and others.

38 One example of such influence was Canadian input into Strategic Air Command plans for overflights of Canadian territory.
39 Department of External Affairs; Internal Memorandum; International Defence Relations; February 12, 1987.
40 Department of National Defence; Memorandum to Assistant Deputy Minister (Policy); Directorate of Strategic Analysis; October 17, 1989.
41 Ibid.
42 Department of External Affairs; Memorandum to Ottawa; Canadian Embassy in Washington; July 2, 1985.
43 Department of External Affairs; Internal Memorandum; International Defence Relations; July 31, 1989.
44 Department of National Defence; Memorandum to Assistant Deputy Minister (Policy); Directorate of Strategic Analysis; October 17, 1989.
45 In 1986, National Defence informed the United States that it would proceed to develop a space-based radar satellite constellation for wide-area surveillance, but it is unclear whether it was ever linked to the missile defence question. Joseph Jockel, *Canada in NORAD 1957-2007: A History* (Kingston, ON: McGill-Queen's University Press, 2007).
46 As a respite, Gorbachev's decisions may be understood as a tactical move that would enable the Soviet Union to recover its strength before re-engaging, which has its roots in Lenin's decision to seek breathing space in 1922, with the enunciation of the New Economic Policy. In so doing, Gorbachev, with his domestic reforms, unleashed political forces that could not be controlled, resulting in the death and breakup of the Soviet Union. Alternatively, Gorbachev can also be understood as representing an entirely new reform movement, reminiscent of the *socialism with a human face,* which had been seen in Czechoslovakia in 1968 under Alexander Dubček prior to the Soviet-led Warsaw Pact intervention. From this perspective, Gorbachev sought to end permanently the Cold War, taking peaceful coexistence as a real possibility. Regardless, the outcome was the same.
47 The Brezhnev Doctrine, enunciated on the occasion of the Soviet-led Warsaw Pact military intervention into Czechoslovakia in August 1968 and removal of Dubček from power and end of the Prague Spring, stated that it was the right and duty of all socialist states to act to prevent counter-revolutionary movements.
48 The only exception in terms of the forward deployment of Russian military forces as the Soviet Union broke into its constituent republics was Kaliningrad, a small piece of Russian territory on the Baltic Sea sandwiched between Poland and Lithuania.
49 The complicated agreement also provided numerous other sub-limits and included significant on-site verification protocols. See Federation of Atomic Scientists, "Strategic Arms Reduction Treaty I," http://www.fas.org.
50 The Lisbon Protocols were signed on May 23, 1992. START I formally came into effect on December 5, 1994.
51 The Helsinki Conference on Security and Cooperation in Europe agreement, another symbol of détente, entailed three baskets: security, economic cooperation, and human rights. For the Soviet Union, it had legitimized the territorial changes in Europe that had occurred at the end of the Second World War and was seen in broader terms as the final peace treaty of the Second World War. For the West, emphasis was placed on the human rights basket as a means to force the Soviet Union and its East European satellites to open up their societies.
52 The long-forgotten treaty remains in effect. Its alliance basis was finally eliminated in 1999 with the new adapted treaty based upon national and territorial holdings. See Jeffrey McCausland, *Conventional Arms Control and European Security,* Adelphi Paper 301 (London: International Institute for Strategic Studies, 1996).

53 The US-led *police* action in Korea was also based upon a Chapter Seven, Article 42, motion. It passed the Security Council only because the Soviet delegate was boycotting its meetings because of the failure to recognize the new People's Republic of China. The Soviets quickly returned to the council, creating deadlock and resulting in the passing of the United for Peace Resolution in the then-US dominated General Assembly. The resolution allows the assembly to take action when or if the council is deadlocked and served as the means to continue UN support for Korea. Since then, the resolution was not acted upon, as most recognized the danger of the UN attempting to act without the support of both superpowers.

54 The other proliferation, which would come to overshadow politically, was intra-state ethnic conflict, as witnessed initially with the breakup of Yugoslavia in June 1990 and then the collapse of Somalia in January 1991.

55 Despite expectations, Iraq did not employ chemical weapons either on the battlefield or in its ballistic missile attacks against Israel and Saudi Arabia. It remains unclear whether the Iraqi decision was because of the lack of the technical ability to marry a chemical warhead to a high-speed ballistic missile or because of the explicit threat made by Secretary of State Baker to Iraqi foreign minister Aziz of US retaliation, with the means left open for Iraqi interpretation, at least publicly. If it was exclusively the latter, the idea of Hussein being undeterrable evaporates.

56 J. Daniel Sherman, "PATRIOT PAC-2 Development and Deployment in the Gulf War," US Army Material Command Study, http://www.dau.mil.

57 The key problem was speed. The PAC-2 was designed to intercept missiles flying between 5,200 and 5,600 feet per second. The Iraqi missiles flew between 6,500 and 7,200 feet per second. Ibid., 38.

58 The modified mobile Scuds or al-Hussein missiles were deployed in the large expanse of the Iraqi western desert within range of major targets in Israel and Saudi Arabia. The attacks did little damage to Israel, as the breakup of the missiles usually meant the conventional warhead did not work, and fears that chemical warheads would be employed proved unfounded. Only one attack into Saudi Arabia inflicted major casualties.

59 See Tim Ripley, *Scud Hunting: Counter-Force Operations against Theatre Ballistic Missiles,* Bailrigg Memorandum no. 18 (Lancaster, UK: Centre for Defence and International Security Operations, 1996).

60 On the Canadian side, see Frank Harvey, *Smoke and Mirrors: Globalized Terrorism and the Illusion of Multilateral Security* (Toronto: University of Toronto Press, 2005).

61 Counting rules for making a reasonably accurate comparison between offensive and defensive costs is extremely difficult. During the SDI debate, offensive costs were always seen as much less than defensive, which underpinned the arms race argument. ICBMs cost less than all the elements required to undertake a successful intercept with a defensive missile, and thus building more ICBMs to defeat an ABM system made sense. However, many of the elements vital for an intercept either had other missile roles to play (early warning, for example) or could serve other military functions. In addition, the costs of counter-measures, decoys, and penetration aids also had to be factored in, and one had to consider overall sunk investment costs from nuclear warheads to basic rocket engines, guidance systems, and the like.

62 See James Fergusson, "From Counter-Proliferation to Non-Proliferation: An Alternative Perspective on Ballistic Missile Defence," *New Approaches to Non-Proliferation* (Toronto: Centre for International and Strategic Studies, York University, Summer 1996).

63 Cueing refers to identifying the target and informing other sensors and radars where to find it during flight in the guidance of interceptor to target.

64 Department of National Defence; Response to External Affairs GPALS Paper; Assistant Deputy Minister (Policy); September 16, 1991.

65 The Standing Consultative Commission in Geneva, consisting of US and Soviet officials, was designed as a forum for discussions on interpretation of the ABM Treaty.

66 Department of External Affairs; Memorandum to the Minister of External Affairs on GPALS; November 8, 1991.

67 Article 15, "Treaty between the United States of America and the Union of Soviet Socialist Republics on the Limitation of Anti-Ballistic Missile Systems," Appendix A in Matthew Bunn, *Foundation for the Future: The ABM Treaty and National Security* (Washington, DC: Arms Control Association, 1990), 163.

68 Alongside the range of strategic nuclear issues that existed during the Cold War and became open for discussion after the Cold War was the issue of a launch-on-warning versus launch-under-attack doctrine. The former was a doctrine to launch a retaliatory strike upon immediate warning of an attack, such that the missiles could escape their silos before a first strike. The latter was a doctrine to absorb a first strike and then retaliate, which naturally required a secure communications system capable of working after an attack. The former raised the spectre of accidental launch, whereas the latter raised the possibility of an opponent's successful first strike, which might target and decapitate the leadership. Both the United States and the Soviet Union had essentially adopted launch-on-warning postures. Alongside this issue was alert status.

69 White House, *Joint U.S. – Russian Statement on a Global Protection System* (Washington, DC: Office of the Press Secretary, June 17, 1992).

70 At the summit, the presidents agreed that START II would cut strategic force levels in half to between 3,000 and 3,500 warheads from START I levels. See Act 4.

71 Also as part of this new environment, Bush announced that all forward-deployed US tactical nuclear forces, except in NATO, would be withdrawn in National Security Directive 64 on November 5, 1991. The document has not yet been formally released. See George Bush Presidential Library, http://bushlibrary.tamu.edu.

72 North Atlantic Treaty Organization, "Rome Declaration on Peace and Cooperation," *NATO Review* 39, 6 (December 1991): 19-22.

73 The Outer Space Treaty contains no explicit definition of where space begins or of a space weapon. By inference with reference to Article 4 on the prohibition on the deployment of weapons of mass destruction in space, the treatment of space analogous to the high seas or international waters, and the references to satellites or objects in space, space may be defined as the point where an object can complete a single orbit, roughly 150 kilometres. Thus, to be a weapon in space, the object must complete at least a single full orbit. Treaty of the Principles Governing the Activities of States in the Exploration and Use of Outer Space, Including the Moon and Other Celestial Bodies, www.state.gov/www/global/arms/treaties/space1.html.

74 Ambassador Mason clearly spelled out the three elements of non-proliferation in Canadian policy: elimination of all WMD, reduction of conventional weapons to the lowest possible level for defence, and prevention of the spread of WMD. Department of External Affairs; Comments on Memorandum to the Minister of External Affairs; Ambassador for Disarmament; October 1, 1992.

75 The Unified Command Plan reviews were instituted after the Second World War. As part of the 1996 Goldwater-Nichols reforms of the US military in the wake of the problems of inter-service cooperation in the invasion of Grenada in 1985, the Joint Chiefs of Staff were required by law to examine every two years the structure of US command within the Unified Command Plan. See Lt. Col. Marcus Fielding, "The United States Unified Command Plan," *Canadian Military Journal*, Autumn 2006: 35-40.

76 Department of National Defence; "Effects of GPALS on CAN-US Cooperation in NORAD," Annex in Memorandum to the Minister of External Affairs; November 8, 1991.
77 Department of External Affairs; Report on PJBD Meeting; March 16, 1992.
78 Canada, *House of Commons Debates,* Vol. 6, 3rd Session, 34th Parliament (March 9, 1992), p. 7837.
79 Ibid. (May 21, 1992), p. 11055.

Act 4: National Missile Defense

1 The JROC reviews and approves requirements for all new weapon systems and consists of the four service vice-chiefs of staff. Jeff Erlich, "U.S. Panel Gives NMD Approval Despite Hurdles," *Defense News,* August 26, 1996.
2 Only one op-ed piece appeared on missile defence immediately following the release of the 1994 White Paper; an improvement somewhat given the failure of public and media attention during the GPALS era. James Fergusson, "The Coming Debate on Ballistic Missile Defence," *Financial Post,* July 15, 1995.
3 Canada, *North American Aerospace Defense Command: 1996 NORAD Agreement and Terms of Reference* (Ottawa: Queen's Printer, 1996), 3.
4 As part of START, the Grand Forks ICBM field was being dismantled. The Minot, or western North Dakota, field was potentially outside the 150-kilometre-radius area that had to contain ICBMs as specified in Article 3(b). "Treaty between the United States of America and the Union of Soviet Socialist Republics on the Limitation of Anti-Ballistic Missile Systems," Appendix A in Matthew Bunn, *Foundation for the Future: The ABM Treaty and National Security* (Washington, DC: Arms Control Association, 1990).
5 The concept underlying THAAD was presented to Congress by Secretary of Defense Cheney on July 2, 1992. Council on Foreign Relations, "Chronology of National Missile Defence Programs," 2007, http://www.cfr.org.
6 Both the air force and the army came to compete for NMD with alternative components for a ground-based system. As missile defence had been assigned to the army in 1957, army officials complained about air force involvement. For details of the different proposals, see Table 4.1 and David J. Mosher, "The Grand Plans," *IEEE Spectrum,* September 1997: 28-39.
7 Department of National Defence; BMD: The Way Ahead – Internal Memorandum; Directorate of Continental Policy; January 12, 1993.
8 Department of External Affairs; Bilateral Security/Defence Relations Memorandum; May 5, 1993.
9 Ibid.
10 Thomas Friedman, "U.S. Formally Rejects 'Star Wars' in ABM Treaty," *New York Times,* July 15, 1993.
11 On January 28, 1991, the first test of the Exo-atmospheric Re-entry Vehicle Interceptor System employing kinetic-kill technology from the 1980s Homing Overlay Experiment was successful. On March 13, 1992, the second test failed.
12 The Goldwater-Nichols Act followed the US intervention in Grenada in 1983, which exposed major gaps in the ability of the three services to work together. The act mandated a series of reforms to ensure jointness among the services, provided more authority to the joint chiefs, and pushed further the US command structure of regional commands supported by force generators. Lt. Col. Marcus Fielding, "The United States Unified Command Plan," *Canadian Military Journal,* Autumn 2006: 35-40.
13 After the gutting of the 1987 White Paper with the 1989 budget, the Mulroney government issued two defence policy statements in 1991 and 1992 respectively. But neither represented a major review of defence policy and were largely ignored.

14 Ivan Head and the Canada 21 Council, *Canada 21: Canada and Common Security in the Twenty First Century* (Toronto: Centre for International Studies, University of Toronto, 1994).
15 Special Joint Committee on Canada's Defence Policy, "Dissenting Report by the Bloc Québécois Members of the Special Joint Committee on Canada's Defence Policy," in *Security in a Changing World* (Ottawa: Parliamentary Publications Directorate, 1994), 78.
16 The Chrétien government, to remove the last *symbolic* vestige of colonialism, changed the name of External Affairs to Foreign Affairs early on in its tenure.
17 Department of National Defence, *1994 Defence White Paper* (Ottawa: Queen's Printer, 1994), 25.
18 The most detailed report on the NMD program at the time is Ballistic Missile Defense Organization, *National Missile Defense,* Annual Report to Congress, Washington, DC, 1997.
19 United States, "Emerging Missile Threats to North America during the Next Fifteen Years," DCI National Intelligence Estimate, President's Summary, http://www.fas.org.
20 United States Congress, *Report of the Commission to Assess the Ballistic Missile Threat to the United States* (Washington, DC: United States Congress, July 15, 1998).
21 Secretary of Defense William Cohen; Press Statement; Washington, DC, Department of Defense; January 21, 1999; p. 1.
22 United States, *Foreign Missile Developments and the Ballistic Missile Threat to the United States through 2015*, Washington, DC, September 9, 1999.
23 Ibid., 2.
24 The fly-by test served to evaluate the ability of the in-flight guidance system to employ dual imaging or phenomenology to identify and track the dummy warhead. William Scott, "Missile Sensor Flyby Boosts NMD Outlook," *Aviation Week and Space Technology,* January 26, 1998: 88.
25 United States, Department of Defense, Office of the Assistant Secretary of Defense (Public Affairs), *National Missile Defense Conducts Successful Intercept Test,* news release, October 2, 1999. See also Bradley Graham, "Missile Defense Plan Scores a Direct Hit," *Washington Post,* October 3, 1999.
26 Theodore Postol became the better-known critic largely because of his appearance on the CBS program *60 Minutes.* George N. Lewis and Theodore Postol, "An Evaluation of the Army Report 'Analysis of Video Tapes to Assess Patriot Effectiveness,' Dated 31 March 1992," September 1992, http://www.fas.org. This view and others critical of the technology are also found in Richard Garvey, "The Wrong Plan," *Bulletin of the Atomic Scientists,* March-April 2000: 36-41.
27 The second intercept test in January, which included the integration of early warning and the X-band guidance radar, failed because of a cooling system malfunction to the on-board terminal infrared guidance sensor in the last five seconds before intercept. United States, Department of Defense, Office of the Assistant Secretary of Defense (Public Affairs), *National Missile Defense Conducts Successful Intercept Test,* news release, October 2, 1999. The third and final test prior to the deployment review and decision on July 8 failed when the second stage did not separate and ignite – a failure of mature technology.
28 This general line was reiterated in the Canadian press as well, even in 2000. Mike Blanchfield, "Shelve Missile Shield: Scientists," *Ottawa Citizen,* June 13, 2000.
29 John Baylis and Kristan Stiddart, "Britain and the Chevaline Project: The Hidden Nuclear Programme, 1967-1982," *Journal of Strategic Studies* 20, 4 (December 2003): 124-55.
30 Department of National Defence, General Larry Welch (ret.) et al., *Report of the Panel on Reducing Risk in Ballistic Missile Defense Flight Test Programs* (Washington, DC: Department of Defense, February 27, 1998). The term "rush to failure" was coined in an article

by Bradley Graham, "Panel Fires at Antimissile Programs: Critique Warns of 'Rush to Failure,'" *Washington Post,* March 22, 1998. See also Paul Mann, "Missile Defense Riddled with Diverse Failures," *Aviation Week and Space Technology,* March 30, 1998, and Lisa Burgess, "BMDO Faces Conflicting Signals," *Defense News,* March 30-April 5, 1998.

31 Department of National Defence, General Larry Welch (ret.) et al., *2nd Report of the Panel on Reducing Risk in Ballistic Missile Defense Flight Test Programs* (Washington, DC: Department of Defense, November 1999).

32 Joseph C. Anselmo, "Pentagon to Spend Big on NMD Testing," *Aviation Week and Space Technology,* September 22, 1997: 88.

33 Congressional Budget Office, *Budgetary and Technical Implications of the Administration's Plan for National Missile Defense* (Washington, DC: Congress of the United States, 2000).

34 In the original bidding for the NMD contract, the companies were required to address all three possible system architectures. James Fergusson, *Canada and Ballistic Missile Defence: Issues, Implications, and Timelines* (Ottawa: Department of Foreign Affairs and International Trade, Nonproliferation, Arms Control, and Disarmament Division, 1998).

35 Actually, the system probably needed more in order to take into account probability estimates of the number of interceptors that might fail anywhere from the launch to release of the EKV or kinetic-kill interceptor warhead.

36 See Act 1.

37 Arguably, there is nothing that explicitly prohibits a national defence as long as it meets Article 3, which forces deployment with the national capital or an ICBM within a 150-kilometre radius. "Treaty between the United States of America and the Union of Soviet Socialist Republics on the Limitation of Anti-Ballistic Missile Systems," Appendix A in Matthew Bunn, *Foundation for the Future: The ABM Treaty and National Security* (Washington, DC: Arms Control Association, 1990).

38 These are found in Article 1, paragraph 2, concerning national defence, Article 3 of the original treaty, and Article 1 of the protocol.

39 The Russians raised the ground-based replacements as a violation, but the United States countered that these radars had been grandfathered in the treaty and already modernized once before without objection.

40 The NMD ground-based radar for the Fort Greely site was tentatively set for deployment at Sheyma Island at the Aleutian chain. In the next iteration, a sea-based X-band radar based in Alaska would be added.

41 Standing Consultative Commission, "Agreed Statement on ABM Treaty Topics," November 1, 1978. Cited in Matthew Bunn, Appendix B in *Foundation for the Future: The ABM Treaty and National Security* (Washington, DC: Arms Control Association, 1990), 168.

42 This interpretation was directly made in testimony by Sidney Graybeal, one of the negotiators of the ABM Treaty, before the Senate Armed Services Committee.

43 Of eight long-range phased array radars, one was in Latvia, two in Ukraine, one in Azerbaijan, and one in Kazakhstan. *Arms Control Reporter: A Chronicle of Treaties, Negotiations, Proposals, Weapons, and Policy* (Brookline, MA: Institute for Defense and Disarmament Studies, 1997), 576.E-TMD.21.

44 David B. Rivkin Jr., Lee A. Casey, and Darin R. Bartram, *The Collapse of the Soviet Union and the End of the 1972 Anti-Ballistic Missile Treaty* (Washington, DC: Heritage Foundation, 1998).

45 Heritage Foundation, *Defending America: A Near- and Long-Term Plan to Deploy Missile Defenses,* report of the Missile Defense Study Team (Washington, DC: Heritage Foundation, 1996). See also Hans Binnendjik and George Stewart, "Toward Missile Defenses from the Sea," *Washington Quarterly* 25, 3 (Summer 2002): 193-207.

46 Aegis refers to the new radar tracking system on the vessels. The specific class of cruiser is the Ticonderoga.

47 The cruisers employed the Standard Missile-2(b) in an air defence role. The follow-on was the Standard Missile-3, which required no significant alterations to the missile launch tubes on the vessel. The Canadian-area air defence Tribal-class destroyers also employed the Standard Missile-2.

48 The S-300 was based upon the SA-12 high-altitude air defence system and came in two variants, the V and P classes. Russia was also seeking to sell the S-300 overseas, and South Korea was considering their purchase. *Arms Control Reporter: A Chronicle of Treaties, Negotiations, Proposals, Weapons, and Policy* (Brookline, MA: Institute for Defense and Disarmament Studies, 1997), 576.E-TMD.27.

49 For a brief discussion and proposals for future cooperation at the time, see K. Scott McMahon, *Pursuit of the Shield: The U.S. Quest for Limited Ballistic Missile Defence* (Lanham, MD: University Press of America, 1997).

50 White House, *Press Briefing by Robert Bell, Senior Director for the NSC for Defense Policy and Arms Control* (Washington, DC: Office of the Press Secretary, March 24, 1997).

51 "1997 ABM Treaty Demarcation Memorandums of Understanding, Agreed, Joint and Unilateral Statements, and Common Understandings," *Arms Control Reporter: A Chronicle of Treaties, Negotiations, Proposals, Weapons, and Policy* (Brookline, MA: Institute for Defense and Disarmament Studies, 1997), 613.D.47-603.D.63.

52 Within the category of theatre systems, the following programs were specifically mentioned: the American Theater High Altitude Area Defense (THAAD), the Navy Theater Wide Ballistic Missile Defence, and the Russian S-300V (also known as the SA-12). "Agreement on Confidence-Building Measures Relative Systems to Counter Ballistic Missiles Other than Strategic Ballistic Missiles," *Arms Control Reporter: A Chronicle of Treaties, Negotiations, Proposals, Weapons, and Policy* (Brookline, MA: Institute for Defense and Disarmament Studies, 1997), 603.D.53.

53 "Statement by the United States of America on Plans with Respect to Systems to Counter Ballistic Missiles Other than Strategic Ballistic Missiles," *Arms Control Reporter: A Chronicle of Treaties, Negotiations, Proposals, Weapons, and Policy* (Brookline, MA: Institute for Defense and Disarmament Studies, 1997), 603.D.63.

54 "Agreement on Confidence-Building Measures Relative Systems to Counter Ballistic Missiles Other than Strategic Ballistic Missiles," *Arms Control Reporter: A Chronicle of Treaties, Negotiations, Proposals, Weapons, and Policy* (Brookline, MA: Institute for Defense and Disarmament Studies, 1997), 603.D.53.

55 See "THAAD and Navy Theater-Wide Violate Treaty," *Arms Control Reporter: A Chronicle of Treaties, Negotiations, Proposals, Weapons, and Policy* (Brookline, MA: Institute for Defense and Disarmament Studies, 1997), 603.C.19.

56 Rodney W. Jones and Nikolai Sokov, "After Helsinki, the Hard Work," *Bulletin of Atomic Scientists* 53, 4 (July-August 1997): 26-31.

57 See, for example, David Hoffman, "Russia Says START II Is Imperiled," *Washington Post*, January 22, 1999.

58 Bill Gertz, "Cohen Sees Russia, U.S. Allies as Hurdles to a Missile Defense," *Washington Times*, October 5, 1999.

59 Executive agreements deal with day-to-day interpretation, implementation, and technical issues and do not require Senate advice and consent.

60 The act passed by a vote of 97 to 3 in the Senate and 317 to 102 in the House on March 17 and 18 respectively. The House bill contained no condition for deployment as did the Senate bill.

61 Bill Gertz, "State Department Cable Argues of a Loophole in Missile Defense Bill," *Washington Times,* March 26, 1999.
62 United States Congress, National Missile Defense Act of 1999, www.cdi.org.
63 Steven Lee Myers, Eric Schmitt, and Marc Lacey, "Russian Resistance Key in Decision to Delay Missile Shield," *New York Times,* September 3, 2000.
64 For example, according to the *Military Balance,* in 1997-98 China possessed seven DF-5 and ten-plus DF-4 single warhead ICBMs. International Institute for Strategic Studies, *The Military Balance 1997/98* (London: Oxford University Press, 1997).
65 See Dean A. Wilkening, *Ballistic Missile Defence and Strategic Stability,* Adelphi Paper 334 (London: International Institute for Strategic Studies, 2000).
66 The resolution passed with eighty in favour, four against, and sixty-eight abstentions. United Nations General Assembly, Resolution 9675, December 1, 1999.
67 For detailed discussion of European attitudes and programs, see James Fergusson, *Ballistic Missile Defence: Implications for the Alliance,* NATO fellowship report, June 2000.
68 North Atlantic Treaty Organization, "Rome Declaration on Peace and Cooperation," *NATO Review* 39, 6 (December 1991): 19-22.
69 North Atlantic Treaty Organization, *Alliance Policy Framework on the Proliferation of Weapons of Mass Destruction* (Istanbul: North Atlantic Council Ministerial, June 9, 1994).
70 For a detailed discussion of the evolution of NATO's missile defence response, see David Martin, "Towards an Alliance Framework for Extended Air Defence/Theatre Missile Defence," *NATO Review* 44, 3 (May 1996): 32-35. See also Gregory Schulte, "Responding to Proliferation: NATO's Role," *NATO Review* 43, 4 (July 1995): 15-19, and North Atlantic Treaty Organization, *NATO's Response to Proliferation of Weapons of Mass Destruction,* Fact Sheet no. 8, April 1997.
71 One of the major political initiatives in this regard is the opening of the Mediterranean Dialogue in 1994, though none of the six Middle East participants includes the so-called rogue states. See North Atlantic Treaty Organization, *The Mediterranean Dialogue,* Fact Sheet no. 16, May 1997.
72 North Atlantic Treaty Organization, *The Alliance's Strategic Concept,* Press Release NAC-S(99)65, April 24, 1999.
73 Luke Hill, "TMD: NATO Starts the Countdown," *Jane's Defence Weekly,* January 3, 2001: 24-27. On June 5, 2001, NATO announced that the consortia led by Science Applications Corporation and Lockheed Martin had been awarded the contracts. North Atlantic Treaty Organization, "NATO's Theatre Missile Defence Programme Reaches New Milestone," Press Release, June 5, 2001.
74 For a detailed discussion, see James Fergusson, "The European Dimension of Ballistic Missile Defence," in *Canada and National Missile Defence,* ed. David Rudd, Jim Hanson, and Jessica Blitt (Toronto: Canadian Institute of Strategic Studies, 2000), 27-38.
75 MEADS was to be based upon the American Corp surface-to-air (SAM) project.
76 See Jeremy Stocker, *Sea-Based Ballistic Missile Defence,* Bailrigg Study 2 (Lancaster, UK: Centre for Defence and International Security Studies, 1999).
77 Besides improved guidance radar, the PAC-3 employs a kinetic-kill intercept, whereas the PAC-2 uses a conventional proximity explosive. The PAC-3 successfully intercepted a test missile in March 1999.
78 In 2000, it became evident that the APAR was too large for Canadian frigates, and Canada withdrew. Given the budget situation, the likelihood that it would be purchased in the foreseeable future was remote.

79 Directorate of Strategic Analysis, Project Report No. 9721 (Ottawa: Department of National Defence, November 1997). The title had been deleted for security reasons upon release, as well as significant sections of the report.
80 In 2002, the term "space-based infrared low" was essentially dropped and restructured into the "space tracking and surveillance system." Missile Defense Agency, "Space Tracking and Surveillance System (STSS)," MDA Facts, http://www.nuclearthreatinitiative.com.
81 Canada, *North American Aerospace Defense Command: 1996 NORAD Agreement and Terms of Reference* (Ottawa: Queen's Printer, 1996), 3.
82 Ibid., 2.
83 Canada, *House of Commons Debates,* Vol. 1, 2nd Session, 35th Parliament (March 11, 1976), p. 498.
84 Ibid., 495.
85 Lloyd Axworthy, "Notes for an Address by the Honourable Lloyd Axworthy, Minister of Foreign Affairs, to Accept the Endicott Peabody Award," Annual Peabody Awards (Boston: Department of Foreign Affairs and International Trade, October 22, 1999).
86 North American Aerospace Defense Command-United States Space Command, *Concept of Operations for Ballistic Missile Defense of North America* (Colorado Springs, CO: North American Aerospace Defense Command – United States Space Command, July 21, 1998).
87 Assigned to USAF Space Command, also commanded by General Horner and his successors up until the decision in the 2001 Unified Command Plan to merge Space Command and Strategic Command, 14th Air Force is tasked operationally with US military satellites.
88 Chief of the Defence Staff (Canada), *Shaping the Future of Canadian Defence: A Strategy for 2020* (Ottawa: Department of National Defence, 2000).
89 The most public recognition was the presence of Canadian Forces personnel with American military units in Iraq, including the position of deputy commander of the US Third Corps upon its deployment, who is now the chief of the defence staff. General Hillier, the most recent chief of the defence staff, also served as a senior commander with the US Army.
90 Although not formally linked at the time, DND had also very recently gone through the costs of environmental cleanup associated with the closure of US installations from the Cold War, such as the base at Argentia, Newfoundland.
91 In the brief nuclear submarine debate, the decision to acquire nuclear-powered submarines was quickly portrayed as Canada acquiring nuclear weapons.
92 J.H. Chapman, P.A. Forsyth, P.A. Lapp, and G.N. Patterson, *Upper Atmosphere and Space Programs in Canada,* Science Secretariat, Privy Council Office Report (Ottawa: Queen's Printer, 1967).
93 The Cold Lake camera was closed in 1981 after roughly two decades of service.
94 In May 2001, the liaison position was raised by Svend Robinson (NDP) in the House and then again by Gilles Duceppe, leader of the Bloc Québécois. On both occasions, Defence Minister Eggleton replied that the position was consistent with Canadian policy in the 1994 White Paper. Canada, *House of Commons Debates,* Vol. 5, 2nd Session, 37th Parliament (May 9, 2001), 3794; Ibid. (May 15, 2001), p. 4070.
95 Kathleen Kenna, "Canada and U.S. Renew NORAD Deal," *Toronto Star*, June 17, 2000.
96 ITAR did not end, however. The current issue for industry is US prohibitions on dual-nationals and immigrant access to ITAR-controlled technology and information, which violates Canada's Charter of Rights provisions.
97 On a critique of civil society, see Denis Stairs, "Foreign Policy Consultations in a Globalizing World: The Case of Canada, the WTO, and the Shenanigans in Seattle," *Policy Matters* 1, 8 (December 2000): 1-44.

98 For the discussion from among the adherents inside and outside Foreign Affairs at the time, see Rob McRae and Don Hubert, eds., *Human Security and the New Diplomacy: Protecting People, Promoting Peace* (Kingston, ON: McGill-Queen's University Press, 2001).
99 The next candidate for the Ottawa Process appeared to be the issue of child soldiers, which also put Canada on a collision course with the United States. For whatever reason, the process beyond some initial conferences with non-governmental organizations never occurred. Some have speculated that the prime minister intervened, possibly after a talk with Clinton. Today it is known as the Oslo Process and is linked to the attempt to replicate the land mines ban in the area of cluster munitions, with negotiations under the rubric of a Convention on Certain Conventional Weapons in the Conference on Disarmament stalled. The most recent negotiations were held in May 2008 in Ireland.
100 The final committee report boldly recommended the "government of Canada argue forcefully within NATO that the present reexamination and update as necessary of the Alliance Strategic Concept should include its nuclear component." Standing Committee on Foreign Affairs and International Trade, *Canada and the Nuclear Challenge: Reducing the Political Value of Nuclear Weapons for the Twenty First Century* (Ottawa: House of Commons, 1998).
101 This was specified in US legislation passed immediately after the establishment of the court. See Jennifer K. Elsea, *U.S. Policy Regarding the International Criminal Court* (Washington, DC: Congressional Research Service, 2006).
102 It appeared to be a favourite title for *Toronto Star* articles, though none ever explained the exact nature of the linkage.
103 United States Space Command, *Vision for 2020* (Colorado Springs, CO: United States Space Command, 1998), and United States Space Command, *The Long Range Plan: Implementing USSPACECOM's Vision for 2020* (Colorado Springs, CO: United States Space Command, 1998).
104 The initial reference was made by the US Air Force chief of staff, General Merrill McPeak, in April 1991. Craig Covault, "Desert Storm Reinforces Military Space Direction," *Aviation Week and Space Technology,* April 8, 1991: 48. Volume 4 of *The Gulf War Airpower Survey* examines space operations but has not been released to the public. For the summary of the survey, see Thomas Keaney and Eliot Cohen, *Gulf War Airpower Survey Summary Report* (Washington, DC: United States Air Force, 1993).
105 It was for this reason that the United States' and Soviet Union's anti-satellite programs were so costly. Both were dependent upon space and could not afford an indiscriminate strike. For a basic discussion, see James Fergusson and Steve James, *Canada, National Security, and Outer Space* (Calgary: Canadian Defence and Foreign Affairs Institute, 2007).
106 Cited in Robert Bothwell, "Canada's Moment: Lester Pearson, Canada, and the World," in *Pearson: The Unlikely Gladiator,* ed. Norman Hillmer (Kingston, ON: McGill-Queen's University Press, 1999), 19-29.
107 In the same article, Hamre warns of the consequences for Canada of being on the outside of missile defence in a speech a month earlier in Calgary: "I believe we are at an important pivot point in our relationship with each other ... That pivot point is going to revolve around the issue of national missile defence." Paul Koring, "Canada Accused of Cold War Thinking in Impasse on Missile Defence: U.S. Official Annoyed with Ottawa's Intransigence on Revamping the Treaty," *Globe and Mail,* May 16, 2000.
108 At the time, the administration wondered, for example, if pouring concrete at a site prohibited by the treaty would be a treaty violation.
109 Jean Chrétien, *My Years as Prime Minister* (Toronto: Knopf, 2007), 302.
110 Greg Seigle, "US NMD Lacks Approval from Canada, NORAD Deputy Says," *Jane's Defence Weekly,* January 26, 2000: 5.

111 Standing Committee on Foreign Affairs and National Defence; Minutes of Proceedings and Evidence; Special Joint Meeting with the Standing Committee on National Defence and Veterans Affairs; March 23, 2000; http://cmte.parl.gc.ca/cmte.
112 Ibid.
113 The remaining rump of Progressive Conservatives was also divided, reflecting the existence of so-called red and blue Tories. Clark, who had returned as leader, not surprisingly as a red Tory, was opposed to all things missile defence.
114 Canada, *House of Commons Debates*, Vol. 4, 2nd Session, 36th Parliament (February 25, 2000), p. 4038.
115 I served as the academic technical advisor on the committee's visit and subsequently testified on the topic of NMD to the committee. Canada, *House of Commons Debates*, Vol. 7, 2nd Session, 36th Parliament (May 9, 2000), p. 6585.
116 Doug Bland calls this one of the ten principles of Canadian defence policy. Doug Bland and Sean Maloney, *Campaigns for International Security* (Kingston, ON: McGill-Queen's University Press, 2003).
117 Kathleen Kenna, "Son of Star Wars Heats Up: Canada Weighs Its Options as Pentagon Presses for a Partner in Controversial Missile-Defence Scheme," *Toronto Star*, May 21, 2000.
118 Sharon Hobson, "Canadian Missile Defence," *Jane's International Defence Review*, March 1, 2002: 17.
119 Jeff Sallot, "Support Missile Project or Else, Ottawa Warned," *Globe and Mail*, March 13, 2000.
120 This reflects my direct experiences at the time as one of the academic participants in the public debate and one of the proponents of this line of reasoning.
121 In the late 1990s, National Defence commissioned several classified polls that reported public support for NORAD at over 80 percent and support for participation in missile defence at over 70 percent.
122 Canada, *House of Commons Debates*, Vol. 5, 2nd Session, 36th Parliament (March 20, 2000), p. 4868.

Act 5: Ground-Based Midcourse Defence

1 The speech raised immediate concerns among the Democrats about the future of the ABM Treaty. Alison Mitchell, "Senate Democrats Square Off with Bush over Missile Plan," *New York Times*, May 3, 2001.
2 The airborne laser has gone through several ground-test firings and air tracking. A successful airborne intercept test occurred in 2010. See Col. Robert McMurry and Lt. Gen. Michael Dunn (ret.), *Airborne Laser (ABL): Recent Developments and Plans for the Future*, Washington Roundtable on Science and Public Policy (Washington, DC: George C. Marshall Institute, 2008).
3 The first and only site was Kwajalien Atoll in the South Pacific, with targets launched from Vandenberg Air Force Base in California.
4 Article 15 permits withdrawal on the grounds of a threat to national security, which is non-negotiable: "Each Party shall, in exercising its national sovereignty, have the right to withdraw from this Treaty if it decides that extraordinary events related to the subject matter of this Treaty have jeopardized its supreme interests. It shall give notice of its decision to the other Party six months prior to withdrawal from the treaty. Such notice shall include a statement of the extraordinary events the notifying Party regards as having jeopardized its supreme interests." "Treaty between the United States of America and the Union of Soviet Socialist Republics on the Limitation of Anti-Ballistic Missile Systems," Appendix A in Matthew Bunn, *Foundation for the Future: The ABM Treaty and National Security* (Washington, DC: Arms Control Association, 1990).

5 The only exception I am aware of is my colleague, Dr. Frank Harvey, who predicted its end several months beforehand.
6 Liu Centre, *The Missile Defence Debate: Guiding Canada's Role* (Vancouver: University of British Columbia, March 2001), 5.
7 The argument was whether the Russian Federation was the legal successor state to the Soviet Union and thus bound to its international legal commitments. See David B. Rifkin Jr., Lee A. Casey, and Darin R. Bartram, *The Collapse of the Soviet Union and the End of the 1972 Anti-Ballistic Missile Treaty* (Washington, DC: Heritage Foundation, 1998).
8 On the Clinton administration's actions, see Eric Schmitt and Steven Lee Myers, "Clinton Lawyers Give a Go-Ahead to Missile Shield," *New York Times*, June 15, 2000.
9 This comment was made by Richard Perle, a former senior arms control official in the Reagan administration, to a missile defence conference. Canadian National Committee of the International Institute for Strategic Studies and the US Committee of the International Institute for Strategic Studies, National Missile Defense conference report, 2001, 7.
10 Peter Baker and William Drozdiak, "Russia Alters Tone, Welcomes Talks on Missile Shield," *Washington Post,* May 3, 2001.
11 In the vacuum of outer space, the blast effects of a nuclear warhead are very limited. Damage is caused by the gamma rays emitted by the detonation, which destroy the electrical systems of satellites, as the Americans found out in 1962 with the test of a warhead in outer space. Also, nuclear warheads can be easily hardened to defeat the impact of these rays.
12 For the state of European programs at the time, see James Fergusson, "The European Dimension of Ballistic Missile Defence," in *Canada and National Missile Defence,* ed. David Rudd, Jim Hanson, and Jessica Blitt (Toronto: Canadian Institute of Strategic Studies, 2000), 27-38.
13 William Drozdiak and Dana Milbank, "Bush Tries to Sell NATO on Missile Plan," *Washington Post,* June 14, 2001.
14 Steven Huldreth, *Missile Defense: The Current Debate* (Washington, DC: Congressional Research Service, December 7, 2004). Interestingly, Canada is not mentioned at all.
15 Canadian Security and Intelligence Service, *Ballistic Missile Proliferation,* Report #2000/09, March 23, 2001, 10.
16 Jan Cienski, "U.S. Calls Ottawa over Missile Shield Plan," *National Post,* May 1, 2001.
17 Article 5 is commonly known as the collective defence clause of the North Atlantic Treaty and is ostensibly equivalent to a declaration of war. NATO's Airborne Warning and Control System (AWACS) capability consisted of six aircraft based at Killenkrichen, Germany.
18 Embassy (in Canada), Condoleezza Rice, "Press Briefing by National Security Advisor Condoleezza Rice," Washington File, Ottawa, US Embassy, November 8, 2001.
19 Celeste A. Wallander, "Russia's Strategic Priorities," *Arms Control Today,* January-February 2002, www.armscontrol.org.
20 Bates Gill, "Can China's Tolerance Last?" *Arms Control Today,* January-February 2003, www.armscontrol.org.
21 Associated Press, *Free from ABM Treaty, Missile Defense Will Take Off,* May 13, 2002.
22 On February 6, 2007, Africa Command was established as independent from EUCOM and CENTCOM.
23 Until July 1, 2005, all operational commands in the Canadian Forces were the responsibility of the deputy chief of the defence staff. On July 1, Canada Command was created ostensibly with the same terms of reference as NORTHCOM, along with Canada Expeditionary Forces Command, with responsibility for all overseas missions. The exact arrangements between Canada Command and NORTHCOM have not yet been worked out.

24 Concerns about this type of cruise missile threat predated 9/11. The first formal briefing to senior officials in the Department of National Defence occurred in 1999. However, it would take 9/11 to give meaning to this threat.
25 US strategic forces that would be the likely target of a Soviet first strike consisted at the end of the Cold War of Minuteman and MX ICBMs deployed in silos across the northwestern United States in North Dakota, Montana, and Wyoming.
26 Sentinel entailed numerous sites such as US and Canadian cities in relatively close proximity, for example Seattle-Vancouver and Detroit-Windsor; perhaps Cleveland-Toronto would also be defended.
27 The first of five initial interceptors was placed in its silo at Fort Greely on July 22, 2004. Five more were deployed by the end of 2004 and ten more by the end of 2005. In addition, four were placed at Vandenberg.
28 The Canadian delegation employed the terms "NMD" and ballistic missile defence "BMD" in their discussions with the United States.
29 Paul Cellucci, *Unquiet Diplomacy* (Toronto: Key Porter, 2005), 135.
30 United Nations Security Council, Resolution 1441, November 8, 2002, http://www.iaea.org; Hans Blix, "An Update on Inspection," United Nations Monitoring, Verification, and Inspection Commission, January 27, 2003, http://www.un.org.
31 At the time, this point was raised by Joseph Jockel in "A Strong Friend Is a Good Defence," *Globe and Mail,* January 14, 2004.
32 Paul Cellucci, *Unquiet Diplomacy* (Toronto: Key Porter, 2005), 153.
33 Canada, *House of Commons Debates,* 2nd Session, 37th Parliament (May 29, 2003). Emphasis mine.
34 Ibid.
35 Both polls were commissioned by National Defence. Jack Aubry, "Canadian Fear of Missile Attack Grows," *National Post,* June 11, 2002.
36 Arms Control Association, "Canada Weighs U.S. Missile Defense Cooperation," *Arms Control Today,* July-August 2003, www.armscontrol.org.
37 Jeff Sallot, "Missile Defence Meeting Comes up Short on Specifics," *Globe and Mail,* May 16, 2001.
38 Allan Thompson, "U.S. Envoy Urges Canada to Weigh Missile Shield," *Toronto Star*, May 16, 2001.
39 Standing protocol placed command in the hands of the deputy when the commander is on duty elsewhere.
40 This and preceding three quotations from Department of National Defence; Letter from Minister Pratt to Secretary of Defense Donald Rumsfeld; Reply from Secretary Rumsfeld to Minister Pratt; January 15, 2004. Emphasis mine. Both this letter and Rumsfeld's response were deleted from the website within the year.
41 See Act 4. North American Aerospace Defense Command-United States Space Command, *Concept of Operations for Ballistic Missile Defence of North America* (Colorado Springs, CO: North American Aerospace Defense Command-United States Space Command, July 21, 1998).
42 Canada, Department of National Defence, "Canadian Participation in the Joint Strike Fighter Program," *Backgrounder,* December 11, 2006.
43 Paul Cellucci, *Unquiet Diplomacy* (Toronto: Key Porter, 2005), 158.
44 Even in the case of a minority government, such as with the NORAD decision in 1958, cabinet alone could decide with no recourse to Parliament. Certainly, the Opposition could force a vote in the House, as had been done in May 2003 and February 2004, but this would not bind the government unless it wished so. Concerning the likely outcome,

both of the previous votes had a majority in favour of Canadian participation, and, even with a divided Liberal caucus in a free vote, it was likely that the vote would be similar, albeit with a reduced majority.

45 Canada, *House of Commons Debates,* Vol. 139, 3rd Session, 37th Parliament (February 24, 2004), p. 1801.

46 See Act 3.

47 Minister of National Defence Bill Graham, press conference transcript, CTV National News, August 5, 2004.

48 Ibid.

49 Janice Gross Stein and Eugene Lang, *The Unexpected War: Canada in Kanduhar* (Toronto: Viking Canada, 2007), 162.

50 From the Opposition day motion by the Bloc Québécois on Thursday, February 19, 2004, through to the announcement of the government decision not to participate a year later, weaponization of space was the dominant rationale for non-participation. The Bloc Québécois and NDP raised the issue in nearly every question on missile defence in the House. For example, see Canada, *House of Commons Debates,* Vol. 140, 1st Session, 38th Parliament (October 8, 2004; November 4, 2004; November 29, 2004; and February 10, 2004).

51 Mel Hurtig, *Rushing to Armageddon: The Shocking Truth about Canada, Ballistic Missile Defence, and Star Wars* (Toronto: McClelland and Stewart), 2004.

52 US Space Command, *The Long Range Plan: Implementing USSPACECOM's Vision for 2020* (Colorado Springs, CO: US Space Command, 1998).

53 For one of the more recent overviews of the weaponization of space, see Wilson Wong, *Weapons in Space,* Silver Dart Canadian Aerospace Studies 3 (Winnipeg: Centre for Defence and Security Studies, 2006).

54 Missile Defense Agency, *Fiscal Year 2004-05 Biennial Budget Submission Estimates* (Washington, DC, 2004), 15-16.

55 Cited by Claude Bachand, Bloc Québécois Member of Parliament (St. Jean), Canada, *House of Commons Debates,* Vol. 139, 3rd Session, 37th Parliament (February 19, 2004), p. 1020.

56 Canadian Broadcasting Corporation, "Slim Majority Oppose Missile Defence: Poll," CBC TV news report, November 5, 2004, http://www.cbc.ca. The findings were echoed in a February Compas survey for the *National Post.* Compas, "Missile Defence: Small Soft, Quebec Majority Opposes It in Practice, While Backing It in Principle, Big Majority Condemns Ottawa's Lack of Public Discussion," February 28, 2005, http://www.compas.ca.

57 *Toronto Star,* "Harper Would Call for Vote on Missile Defence," June 9, 2004.

58 In speculative media discussions about the visit, references were made to the heckling of President Reagan during his speech to the Joint Houses of Parliament.

59 Paul Cellucci, *Unquiet Diplomacy* (Toronto: Key Porter, 2005), 161.

60 White House, "President Discusses Strong Relationship with Canada: Halifax, Nova Scotia," Press Release, Office of the Press Secretary, December 1, 2004.

61 Stein and Lang identify the meeting in early December but suggest that many issues were under consideration for which the prime minister could not obtain answers. Most within the list they provide in fact did have clear answers. For the prime minister, cost and protection were the only issues at play. Janice Gross Stein and Eugene Lang, *The Unexpected War: Canada in Kanduhar* (Toronto: Viking Canada, 2007), 164.

62 Canada, *House of Commons Debates,* Vol. 40, 38th Parliament, First Session (February 24, 2005).

63 *The Economist,* "The Uncertain Leadership of Canada's Paul Martin," February 17, 2005.

64 Canadian Broadcasting Corporation, "U.S. Contractor Prefers Goose Bay for Missile Defence Radar," CBC TV news report, June 17, 2005, http://www.cbc.ca.

65 Janice Gross Stein and Eugene Lang, *The Unexpected War: Canada in Kanduhar* (Toronto: Viking Canada, 2007), 175.

Epilogue

1 Since the first intercept test in February 1999, eight of thirteen have been successful. One additional test did not occur because of the failure of the target to launch on May 25, 2007. Missile Defense Agency, "Ballistic Missile Defense Flight Test Record," fact sheet, March 27, 2009.

2 Andrew Scutro, "CNO Announces New Missile Defense Command," *Air Force Times*, March 6, 2009, www.airforcetimes.com.

3 Jenny Shin, Missile Defence Update no. 2 (Washington, DC: Center for Defence Information, March 13, 2009), http://www.cdi.org.

4 Missile Defense Agency, "ABL Successfully Fires Complete Weapon System from Aircraft," News Release, November 26, 2008, http://www.mda.mil.

5 William J. Broad, "North Korean Missiles Test Was a Failure, Experts Say," *New York Times*, April 5, 2009, http://www.nytimes.com.

6 Associated Press, "Officials: US Tracking Suspicious Ship from N Korea" Yahoo! News, June 18, 2009, http://news.yahoo.com.

7 Eli Lake, "Iran's Space Launch Turns Clock Back," *Washington Times*, February 4, 2009, http://www.washingtontimes.com.

8 Gabriela Baczynska and David Alexander, "Polish, U.S. Missile Shield Irks Moscow," *National Post*, August 15, 2008.

9 North Atlantic Treaty Organization, Bucharest Summit Declaration, April 3, 2008, paragraph 37, p. 7.

10 Martin Sieff, "Russian Threat to Withdraw from INF Not Bluff," *Space War*, February 21, 2007, http://www.spacewar.com.

11 Radio Free Europe/Radio Liberty, "U.S./Russia: Missile Expert Assesses Azerbaijan Radar Proposal," June 8, 2007, http://www.rferl.org.

12 North Atlantic Treaty Organization, *Bucharest Summit Declaration*, April 3, 2008, paragraph 38, p. 7.

13 John McCain Campaign, "Effective Missile Defense," http://www.johnmccain.com.

14 Barack Obama Campaign, "National Missile Defence," http://www.barackobama.com.

15 Michael O'Hanlon, "Obama Administration's Sound Thinking on Missile Defense," Brookings Institution, June 24, 2009, http://www.brookings.edu.

16 Ewen MacAskill, "Obama Offers to Drop Missile Project if Russia Helps Deal with Iran," *Guardian*, March 3, 2009, http://www.guardian.co.uk.

17 Martin Bucher, "NATO and Missile Defence," *NATO Monitor*, April 4, 2009.

18 White House, "A 'Phased and Adaptive Approach' for Missile Defense in Europe," *Fact Sheet on Missile Defense* (Washington, DC: Office of the Press Secretary, Sepetember 17, 2009), www.whitehouse.gov.

19 Jason Wood, "Obama's Position on Space Treaty Impractical and Dangerous," *World Politics Review*, February 10, 2009, http://www.worldpoliticsreview.com.

20 See Dwight Mason, "The 2006 NORAD Renewal Agreement," *Frontline Defence* 3, 3 (2006), http://www.frontline-canada.com.

21 Canada, "Defence," in *Canada's International Policy Statement: A Role of Pride and Influence in the World* (Ottawa: Government of Canada, 2005), 1-32.

22 One can speculate that, if Canada had committed to some form of participation prior to the ABM Treaty, its prohibitions on third-party involvement might have been much different.

23 For a full discussion, see James Fergusson, "A Strange Decision in a Strange Land: The Irrelevance of Ottawa's Missile Defence 'No,'" in *The Dilemmas of American Strategic Primacy: Implications for the Future of Canadian-American Cooperation,* ed. David S. McDonough and Douglas A. Ross (Toronto: Royal Canadian Military Institute, 2005), 1-18.

24 For example, in order for NORAD to undertake its ballistic missile early warning role, Canadian officers had to have access to intelligence about adversarial ballistic missile capabilities – intelligence that Canada cannot acquire on its own.

25 Canada, *House of Commons Debates,* Vol. 8, 1st Session, 28th Parliament (March 17, 1969), p. 6683.

26 Cited in Kim Richard Nossal, *The Politics of Canadian Foreign Policy* (Scarborough, ON: Prentice Hall, 1989), 141.

27 James Fergusson, "Thinking about a 'Known Unknown': U.S. Strategy and the Past, Present, and Future of Strategic Defence," *International Journal,* Autumn 2008: 823-46.

Bibliography

Government Document Sources

Library and Archives Canada, Ottawa

Department of External Affairs and Department of National Defence. Memorandums to Cabinet. RG 25, 1967-75.

Department of National Defence. Reports, memorandums, and other documents. RG 24, 1954-64.

Privy Council Office. Cabinet Conclusions, 1965-70.

Special Joint Committee of the Senate and the House of Commons on Canada's International Relations. *Independence and Internationalism*. Ottawa: Queen's Printer, 1986.

Special Joint Committee on Canada's Defence Policy. "Dissenting Report by the Bloc Québécois Members of the Special Joint Committee on Canada's Defence Policy." In *Security in a Changing World.* Ottawa: Parliamentary Publications Directorate, 1994.

Standing Committee on External Affairs. Minutes of Proceedings and Evidence. No. 14. Ottawa, March 7, 1968.

Standing Committee on External Affairs and National Defence. Minutes, 1969.

–. *NORAD 1986*. Ottawa: House of Commons, 1986.

Standing Committee on Foreign Affairs and International Trade. *Canada and the Nuclear Challenge: Reducing the Political Value of Nuclear Weapons for the Twenty First Century.* Ottawa: House of Commons, 1998.

–. Minutes of Proceedings and Evidence, Special Joint Meeting with the Standing Committee on National Defence and Veterans Affairs, March 23, 2000. http://cmte.parl.gc.ca/cmte.

Canada

Canada. *Challenge and Commitment: A Defence Policy for Canada.* Ottawa: Queen's Printer, 1987.

–."Defence." In *Canada's International Policy Statement: A Role of Pride and Influence in the World,* 32. Ottawa: Government of Canada, 2005.

–. *Defence in the 70s: White Paper on Defence*. Ottawa: Queen's Printer, 1971.

–. *House of Commons Debates*. Selected from 1967-2005.

–. *1994 Defence White Paper*. Ottawa: Queen's Printer, 1994.

–. *North American Aerospace Defence Command: 1996 NORAD Agreement and Terms of Reference*. Ottawa: Queen's Printer, 1996.

–. *White Paper on Defence*. Ottawa: Queen's Printer, 1964.

Canadian Security and Intelligence Service. *Ballistic Missile Proliferation*. Report no. 2000/09. Ottawa: Government of Canada, March 23, 2001.

Chapman, J.H., P.A. Forsyth, P.A. Lapp, and G.N. Patterson. *Upper Atmosphere and Space Programs in Canada*. Science Secretariat, Privy Council Office Report. Ottawa: Queen's Printer, February 1967.

Chief of the Defence Staff. *Shaping the Future of Canadian Defence: A Strategy for 2020.* Ottawa: Department of National Defence, 2000.

Department of External Affairs. Letters, memorandums, and other documents, 1963-93.

–. International Security and Arms Control Bureau. *Outer Space: A Review of Canadian Activities and Policies in the Context of Canada's Approach to Arms Control and Disarmament.* Discussion Paper. Ottawa: External Affairs, December 13, 1984.

Department of National Defence. Reports, memorandums, and other documents, 1974-93.

–. *Ballistic Missile Defence in the 1990s: Implications for Canada.* Operational Research Analysis Report PR 63. Ottawa: Department of National Defence, 1993.

–. "Canada and the United States Amend the NORAD Agreement." News Release. August 5, 2004.

–. "Canadian Participation in the Joint Strike Fighter Program." *Backgrounder.* December 11, 2006.

–. *The State of the U.S. Missile Defence Program.* Issue Brief. Ottawa: Directorate of Strategic Analysis, Department of National Defence, September 25, 2006.

–. *The United States Strategic Defense Initiative: Implications for Deterrence, Arms Control, Adversaries, and Allies.* Operational Research and Analysis Establishment (ORAE) Report. Ottawa: Department of National Defence, 1984.

Privy Council Office. Memorandums, notes, Cabinet Conclusions, and other documents, 1984-85.

–. Department of Regional and Industrial Expansion. *Economic Implications of Canadian Participation in Strategic Defense Initiative: Memorandum for Arthur Kroeger.* Electronics and Aerospace Branch, Research and Development, May 17, 1985.

United States

Ballistic Missile Defense Organization. *National Missile Defense.* Annual Report to Congress. Washington, DC, 1997.

–. "National Missile Defense Program Evolution." *Fact Sheet.* Washington, DC, July 1997.

Cohen, William. *Press Statement.* Washington, DC: Department of Defense, January 21, 1999.

Congress of the United States. *Ballistic Missile Defense Technologies.* Washington, DC: 104th Congress of the United States, September 1985.

–. *Report of the Commission to Assess the Ballistic Missile Threat to the United States.* Washington, DC: 104th Congress of the United States, July 15, 1998.

–. Office of Technology Assessment. *Anti-Satellite Weapons, Countermeasures, and Arms Control.* Washington, DC: 104th Congress of the United States, September 1985.

Congressional Budget Office. *Alternatives for Boost-Phase Missile Defence.* Washington, DC: Congress of the United States, July 2004.

–. *Budgetary and Technical Implications of the Administration's Plans for National Missile Defense.* Washington, DC: Congress of the United States, April 2000.

Department of Defense. "DOD Announces Merger of U.S. Space and Strategic Commands." News Release, June 26, 2002.

–. "DOD Establishes Missile Defense Agency." News Release, January 4, 2002.

–. "Emerging Missile Threats to North America during the Next Fifteen Years." DCI National Intelligence Estimate. President's Summary. http://www.fas.org.

–. *Foreign Missile Developments and the Ballistic Missile Threat to the United States through 2015.* Washington, DC, September 9, 1999.

–. *Missile Defense Policy.* Office of Missile Defense Policy, November 15, 2001.

–. *National Missile Defense Conducts Successful Intercept Test.* News Release, October 2, 1999.
–. Office of the Assistant Secretary of Defense (Public Affairs). *National Missile Defense Conducts Intercept Test.* News Release, January 18, 2000.
–. Office of the Deputy Secretary of Defense. News Conference. Transcript, March 14, 1969.
–. Welch, General Larry (ret.), et al. *2nd Report of the Panel on Reducing Risk in Ballistic Missile Defense Flight Test Programs.* Washington, DC: Department of Defense, November 1999.
–. Welch, General Larry (ret.) et al. *Defense Acquisitions: Despite Restructuring SBIRS High Program Remains at Risk of Cost and Schedule Overruns.* Report to the Subcommittee on Strategic Forces, Senate Committee on Armed Services, October 2003.
–. Welch, General Larry (ret.) et al. *Report of the Panel on Reducing Risk in Ballistic Missile Defense Flight Test Programs.* Washington, DC: Department of Defense, February 27, 1998.
Embassy (in Canada). "Wolfowitz Says US Will Move beyond ABM Treaty." Ottawa: Information Resource Centre, July 12, 2001. http://www.usembassycanada.gov.
–. Rice, Condoleezza. "Press Briefing by National Security Advisor Condoleezza Rice." Washington File. Ottawa: US Embassy, November 8, 2001.
General Accounting Office. *National Missile Defence: Even with Increased Funding Technical and Schedule Risks Are High.* Washington, DC, June 23, 1998.
–. *Schedule and Technical Risks Represent Significant Technological Challenges.* Washington, DC, December 1997.
–. *Strategic Defense Initiative Program: Extent of Foreign Participation.* Washington, DC, February 7, 1990.
Missile Defense Agency. "ABL Successfully Fires Complete Weapon System from Aircraft." News Release. November 26, 2008. http://www.mda.mil.
–. "Ballistic Missile Defense Flight Test Record." *Fact Sheet.* Washington, DC, March 27, 2009.
–. *Ballistic Missile Defense Worldwide.* Washington, DC: Department of Defense, 2008. www.mda.mil.
–. *Fiscal Year (FY) 2004/FY 2005 Biennial Budget Estimates Submission Press Release.* Washington, DC: Department of Defense, 2003.
–. "Space Tracking and Surveillance System (STSS)." *Fact Sheet.* Washington, DC, January 11, 2010. www.mda.mil/system/sensors.html.
Office of the President. "Arms Control for Anti-Satellite Systems: Presidential Directive-33." Washington, DC, March 10, 1978. http://www.jimmycarterlibrary.org.
–. "Nuclear Weapons Employment Policy: Presidential Directive-59." Washington, DC, July 25, 1980. http://www.jimmycarterlibrary.org.
Perry, William. Secretary of Defense Press Briefing. Washington, DC, February 16, 1996.
United States Air Force. "Defense Support Program Satellites." *Fact Sheet.* January 2008. http://www.af.mil.
United States Space Command. *Long Range Plan: Implementing USSPACECOM's Vision for 2020.* Colorado Springs, CO: United States Space Command, 1998.
–. *Vision for 2020.* Colorado Springs, CO: United States Space Command, 1998.
White House. "A 'Phased and Adaptive Approach' for Missile Defense in Europe." *Fact Sheet on Missile Defense.* Washington, DC: Office of the Press Secretary, September 17, 2009. www.whitehouse.gov.
–. *Joint U.S.-Russian Statement on a Global Protection System.* Washington, DC: Office of the Press Secretary, June 17, 1992.

–. "President Discusses Strong Relationship with Canada: Halifax, Nova Scotia." Press Release. Office of the Press Secretary, December 1, 2004.
–. *Press Briefing by Robert Bell, Senior Director for the NSC for Defense Policy and Arms Control.* Washington, DC: Office of the Press Secretary, March 24, 1997.

Treaties

1997 ABM Treaty Demarcation Memorandums of Understanding, Agreed, Joint and Unilateral Statements, and Common Understandings. *Arms Control Reporter: A Chronicle of Treaties, Negotiations, Proposals, Weapons, and Policy.* Brookline, MA: Institute for Defense and Disarmament Studies, 1997.
Treaty between the United States of America and the Union of Soviet Socialist Republics on the Limitation of Anti-Ballistic Missile Systems; Agreed Statements, Common Understandings, and Unilateral Statements Regarding the Treaty between the United States of America and the Union of Soviet Socialist Republics on the Limitation of Anti-Ballistic Missile Systems." Appendix A in Matthew Bunn, *Foundation for the Future: The ABM Treaty and National Security.* Washington, DC: Arms Control Association, 1990. Also available at http://www.acq.osd.mil.
"Treaty on Principles Governing the Activities of States in the Exploration and Use of Outer Space, Including the Moon and Other Celestial Bodies." January 27, 1967. *United Nations Treaty Series.*

Other Sources

American Physical Society Boost-Phase Study Group. *Boost-Phase Intercept Systems for National Missile Defense: Scientific and Technical Issues.* Washington DC: American Physical Society, July 2003.
Andrew, Arthur. *The Rise and Fall of a Middle Power: Canadian Diplomacy from King to Mulroney.* Toronto: James Lorimer, 1993.
Arms Control Association. "ABM Treaty Withdrawal: Neither Necessary Nor Prudent." *Arms Control Today,* January-February 2003. www.armscontrol.org.
–. "Canada Weighs U.S. Missile Defense Cooperation." *Arms Control Today,* July-August 2003. www.armscontrol.org.
Associated Press. "Bonn Offered Role in Star Wars." *Globe and Mail,* February 12, 1985.
–. "Officials: US Tracking Suspicious Ship from N Korea." *Yahoo! News,* June 18, 2009. news.yahoo.com.
Aubry, Jack. "Canadian Fear of Missile Attack Grows." *National Post,* June 11, 2002.
Axworthy, Lloyd. "Notes for an Address by the Honourable Lloyd Axworthy, Minister of Foreign Affairs, to Accept the Endicott Peabody Award." Annual Peabody Awards, Boston, October 22, 1999. http://w01.international.gc.ca.
Baczynska, Gabriela, and David Alexander. "Polish, U.S. Missile Shield Irks Moscow." *National Post,* August 15, 2008. http://www.nationalpost.com.
Baker, Peter, and William Drozdiak. "Russia Alters Tone, Welcomes Talks on Missile Shield." *Washington Post,* May 3, 2001.
Barack Obama Campaign. "National Missile Defence." http://www.barackobama.com.
Baylis, John, and Kristan Stiddart. "Britain and the Chevaline Project: The Hidden Nuclear Programme, 1967-1982." *Journal of Strategic Studies* 20, 4 (December 2003): 124-55.
Best, Jack. "Clark and Coates Disagree over 'Star Wars.'" *Ottawa Citizen,* January 3, 1985.
Binational Planning Group. *The Final Report on Canada and the United States (CANUS) Enhanced Military Cooperation.* Colorado Springs, CO: Binational Planning Group, March 13, 2006.

Binnendjik, Hans, and George Stewart. "Toward Missile Defenses from the Sea." *Washington Quarterly* 25, 3 (Summer 2002): 193-207.
Binx, Hans. "An Update on Inspection." United Nations Monitoring, Verification, and Inspection Commission. January 27, 2003. http://www.un.org.
Blanchfield, Mike. "Shelve Missile Shield: Scientists." *Ottawa Citizen,* June 13, 2000.
Bland, Doug, and Sean Maloney. *Campaigns for International Security*. Kingston, ON: McGill-Queen's University Press, 2003.
Boese, Wade. "Canada Draws Line on Missile Defense." *Arms Control Today,* January-February 2005. www.armscontrol.org.
–. "Russia Considers Missile Defense." *Arms Control Today,* March 2003. www.armscontrol.org.
Bothwell, Robert. "Canada's Moment: Lester Pearson, Canada, and the World." In *Pearson: The Unlikely Gladiator,* edited by Norman Hillmer, 19-29. Kingston, ON: McGill-Queen's University Press, 1999.
Boyer, Peter J. "When Missiles Collide: How President Clinton Set His Own Star Wars Trap." *New Yorker,* September 8, 2000.
Bradley, Graham. "Missile Defense Plan Scores a Direct Hit." *Washington Post,* October, 3, 1999.
Brennon, Donald, ed. *Arms Control, Disarmament, and National Security*. New York: Braziller, 1961.
British Broadcasting Corporation. "Russia 'Open' to ABM Changes." News Report. June 29, 2001. http://news.bbc.co.uk.
Broad, William J. "North Korean Missiles Test Was a Failure, Experts Say." *New York Times,* April 5, 2009. http://www.nytimes.com.
Brodie, Bernard, ed. *The Absolute Weapon: Atomic Power and World Order.* New York: Harcourt, Brace, 1946.
Brown, Neville. *The Fundamental Issues Study within the British BMD Review*. Oxford: Mansfield College, 1998.
Bucher, Martin. "NATO and Missile Defence." *NATO Monitor,* April 4, 2009. http://natomonitor.blogspot.com.
Bunn, Matthew. *Foundation for the Future: The ABM Treaty and National Security*. Washington, DC: Arms Control Association, 1990.
Burgess, Lisa. "BMDO Faces Conflicting Signals." *Defense News,* March 30-April 5, 1998.
Burr, William. "The Nixon Administration, the 'Horror Strategy,' and the Search for Limited Nuclear Options." *Journal of Cold War Studies* 7, 3 (Summer 2005): 34-87.
–. "The Nixon Administration, the SIOP, and the Search for Limited Nuclear Options." National Security Archive Electronic Briefing Book no. 173. November 23, 2005. http://www.gwu.edu.
Buteux, Paul. *The Politics of Nuclear Consultation in NATO, 1965-1980.* Cambridge, UK: Cambridge University Press, 1983.
Canadian Broadcasting Corporation. "Slim Majority Oppose Missile Defence: Poll." CBC TV News Report, November 5, 2004. http://www.cbc.ca
–. "U.S. Contractor Prefers Goose Bay for Missile Defence Radar." CBC TV News Report. June 17, 2005. www.cbc.ca.
Canadian National Committee of the International Institute for Strategic Studies and the US Committee of the International Institute for Strategic Studies. *National Missile Defence*. Conference Report, Toronto, 2001.
Canadian Press. "PC's Deny Link between Star Wars, Modernization of Northern Radar." *Ottawa Citizen,* March 7, 1985.

Cannon, Lou. "Reagan's Big Idea: How the Gipper Conceived Star Wars." *National Review,* February 22, 1999.
Cellucci, Paul. *Unquiet Diplomacy*. Toronto: Key Porter Books, 2005.
Chayes, Abram, and Antonia Handler Chayes. "Testing and Development of 'Exotic' Systems under the ABM Treaty: The Great Reinterpretation Caper." *Harvard Law Review* 99, 8: 1965-96.
Chayes, Abram, and Jerome Wiesner, eds. *ABM: An Evaluation of the Decision to Deploy an Antiballistic Missile System.* New York: Signet Books, 1969.
Chayes, Abram, Jerome B. Wiesner, George W. Rathjens, and Steven Weinberg. "Overview." In *ABM: An Evaluation of the Decision to Deploy an Antiballistic Missile System,* edited by Abram Chayes and Jerome Wiesner. New York: Signet Books, 1969.
Cienski, Jan. "U.S. Calls Ottawa over Missile Shield Plan." *National Post,* May 1, 2001.
Clearwater, John. *Canadian Nuclear Weapons*. Toronto: Dundurn Press, 1998.
Compas. "Missile Defence: Small Soft, Quebec Majority Opposes It in Practice, while Backing It in Principle, Big Majority Condemns Ottawa's Lack of Public Discussion." Survey for the *National Post,* February 28, 2005. http://www.compas.ca.
Council on Foreign Relations. "Chronology of National Missile Defence Programs." 2007. http://www.cfr.org.
Covault, Craig. "Chinese Test Anti-Satellite Weapon." *Aviation Week and Space Technology,* January 17, 2007.
–. "Desert Storm Reinforces Military Space Direction." *Aviation Week and Space Technology,* April 8, 1991.
Coyle, Phillip. "Rhetoric or Reality? Missile Defense under Bush." *Arms Control Today,* May 2002. www.armscontrol.org.
Denholm Crosby, Ann. *Dilemmas in Defence Decision-Making: Constructing Canada's Role in NORAD, 1958-1996*. London: Macmillan, 1998.
–. "The Print Media's Shaping of the Security Discourse: Cruise Missile Testing." *International Journal* 52, 1 (Winter 1996-97): 89-117.
Denoon, David B.H. *Ballistic Missile Defense in the Post-Cold War Era.* Boulder, CO: Westview Press, 1995.
Donaghy, Greg. *Tolerant Allies: Canada and the United States, 1963-1968*. Montreal: McGill-Queen's University Press, 2002.
Dornheim, Michael A. "Missile Defense Soon, but Will It Work?" *Aviation Week and Space Technology,* February 24, 1997.
Drozdiak, William, and Dana Milbank. "Bush Tries to Sell NATO on Missile Plan." *Washington Post,* June 14, 2001.
Edgar, Alastair, and David Haglund. *The Canadian Defence Industry in the New Global Environment*. Kingston, ON: McGill-Queen's University Press, 1995.
Editorial. "Time to Stand Up." *Globe and Mail,* May 3, 2001.
Elsea, Jennifer K. *U.S. Policy Regarding the International Criminal Court*. Washington, DC: Congressional Research Service, 2006.
Erlich, Jeff. "U.S. Panel Gives NMD Approval Despite Hurdles." *Defense News,* August 26, 1996.
Ewing, Humphrey Crum, Robin Ranger, and David Bosdet. *Ballistic Missiles: The Approaching Threat.* Bailrigg Memorandum no. 9. Lancaster, UK: Centre for Defence and International Security Studies, 1995.
Ewing, Humphrey Crum, Robin Ranger, David Bosdet, and David Wienck. *Cruise Missiles: Precision and Countermeasures.* Bailrigg Memorandum no. 10. Lancaster, UK: Centre for Defence and International Security Studies, 1995.

Federation of Atomic Scientists. "Strategic Arms Reduction Treaty I." 2007. http://www.fas.org.

Fergusson, James. *Ballistic Missile Defence and Proliferation*. Report to the Arms Control, Disarmament, and Non-Proliferation Directorate, Department of Foreign Affairs and International Trade. Ottawa: Government of Canada, 1995.

–. *Ballistic Missile Defence: Implications for the Alliance.* NATO Fellowship Report, June 2000.

–. *Canada and Ballistic Missile Defence*. Forum Report 26-27 November. Occasional Paper 38. Winnipeg: Centre for Defence and Security Studies, 1998.

–. *Canada and Ballistic Missile Defence: Issues, Implications, and Timelines*. Report to the Nonproliferation, Arms Control, and Disarmament Division, Department of Foreign Affairs and International Trade. Ottawa: Department of Foreign Affairs, 1998.

–. "Canada and Ballistic Missile Defence: What We Know, Don't Know, and Can't Know!" *Breakfast on the Hill Seminar Series*. Ottawa: Canadian Federation for the Humanities and Social Sciences, November 4, 2004. http://www.fedcan.ca.

–. "The Coming Debate on Ballistic Missile Defence." *Financial Post,* July 15, 1995.

–. "The European Dimension of Ballistic Missile Defence." In *Canada and National Missile Defence,* edited by David Rudd, Jim Hanson, and Jessica Blitt, 27-38. Toronto: Canadian Institute of Strategic Studies, 2000.

–. "From Counter-Proliferation to Non-Proliferation: An Alternative Perspective on Ballistic Missile Defence." *New Approaches to Non-Proliferation.* Toronto: Centre for International and Strategic Studies, York University, Summer 1996.

–. "National Missile Defense, Homeland Defence, and Outer Space: Policy Dilemmas in the Canada-US Relationship." In *The Axworthy Legacy: Canada among Nations 2001,* edited by Fen Osler Hampson, Norman Hillmer, and Maureen Appel Molot, 233-52. Don Mills, ON: Oxford University Press, 2001.

–. "Not Home Alone: Canada and Ballistic Missile Defence." *International Journal,* Autumn 2001: 678-85.

–. "Shall We Dance? The Missile Defence Decision, NORAD Renewal, and the Future of Canada-US Defence Relations." *Canadian Military Journal,* Summer 2005: 678-85.

–. "A Strange Decision in a Strange Land: The Irrelevance of Ottawa's Missile Defence 'No.'" In *The Dilemmas of American Strategic Primacy: Implications for the Future of Canadian-American Cooperation,* edited by David S. McDonough and Douglas A. Ross, 1-18. Toronto: Royal Canadian Military Institute, 2005.

–. *Strategic Stability Reconsidered*. International Security Research and Outreach Programme, Department of Foreign Affairs and International Trade, September 2001.

–. "Thinking about a 'Known Unknown': US Strategy and the Past, Present, and Future of Strategic Defence." *International Journal,* Fall 2008: 823-46.

Fergusson, James, and Steve James. *Canada, National Security, and Outer Space*. Calgary: Canadian Defence and Foreign Affairs Institute, 2007.

Fergusson, James, James Mackintosh, Patrick McDonnell, and Shane Levesque. *Missile Defence: Options for Canadian Industry*. Report prepared for the Director-General of International and Industry Programs. Ottawa: Department of National Defence, July 2001.

Fernandez, Adolfo J. *Military Role in Space: A Primer*. Report for Congress. Washington, DC: Congressional Research Service, September 23, 2004.

Fielding, Lt. Col. Marcus. "The United States Unified Command Plan." *Canadian Military Journal,* Autumn 2006: 35-40.

Fitzgerald, Francis. *Way Out There in the Blue: Reagan, Star Wars, and the End of the Cold War*. New York: Simon and Schuster, 2000.

Forden, Geoffrey E. "The Airborne Laser." *IEEE Spectrum,* September 1997: 40-49.
Foxwell, David, and Joris Janssen Lok. "Naval TBM Defense Matures: Area and Theater-Wide Naval Theater Ballistic Missile Defence." *Jane's International Defence Review,* January 1, 1998: 28-30, 32-34.
Freedman, Lawrence. *The Evolution of Nuclear Strategy.* 2nd ed. London: Macmillan, 1989.
Friedman, Thomas. "U.S. Formally Rejects 'Star Wars' in ABM Treaty." *New York Times,* July 15, 1993.
Garvey, Richard. "The Wrong Plan." *Bulletin of the Atomic Scientists,* March-April 2000: 36-41.
Gershwin, Lawrence K. "Threats to U.S. Interests from Weapons of Mass Destruction." *Comparative Strategy* 12 (1993), 7-13.
Gertz, Bill. "Arms Control Freaks: The Administration Likes Negotiations Better than It Does Defense." *National Review,* February 22, 1999.
–. "Cohen Sees Russia, U.S. Allies as Hurdles to a Missile Defense." *Washington Times,* October 5, 1999.
–. "Cohen Welcomes Test of Missile Defense." *Washington Times,* October 4, 1999.
–. "Missile Defense Survives Veto Threat." *Washington Times,* March 17, 1999.
–. "Senate Panel OKs Bill on Missile Defense." *Washington Times,* February 10, 1999.
–. "State Department Cable Argues of a Loophole in Missile Defense Bill." *Washington Times,* March 26, 1999.
Gill, Bates. "Can China's Tolerance Last?" *Arms Control Today,* January-February 2003. www.armscontrol.org.
Glasser, Susan B. "Putin Is Upbeat on U.S. Ties: Missile Defense Aside, He Sees Partner in Bush." *Washington Post,* June 19, 2001.
Gordon, Michael R. "Allies' Mood on 'Star Wars' Shifts." *New York Times,* February 5, 2001.
Graham, Bill (Minister of National Defence). Press Conference Transcript. CTV National News, August 5, 2004.
Graham, Bradley. "Missile Defense Plan Scores a Direct Hit." *Washington Post,* October 3, 1999.
–. "Panel Fires at Antimissile Programs: Critique Warns of 'Rush to Failure.'" *Washington Post,* March 22, 1998.
Gray, Colin S. *Modern Strategy.* London: Oxford University Press, 1999.
Grover, Bernie. *The National Missile Defense Program: An Assessment of Market Opportunities for Canadian Industry.* Report for the Canadian Defence Industries Association. Ottawa: Canadian Defence Industries Association, August 2000.
Hadley, Stephen. "Global Protection System: Concept and Progress." *Comparative Strategy* 12 (1993): 2-6.
Haglund, David. "Yesterday's Issue? National Missile Defence, Canada, and the Allies." *International Journal,* Autumn 2001: 686-98.
Haglund, David, and Joel Sokolsky, ed. *The US-Canada Security Relationship: The Politics, Strategy, and Technology of Defense.* Boulder, CO: Westview Press, 1989.
Halberstam, David. *The Best and the Brightest.* New York: Random House, 1972.
Harvey, Frank. *Smoke and Mirrors: Globalized Terrorism and the Illusion of Multilateral Security.* Toronto: University of Toronto Press, 2005.
Haydon, Peter. *The Cuban Missile Crisis Reconsidered: Canadian Involvement Reconsidered.* Toronto: Canadian Institute for Strategic Studies, 1993.
Hays, Peter L. *United States Military Space: Into the Twenty-First Century.* Occasional Paper 42. Colorado Springs, CO: USAF Institute for National Security Studies, USAF Academy, 2002.

Head, Ivan, and the Canada 21 Council. *Canada 21: Canada and Common Security in the Twenty First Century*. Toronto: Centre for International Studies, University of Toronto, 1994.

Hellyer, Paul. *Damn the Torpedoes: My Fight to Unify Canada's Armed Forces*. Toronto: McClelland and Stewart, 1990.

Helprin, Mark. "The War of Lights: As Offensive Weapons Evolve, So Must Defenses." *National Review,* February 22, 1999.

Heritage Foundation. *Defending America: A Near- and Long-Term Plan to Deploy Missile Defenses*. Report of the Missile Defense Study Team. Washington, DC: Heritage Foundation, 1996.

–. *Defending America: A Plan to Meet the Urgent Missile Threat*. Commission on Missile Defenses. Washington, DC: Heritage Foundation, 1999.

Herken, Gregory. *The Counsels of War*. New York: Oxford University Press, 1985.

Hertzman, Lewis, John Warnock, and Thomas Hockin. *Alliances and Illusions: Canada and the NATO-NORAD Question*. Edmonton: Hurtig, 1969.

Hill, Luke. "TMD: NATO Starts the Countdown." *Jane's Defence Weekly,* January 3, 2001: 24-27.

Hobson, Sharon. "Canadian Missile Defence." *Jane's International Defence Review,* March 1, 2002: 16-20.

Hoffman, David. "Russia Says START II Is Imperiled." *Washington Post,* January 22, 1999.

–. "Russia's Myopic Missile Defense." *Washington Post,* February 10, 1999.

Holloway, Steven Kendall. *Canadian Foreign Policy: Defining the National Interest*. Peterborough, ON: Broadview Press, 2006.

Huldreth, Steven. *Missile Defense: The Current Debate*. Washington, DC: Congressional Research Service, December 7, 2004.

Hurtig, Mel. *Rushing to Armageddon: The Shocking Truth about Canada, Ballistic Missile Defence, and Star Wars*. Toronto: McClelland and Stewart, 2004.

Ing, Greg, and Anne McIlroy. "Atomic Energy Lab Working on U.S. Star Wars Research." *Ottawa Citizen,* July 13, 1987.

Institute for Defense and Disarmament Studies. *Arms Control Reporter: A Chronicle of Treaties, Negotiations, Proposals, Weapons, and Policy*. Brookline, MA: Institute for Defense and Disarmament Studies, 1986.

–. *Arms Control Reporter: A Chronicle of Treaties, Negotiations, Proposals, Weapons, and Policy*. Brookline, MA: Institute for Defense and Disarmament Studies, 1997.

–. *Arms Control Reporter: A Chronicle of Treaties, Negotiations, Proposals, Weapons, and Policy*. Vol. 25. New York: Institute for Defense and Disarmament Studies, 2006.

Institute for Foreign Policy Analysis. *Missile Defense, the Space Relationship, and the Twenty First Century*. Washington, DC: Institute for Foreign Policy Analysis, 2007.

–. *National Missile Defense: Policy Issues and Technological Challenges*. Washington, DC, July 5, 2000.

Jockel, Joseph T. *Canada in NORAD 1957-2007: A History*. Kingston: McGill-Queen's University Press, 2007.

–. *Four US Military Commands: NORTHCOM, NORAD, SPACECOM, STRATCOM – The Canadian Opportunity*. Working Paper Series 2003-03. Montreal: Institute for Research on Public Policy, 2003.

–. *No Boundary Upstairs: Canada, the United States, and the Origins of North American Air Defence, 1945-1958*. Vancouver: UBC Press, 1987.

–. "A Strong Friend Is a Good Defence." *Globe and Mail,* January 14, 2004.

–. "US National Missile Defense, Canada, and the Future of NORAD." In *Vanishing Borders: Canada among Nations 2000,* edited by Maureen Appel Molot and Fen Osler Hampson, 73-92. Don Mills, ON: Oxford University Press, 2000.

Jockel, Joseph J., and Joel J. Sokolsky. *The End of the Canada-U.S. Relationship.* Occasional Paper no. 53. Kingston, ON: Centre for International Relations, 1996.

John McCain Campaign. "Effective Missile Defense." http://www.johnmccain.com.

Jones, Rodney W., and Nikolai Sokov. "After Helsinki, the Hard Work." *Bulletin of Atomic Scientists* 53, 4 (July-August 1997): 26-31.

Kaplan, Fred. *The Wizards of Armageddon.* New York: Simon and Schuster, 1987.

Karp, Aaron. *Ballistic Missile Proliferation: The Politics and Technics.* London: Oxford University Press, 1996.

Keaney, Thomas J., and Eliot Cohen. *Gulf War Airpower Survey Summary Report.* Washington, DC: United States Air Force, 1993.

Keeny, Spurgeon M. Jr., Steve Fetter, Joseph Cirincione, and Jack Mendelsohn "Evaluating the Criteria for NMD Deployment." *Arms Control Today,* April 2000. www.armscontrol.org.

Kenna, Kathleen. "Canada and U.S. Renew NORAD Deal." *Toronto Star,* June 17, 2000.

–. "Son of Star Wars Heats Up: Canada Weighs Its Options as Pentagon Presses for a Partner in Controversial Missile-Defence Scheme." *Toronto Star,* May 21, 2000.

Kirkpatrick, Jeane. "Target America: The United States Is in the Sights of Rogue States." *National Review,* February 22, 1999.

Kissinger, Henry. *Nuclear Weapons and Foreign Policy.* New York: Doubleday, 1958.

Koring, Paul. "Canada Accused of Cold War Thinking in Impasse on Missile Defence: U.S. Official Annoyed with Ottawa's Intransigence on Revamping the Treaty." *Globe and Mail,* May 16, 2000.

Koring, Paul, and Daniel LeBlanc. "Canada Aims to Join 'America Command.'" *Globe and Mail,* January 29, 2002.

Kueter, Jeff. "The Importance of Boost Phase Missile Defense." *Policy Outlook.* Washington, DC: George C. Marshall Institute, 2007.

Lagasse, Philippe. *The SORT Debate: Implications for Canada.* Working Paper Series 2003-01. Montreal: Institute for Research on Public Policy, 2003.

Lake, Eli. "Iran's Space Launch Turns Clock Back." *Washington Times,* February 4, 2009.

Lambeth, Benjamin. *Mastering the Ultimate High Ground: Next Steps in the Military Uses of Space.* Santa Monica, CA: Rand Press, 2003.

Legault, Albert, and Michel Fortmann. *A Diplomacy of Hope: Canada and Disarmament 1945-1988.* Kingston, ON: McGill-Queen's University Press, 1992.

Lehman II, Ronald F. "Changing Realities." *Comparative Strategy* 12 (1993): 47-51.

Lewis, George N., and Lisbeth Gronlund. *An Assessment of the Missile Defense Agency's "Endgame Success" Argument.* Cambridge, MA: Union of Concerned Scientists, December 2, 2002.

Lewis, George N., and Theodore A. Postol. "An Evaluation of the Army Report 'Analysis of Video Tapes to Assess Patriot Effectiveness,' Dated 31 March 1992." http://www.fas.org.

–. "Portrait of a Bad Idea." *Bulletin of Atomic Scientists* 53, 4 (July-August 1997): 18-25.

Lewis, John Wilson, and Xue Litai. *China Builds the Bomb.* Stanford, CA: Stanford University Press, 1988.

Lindsay, James M., and Michael E. O'Hanlon. *Defending America: The Case for Limited National Missile Defense.* Washington, DC: Brookings Institution, 2001.

–. "Missile Defense after the ABM Treaty." *Washington Quarterly* 25, 3 (Summer 2002): 163-75.

Lindsey, George R. "Ballistic Missile Defence in the 1990s." *Canadian Defence Quarterly,* September 1995: 6-11.

–. "Defence against Ballistic Missiles." *RCAF Staff College Journal,* 5th anniversary ed., 1960.

–. *The Information Requirements for Aerospace Defence: The Limits Imposed by Geometry and Technology*. Bailrigg Memorandum no. 27. Lancaster, UK: Centre for Defence and International Security Studies, 1997.

Lindsey, George, and Arnold Simoni. *Prospects for a Multilateral Missile Defence Regime: A Research Report.* Toronto: York Centre for International and Strategic Studies, 1993.

Liu Centre. *The Missile Defence Debate: Guiding Canada's Role.* Vancouver: University of British Columbia, March 2001.

Lok, Joris Janssen. "NATO Exercises Prop Up TMD Pillars." *Jane's International Defence Review,* July 1, 1998: 58-59.

Lyon, Peyton. "Defence and Foreign Policy." In *Special Studies Prepared for the Special Committee of the House of Commons on Matters Relating to Defence.* Supplement. Canada, House of Commons. Ottawa: Queen's Printer, 1965.

MacAskill, Ewen. "Obama Offers to Drop Missile Project If Russia Helps Deal with Iran." *Guardian,* March 3, 2009. www.guardian.co.uk.

MacDonald, Marci. *Yankee Doodle Dandy: Brian Mulroney and the American Agenda.* Toronto: Stoddart, 1995.

MacGuigan, Mark. *An Inside Look at External Affairs during the Trudeau Years.* Calgary: University of Calgary Press, 2002.

MacIsaac, David. "Voices from the Central Blue: The Air Power Theorists." In *Makers of Modern Strategy from Machiavelli to the Nuclear Age,* edited by Peter Paret, 612-47. Princeton, NJ: Princeton University Press, 1986.

Mann, Paul. "Missile Defense Riddled with Diverse Failures." *Aviation Week and Space Technology,* March 30, 1998.

Martin, David. "Towards an Alliance Framework for Extended Air Defence/Theatre Missile Defence." *NATO Review* 44, 3 (May 1996): 32-35.

Martin, Lawrence. *Pledge of Allegiance: The Americanization of Canada in the Mulroney Years.* Toronto: McClelland and Stewart, 1993.

–. *Presidents and Prime Ministers: Washington and Ottawa Face to Face; The Myth of Bilateral Bliss.* Toronto: Doubleday, 1982.

Mason, Dwight. "Canada Alert: Canada and the U.S. Missile Defense System." *Hemispheric Focus* 12, 1 (January 9, 2004). http://csis.org.

McCausland, Jeffrey. *Conventional Arms Control and European Security*. Adelphi Paper 301. London: International Institute for Strategic Studies, 1996.

McFate, Patricia Bliss, and Sidney N. Graybeal. "The Relationship of the ABM Treaty to Strategic Stability and Outer Space." Unpublished paper, February 28, 1998.

McKenzie, Colin. "Russia-US Reach Deal on Arms Cuts." *Globe and Mail,* June 17, 1992.

McLauglin, Kevin. "Would Space-Based Defenses Improve Security?" *Washington Quarterly* 25, 3 (Summer 2002): 177-91.

McMahon, K. Scott. *Pursuit of the Shield: The U.S. Quest for Limited Ballistic Missile Defence.* Lanham, MD: University Press of America, 1997.

McMurry, Col. Robert, and Lt. Gen. Michael Dunn (ret.). *Airborne Laser (ABL): Recent Developments and Plans for the Future.* Washington Roundtable on Science and Public Policy. Washington, DC: George C. Marshall Institute, 2008.

McRae, Rob, and Don Hubert, eds. *Human Security and the New Diplomacy: Protecting People, Promoting Peace.* Kingston, ON: McGill-Queen's University Press, 2001.

Michaud, Nelson, and Kim Richard Nossal, eds. *Diplomatic Overtures: The Conservative Era in Canadian Foreign Policy 1984-1993.* Vancouver: UBC Press, 2001.

Mitchell, Alison. "Senate Democrats Square Off with Bush over Missile Plan." *New York Times,* May 3, 2001.
Montgomery, Charlotte. "Clark Denies Receipt of Invitation by US: Unaware of Star Wars Move." *Globe and Mail,* March 27, 1985.
–. "Star Wars Invitation Extended by US." *Globe and Mail,* March 26, 1985.
Mosher, David J. "The Grand Plans." *IEEE Spectrum,* September 1997: 28-39.
Mowthorpe, Matthew. *The Militarization and Weaponization of Space.* Lanham, MD: Lexington Books, 2004.
"MPs to Poll Public on Missile Shield." *Winnipeg Sun,* May 3, 2001.
Mulroney, Brian. *Memoirs 1939-1993.* Toronto: Douglas Gibson Books, 2007.
Myers, Steven Lee, Eric Schmitt, and Marc Lacey. "Russian Resistance Key in Decision to Delay Missile Shield." *New York Times,* September 3, 2000.
Nash, Knowlton. *Kennedy and Diefenbaker: The Feud that Helped Topple a Government.* Toronto: McClelland and Stewart, 1990.
Nitze, Paul. "The Objective of Arms Control." Lecture transcript, Alistair Buchan Memorial Lecture Series. Washington, DC: State Department, March 28, 1985.
North American Aerospace Defense Command – United States Space Command. *Concept of Operations for Ballistic Missile Defence of North America.* Colorado Springs, CO: North American Aerospace Defense Command – United States Space Command, July 21, 1998.
North Atlantic Treaty Organization. *Alliance Policy Framework on the Proliferation of Weapons of Mass Destruction.* Istanbul: North Atlantic Council Ministerial, June 9, 1994.
–. "The Alliance's Strategic Concept." *The Reader's Guide to the NATO Summit.* Washington, DC: NATO, 1999.
–. *Bucharest Summit Declaration,* April 3, 2008.
–. *The Future Tasks of the Alliance.* Brussels, December 14, 1967.
–. "NATO's Theatre Missile Defence Programme Reaches New Milestone." Press Release. June 5, 2001.
–. "Rome Declaration on Peace and Cooperation." *NATO Review* 39, 6 (December, 1991): 19-22.
–. Missile Defence Ad Hoc Group. *NATO TBMD Programme Options.* Report to the North Atlantic Council. Conference of National Armaments Directors. Brussels: NATO, 1997.
Nossal, Kim Richard. "The Mulroney Years: Transformation and Tumult." *Policy Options,* June-July 2003: 76-81.
–. *The Politics of Canadian Foreign Policy.* Scarborough, ON: Prentice Hall, 1989.
O'Brien, Kevin A. *Canada and Aerospace Defence: NORAD, Global Warning, and Theatre Missile Defence in the Evolving International Security Environment.* Toronto: Canadian Institute of Strategic Studies, 1995.
O'Hanlon, Michael. "Obama Administration's Sound Thinking on Missile Defense." Washington, DC: Brookings Institution, June 24, 2009. www.brookings.edu.
–. *Technological Change and the Future of Warfare.* Washington, DC: Brookings Institution, 2000.
Overy, Richard. *The Air War 1939-1945.* London: Stein and Day, 1981.
Paul, T.V., Richard J. Harknett, and James J. Wirtz, eds. *The Absolute Weapon Revisited: Nuclear Arms and the Emerging International Order.* Ann Arbor: University of Michigan Press, 1998.
Peterson, M.J. "The Use of Analogies in Developing Space Law." *International Organization* 51 (Spring 1997): 245-74.

Podvig, Pavel, ed. *Russian Strategic Nuclear Forces*. Cambridge, MA: MIT Press, 2001.

Polyani, John. "Canada Must Oppose U.S. Missile Defence Plan." *Toronto Star*, May 3, 2001.

Porth, Jacquelyn S. "Rumsfeld Says U.S. Expects Limited Missile Defense Shield by End of 2004 – Developing Missile Defenses Promotes Closer Relations with Allies, He Says." Washington File. Ottawa: US Embassy, August 20, 2004.

Postol, Theodore. "The Lessons of the Gulf War Experience with PATRIOT." *International Security*, Winter 1991-92: 119-71.

Preston, Robert, Dana J. Johnson, Sean J.A. Edwards, Michael D. Miller, and Calvin Shipbaugh. *Space Weapons, Earth Wars*. Project Air Force. Santa Monica, CA: Rand Press, 2002.

Pugliese, David. "Missile Defence System to 'Preserve' U.S. Dominance." *Ottawa Citizen*, May 24, 2000.

Questor, George. *Deterrence before Hiroshima*. New York: Transaction, 1986.

Radio Free Europe/Radio Liberty. "U.S./Russia: Missile Expert Assesses Azerbaijan Radar Proposal, June 8, 2007. www.rferl.org.

Ranger, Robin. *Arms and Politics 1958-1978: Arms Control in a Changing Political Context*. Toronto: Macmillan, 1979.

–, ed. *The Devil's Brew I: Chemical and Biological Weapons and Their Delivery Systems*. Bailrigg Memorandum no. 16. Lancaster, UK: Centre for Defence and International Security Studies, 1996.

–, ed. *Extended Air Defence and the Long-Range Missile Threat*. Bailrigg Memorandum no. 30. Lancaster, UK: Centre for Defence and International Security Studies, 1997.

Ranger, Robin, with Jeremy Stocker and David Wiencek, eds. *Theatre Missile Defence*. Bailrigg Study no. 1. Lancaster, UK: Centre for Defence and International Studies, 1998.

Ranger, Robin, and David Wiencek, eds. *The Devil's Brew II: Weapons of Mass Destruction and International Security*. Bailrigg Memorandum no. 17. Lancaster, UK: Centre for Defence and International Security Studies, 1997.

Reagan, Ronald. *The Reagan Diaries*. Edited by Douglas Brinkley. New York: Harper Perrennial, 2007.

Regehr, Ernie. "BMD, NORAD, and Canada-US Security Relations." *Ploughshares Monitor* 25, 1 (Spring 2004): www.ploughshares.ca.

–. "A Question of Defense: How American Allies Are Responding to the US Missile Defense Program." *Comparative Strategy* 23, 2 (2004): 143-72.

Richter, Andrew. *Avoiding Armageddon: Canadian Military Strategy and Nuclear Weapons, 1950-1963*. Vancouver: UBC Press, 2002.

Ripley, Tim. *Scud Hunting: Counter-Force Operations against Theatre Ballistic Missiles*. Bailrigg Memorandum no. 18. Lancaster, UK: Centre for Defence and International Security Studies, 1996.

Rivkin David B. Jr., Lee A. Casey, and Darin R. Bartram. *The Collapse of the Soviet Union and the End of the 1972 Anti-Ballistic Missile Treaty*. Washington, DC: Heritage Foundation, 1998.

Robinson, Bill. "Return of SDI: Ballistic Missile Defence for North America." *Ploughshares Monitor*, 18, 4 (December 1997). www.ploughshares.ca.

Ross, Douglas. "Foreign Policy Challenges for Paul Martin." *International Journal* 58, 4 (Autumn 2003): 533-71.

–. "SDI and Canadian-American Relations." In *America's Alliances and Canada-American Relations*, edited by Lauren McKinsey and Kim Richard Nossal, 137-54. Toronto: Summerhill Press, 1988.

Roussel, Stephané. *The North American Democratic Peace: Absence of War and Security Institution-Building in Canada-US Relations.* Kingston, ON: McGill-Queen's University Press, 2004.

Rowan, Geoffrey. "Patriot Missile Rising to Occasion." *Globe and Mail,* January 22, 1991.

Rowen, Henry S. "The Evolution of Strategic Nuclear Doctrine." In *Strategic Thought in the Nuclear Age,* edited by Laurence Martin, 131-56. Baltimore: Johns Hopkins University Press, 1979.

Rudd, David, Jim Hanson, and Jessica Blitt, eds. *Canada and National Missile Defence.* Toronto: Canadian Institute of Strategic Studies, 2000.

Sallot, Jeff. "Missile Defence Meeting Comes up Short on Specifics." *Globe and Mail,* May 16, 2001.

–. "No Hurry to Endorse Missile Plan." *Globe and Mail,* May 2, 2001.

–. "Support Missile Project or Else, Ottawa Warned." *Globe and Mail,* March 13, 2000.

Schelling, Thomas. *The Strategy of Conflict.* New York: Oxford University Press, 1963.

Schmitt, Eric, and Steven Lee Myers. "Clinton Lawyers Give a Go-Ahead to Missile Shield." *New York Times,* June 15, 2000.

Schneider, William Jr. "The Future of the ABM Treaty." *Comparative Strategy* 12 (1993): 53-56.

Scott, William. "Missile Sensor Flyby Boosts NMD Outlook." *Aviation Week and Space Technology,* January 26, 1998.

Scutro, Andrew. "CNO Announces New Missile Defense Command." *Air Force Times,* March 6, 2009. www.airforcetimes.com.

Seigle, Greg. "US NMD Lacks Approval from Canada, NORAD Deputy Says." *Jane's Defence Weekly,* January 26, 2000: 5.

Sherman, J. Daniel. "PATRIOT PAC-2 Development and Deployment in the Gulf War." US Army Material Command Study. www.dau.mil.

Shin, Jenny. Missile Defence Update no. 2. Washington, DC: Center for Defence Information, March 13, 2009. www.cdi.org.

Sieff, Martin. "Russian Threat to Withdraw from INF Not Bluff." *Space War,* February 21, 2007. www.spacewar.com.

Simon, Jeffrey, ed. *Security Implications of SDI: Will We Be More Secure in 2010?* Washington, DC: National Defense University Press, 1990.

Smith, David J. "The Missile Defense Act of 1991." *Comparative Strategy* 12 (1993): 71-73.

Smith, James, ed. *Nuclear Deterrence and Defense: Strategic Considerations.* Colorado Springs, CO: USAF Institute for National Security Studies, 2001.

Sokolsky, Joel. "The Bilateral Defence Relationship with the United States." In *Canada's International Security Policy,* edited by David Dewitt and David Leyton-Brown, 171-98. Scarborough, ON: Prentice Hall, 1995.

–. "The Bilateral Security Relationship: Will 'National' Missile Defence Involve Canada?" *American Review of Canadian Studies,* Summer 2000: 227-54.

–. "Changing Strategies, Technologies, and Organization: The Continuing Debate on NORAD and the Strategic Defense Initiative." *Canadian Journal of Political Science* 19, 4 (December 1986): 751-74.

Soofer, Robert. "SDI Structural Impediments." *Comparative Strategy* 12 (1993): 75-77.

Stairs, Denis. "Foreign Policy Consultations in a Globalizing World: The Case of Canada, the WTO, and the Shenanigans in Seattle." *Policy Matters* 1, 8 (December 2000): 1-44.

Stares, Paul B. *The Militarization of Space: US Policy, 1945-1984.* Ithaca, NY: Cornell University Press, 1985.

Stein, Janice Gross, and Eugene Lang. *The Unexpected War: Canada in Kanduhar*. Toronto: Viking Canada, 2007.

Steinbrunner, John. "National Missile Defense: Collision in Progress." *Arms Control Today*, November 1999. www.armscontrol.org.

Stocker, Jeremy. *Sea-Based Ballistic Missile Defence*. Bailrigg Study no. 2. Lancaster, UK: Centre for Defence and International Security Studies, 1999.

Stromseth, Jane E. *The Origins of Flexible Response: NATO's Debate over Strategy in the 1960s*. New York: St. Martin's Press, 1988.

Talbott, Strobe. *The Master of the Game: Paul Nitze and the Nuclear Peace*. New York: Alfred A. Knopf, 1988.

Thatcher, Margaret. "Deterrence Is Not Enough: Security Requirements for the 21st Century." Speech to the Ballistic Missile Defense: New Requirements for a New Century Conference, Institute for Public Policy, Washington, DC, December 3, 1998.

The Economist. "The Uncertain Leadership of Canada's Paul Martin." February 17, 2005.

Thompson, Allan. "U.S. Envoy Urges Canada to Weigh Missile Shield." *Toronto Star*, May 16, 2001.

Toronto Star. "Harper Would Call for Vote on Missile Defence." June 9, 2004.

Turner, Stansfield. *Caging the Nuclear Genie: An American Challenge for Global Security*. Boulder, CO: Westview Press, 1997.

United Nations Security Council. Resolution 1441, November 8, 2002. http://www.iaea.org.

Wallander, Celeste A. "Russia's Strategic Priorities." *Arms Control Today*, January-February 2002. www.armscontrol.org.

Wallop, Senator Malcolm. "What SDI Consensus?" *Comparative Strategy* 12 (1993): 67-70.

Warnock, John. "Canada and North American Defence." In *Alliances and Illusions: Canada and the NATO-NORAD Question*, edited by Lewis Hertzman, John Warnock, and Thomas Hockin, 43-94. Edmonton: Hurtig, 1969.

Weston, William. "Canada and Ballistic Missile Defence: A Challenge for the 21st Century." *Strategic Datalink*. Toronto: Canadian Institute for Strategic Studies, 1996.

Whiting, Allen S. "The Chinese Nuclear Threat." In *ABM: An Evaluation of the Decision to Deploy an Antiballistic Missile System*, edited by Abram Chayes and Jerome Wiesner, 160-70. New York: Signet Books, 1969.

Whittell, Giles, and Ben MacIntrye. "Putin Threat to New Arms Race." *Times of London*, June 20, 2001.

Wiencek, David. *Dangerous Arsenals: Missile Threats in and from Asia*. Bailrigg Memorandum no. 22. Lancaster, UK: Centre for Defence and International Security Operations, 1997.

Wilkening, Dean A. *Ballistic Missile Defence and Strategic Stability*. Adelphi Paper 334. London: International Institute for Strategic Studies, 2000.

Wong, Wilson. *Weapons in Space*. Silver Dart Canadian Aerospace Studies 3. Winnipeg: Centre for Defence and Security Studies, 2006.

Wood, Jason. "Obama's Position on Space Treaty Impractical and Dangerous." *World Politics Review*, February 10, 2009. www.worldpoliticsreview.com.

Woolf, Amy. *U.S. Strategic Nuclear Forces: Background, Development, and Issues*. Washington, DC: Congressional Research Service, September 5, 2007.

Worden, Simon P. *SDI and the Alternatives*. Washington, DC: National Defense University Press, 1991.

Yost, David. "Ballistic Missile Defense and the Western Alliance." *International Security* 7, 2 (Autumn 1982): 143-74.

–. *Soviet Ballistic Missile Defence and the Western Alliance.* Cambridge, MA: Harvard University Press, 1988.

–. *The US and Nuclear Deterrence in Europe.* Adelphi Paper 326. London: International Institute for Strategic Studies, 1999.

Index

Note: "ABM" stands for anti-ballistic missile defence; "GMD" stands for Ground-based Midcourse Defense; GPALS stands for Global Protection against Limited Strikes; "NMD" stands for National Missile Defense; "SDI" stands for Strategic Defense Initiative;

Printed and bound in Canada by Friesens

Set in Minion and Helvetica Condensed by Artegraphica Design Co. Ltd.

Copy editor: Judy Phillips

Proofreader: Dallas Harrison